ADVANCES IN
ATOMIC, MOLECULAR, AND OPTICAL PHYSICS

Edited by

E. Arimondo
PHYSICS DEPARTMENT
UNIVERSITY OF PISA
PISA, ITALY

P.R. Berman
PHYSICS DEPARTMENT
UNIVERSITY OF MICHIGAN
ANN ARBOR, MI, USA

and

C.C. Lin
DEPARTMENT OF PHYSICS
UNIVERSITY OF WISCONSIN
MADISON, WI, USA

Volume 56

AMSTERDAM • BOSTON • HEIDELBERG • LONDON • NEW YORK
OXFORD • PARIS • SAN DIEGO • SAN FRANCISCO • SINGAPORE
SYDNEY • TOKYO

Academic Press is an imprint of Elsevier

Academic Press is an imprint of Elsevier
32 Jamestown Road, London NW1 7BY, UK
Radarweg 29, PO Box 211, 1000 AE Amsterdam, The Netherlands
Linacre House, Jordan Hill, Oxford OX2 8DP, UK
30 Corporate Drive, Suite 400, Burlington, MA 01803, USA
525 B Street, Suite 1900, San Diego, CA 92101-4495, USA

First edition 2008

Copyright © 2008 Elsevier Inc. All rights reserved

No part of this publication may be reproduced, stored in a retrieval system or transmitted in any form or by any means electronic, mechanical, photocopying, recording or otherwise without the prior written permission of the publisher

Permissions may be sought directly from Elsevier's Science & Technology Rights Department in Oxford, UK: phone (+44) (0) 1865 843830; fax (+44) (0) 1865 853333; email: permissions@elsevier.com. Alternatively you can submit your request online by visiting the Elsevier web site at http://elsevier.com/locate/permissions, and selecting: *Obtaining permission to use Elsevier material*

Notice
No responsibility is assumed by the publisher for any injury and/or damage to persons or property as a matter of products liability, negligence or otherwise, or from any use or operation of any methods, products, instructions or ideas contained in the material herein. Because of rapid advances in the medical sciences, in particular, independent verification of diagnoses and drug dosages should be made

ISBN: 978-0-12-374290-2

ISSN: 1049-250X

For information on all Academic Press publications
visit our website at elsevierdirect.com

Printed and bound in USA

08 09 10 11 12   10 9 8 7 6 5 4 3 2 1

Working together to grow libraries in developing countries

www.elsevier.com | www.bookaid.org | www.sabre.org

ELSEVIER   BOOK AID International   Sabre Foundation

Contents

CONTRIBUTORS ix

1. Ionizing Collisions by Positrons and Positronium Impact on the Inert Atoms

G. Laricchia, S. Armitage, Á. Kövér and D.J. Murtagh

1. Introduction	2
2. Ion Production by Positron Impact	4
3. Positronium Ionizing Collisions	27
4. Conclusions and Outlook	36
5. Acknowledgments	38
6. References	39

2. Interactions Between Thermal Ground or Excited Atoms in the Vapor Phase: Many-Body Dipole–Dipole Effects, Molecular Dissociation, and Photoassociation Probed By Laser Spectroscopy

J.G. Eden, B.J. Ricconi, Y. Xiao, F. Shen and A.A. Senin

1. Introduction	51
2. Detection of Quantum Beating in Atoms and Molecules by Wavepacket Interferometry and a Coherent Nonlinear Optical Process	54
3. Many-Body, Dipole–Dipole Interactions Among Excited Alkali Atoms	70
4. Observation of Molecular Dissociation and Nascent Product State Distributions by the Dipole–Dipole Interaction	82
5. Coherent Control of Rb_2 Predissociation	95
6. Photoassociation of Rare-Gas–Halogen Atomic Pairs at Ambient Temperature	100
7. Application of Photoassociation to High Intensity Discharge Lighting	110
8. Summary and Conclusions	112
9. Acknowledgments	114
10. References	114

3. Bose–Einstein Condensates in Disordered Potentials

Leonardo Fallani, Chiara Fort and Massimo Inguscio

1. Introduction 120
2. How to Produce a Disordered Potential 121
3. Weakly-interacting Regime 127
4. Strongly-interacting Regime 145
5. Conclusions 154
6. Acknowledgments 154
7. References 155

4. Dipole–Dipole Interactions of Rydberg Atoms

Thomas F. Gallagher and Pierre Pillet

1. Introduction 162
2. Principles of Resonant Dipole–Dipole Collisions 166
3. Verification of the Predictions 174
4. Manipulating Resonant Collisions 181
5. Dipole–dipole Interactions in the Frozen Rydberg Gas ... 189
6. Line Broadening and Blockades 198
7. Mechanical Effects 204
8. Conclusion 215
9. Acknowledgments 216
10. References 216

5. Strong-Field Control of X-Ray Processes

Robin Santra, Robert W. Dunford, Elliot P. Kanter, Bertold Krässig, Stephen H. Southworth and Linda Young

1. Motivation 220
2. Basic concepts 222
3. Theoretical aspects 227
4. General experimental considerations 240
5. Case studies 242
6. Summary and outlook 253
7. Acknowledgments 255
8. References 255

6. Optical Trapping Takes Shape: The Use of Structured Light Fields

K. Dholakia and W.M. Lee

1. Introduction 262
2. Single beam optical tweezers 264
3. Applications within biophysics and the colloidal sciences 274
4. Optical trapping with structured light fields and its applications 282
5. Optical binding 323
6. Conclusions 324
7. Acknowledgments 324
8. References 325

INDEX 339

CONTENTS OF VOLUMES IN THIS SERIAL 341

CONTRIBUTORS

Numbers in parentheses indicate the pages on which the author's contributions begin.

G. LARICCHIA (1), Department of Physics and Astronomy, UCL (University College London), London WC1E 6BT, United Kingdom

S. ARMITAGE (1), Department of Physics and Astronomy, UCL (University College London), London WC1E 6BT, United Kingdom

Á. KÖVÉR (1), Institute of Nuclear Research, Debrecen, Hungary (ATOMKI)

D.J. MURTAGH (1), Department of Physics and Astronomy, UCL (University College London), London WC1E 6BT, United Kingdom

J.G. EDEN (49), Laboratory for Optical Physics and Engineering, Department of Electrical and Computer Engineering, University of Illinois, Urbana, IL 61801, United States

B.J. RICCONI (49), Laboratory for Optical Physics and Engineering, Department of Electrical and Computer Engineering, University of Illinois, Urbana, IL 61801, United States

Y. XIAO (49), Laboratory for Optical Physics and Engineering, Department of Electrical and Computer Engineering, University of Illinois, Urbana, IL 61801, United States

F. SHEN (49), Laboratory for Optical Physics and Engineering, Department of Electrical and Computer Engineering, University of Illinois, Urbana, IL 61801, United States

A.A. SENIN (49), Laboratory for Optical Physics and Engineering, Department of Electrical and Computer Engineering, University of Illinois, Urbana, IL 61801, United States

LEONARDO FALLANI (119), LENS European Laboratory for Nonlinear Spectroscopy, Italy;
Dipartimento di Fisica, Università di Firenze, Via Nello Carrara 1, 50019 Sesto Fiorentino (FI), Italy

CHIARA FORT (119), LENS European Laboratory for Nonlinear Spectroscopy, Italy;
Dipartimento di Fisica, Università di Firenze, Via Nello Carrara 1, 50019 Sesto Fiorentino (FI), Italy

MASSIMO INGUSCIO (119), LENS European Laboratory for Nonlinear Spectroscopy, Italy;
Dipartimento di Fisica, Università di Firenze, Via Nello Carrara 1, 50019 Sesto Fiorentino (FI), Italy

THOMAS F. GALLAGHER (161), Department of Physics, University of Virginia, Charlottesville, VA 22903, USA;
Laboratoire Aimé Cotton, CNRS, Univ Paris-Sud, 91405 Orsay cedex, France

PIERRE PILLET (161), Laboratoire Aimé Cotton, CNRS, Univ Paris-Sud, 91405 Orsay cedex, France

ROBIN SANTRA (219), Argonne National Laboratory, Argonne, Illinois 60439, USA

ROBERT W. DUNFORD (219), Argonne National Laboratory, Argonne, Illinois 60439, USA

ELLIOT P. KANTER (219), Argonne National Laboratory, Argonne, Illinois 60439, USA

BERTOLD KRÄSSIG (219), Argonne National Laboratory, Argonne, Illinois 60439, USA

STEPHEN H. SOUTHWORTH (219), Argonne National Laboratory, Argonne, Illinois 60439, USA

LINDA YOUNG (219), Argonne National Laboratory, Argonne, Illinois 60439, USA

K. DHOLAKIA (261), SUPA, School of Physics and Astronomy, University of St Andrews, North Haugh, Fife, Scotland KY16 9SS

W.M. LEE (261), SUPA, School of Physics and Astronomy, University of St Andrews, North Haugh, Fife, Scotland KY16 9SS

CHAPTER 1

Ionizing Collisions by Positrons and Positronium Impact on the Inert Atoms

G. Laricchia[a], S. Armitage[a], Á. Kövér[b] and D.J. Murtagh[a]

[a] *Department of Physics and Astronomy, UCL (University College London), London WC1E 6BT, United Kingdom*
[b] *Institute of Nuclear Research, Debrecen, Hungary (ATOMKI)*

Contents		
1.	Introduction	2
2.	Ion Production by Positron Impact	4
	2.1 Experimental Methods for Integral Cross-Sections	6
	2.2 Total Ionization Cross-Sections	9
	2.3 Direct Ionization	11
	2.4 Positronium Formation	21
3.	Positronium Ionizing Collisions	27
	3.1 Background	27
	3.2 Experimental Methods	31
	3.3 Results	32
4.	Conclusions and Outlook	36
	Acknowledgments	38
	References	39

Abstract   Recent progress in the study of positron and positronium in-duced ionization is reviewed within the context of the inert atoms. The focus is on experimental techniques and results, which comprise both integral and differential cross-sections. Measurements for positronium formation and direct ioniza-tion by positron impact, accompanied by singly- or multiply-charged ions, are compared and discussed. First data, inte-gral and differential, on the fragmentation of positronium in collision with helium and xenon are also presented. Comparisons with theories and other projectiles are made where possible and future prospects are considered.

1. INTRODUCTION

Almost two decades ago in this journal, Kauppila and Stein (1990) presented an interesting comparison between the total cross-sections of electrons (e^-) and positrons (e^+) scattering from a variety of atoms and molecules. Since then, a considerable amount of experimental data have been accumulated for specific positron-driven inelastic processes as well as for the controlled scattering of positronium (Ps), the bound state of a positron and an electron whose basic properties are summarized in Table 1. These experimental advances have been accompanied by concomitant progress in theoretical methods so that considerable success has been achieved in the detailed description of positrons interacting with simple atoms. This is exemplified for atomic hydrogen in Figure 1 where a high degree of consistency may be observed between theoretical and experimental cross-sections as well as between the sum of the various partial cross-sections and the total cross-section (Q_t).

These types of studies contribute to the array of investigations with positrons and positronium, which range from searches of CPT-violating

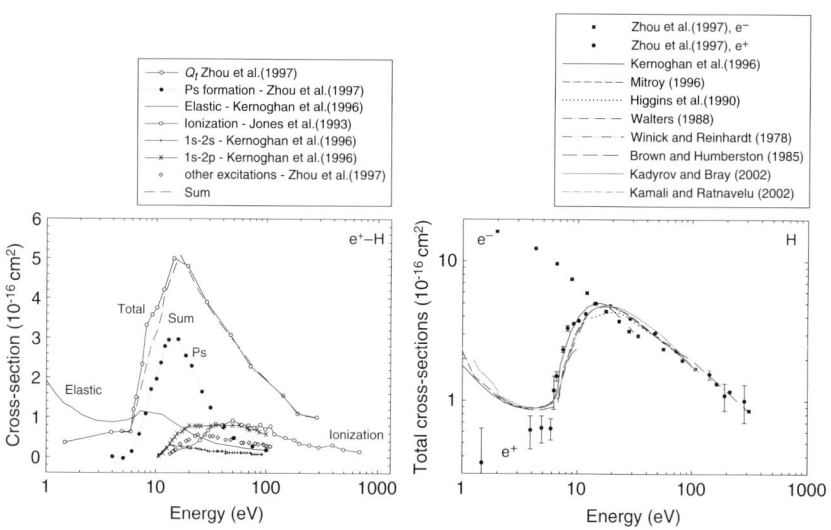

FIGURE 1 (lhs): Total and partial cross-sections for e^+–H scattering. (rhs): comparison of total cross-sections for $e^\pm$–H. (adapted from Stein et al. (1998) and Zhou et al. (1997))

TABLE 1 Properties of ground-state positronium.

Name	State	Lifetime (s)	N_γ	E_γ
ortho-Ps	$^3S_1$	142×10^{-9}	odd, dominantly 3[a]	0–511 keV
para-Ps	$^1S_0$	125×10^{-12}	even, dominantly 2[a]	511 keV

N_γ is the number of annihilation photons which, by charge-parity conservation, must satisfy $(-1)^{N_\gamma} = (-1)^{L+S}$ where L, S refer to the orbital and spin angular momenta of Ps (Yang, 1950). E_γ is the energy of the annihilation photons if the kinetic energy of Ps is negligible.
[a] The annihilation probability is proportional to α^m where $\alpha \approx 1/137$ is the fine structure constant and m corresponds, in this case, to the number of photons emitted; the $m = 1$ mode is reduced by the additional requirement of the presence of a third body to absorb any excess momentum.

TABLE 2 Dominant interactions in electron, positron and positronium scattering from atoms.

Interaction	e^-	e^+	Ps
Static	Negative	Positive	Zero
Polarization[a]	Negative	Negative	Zero
Exchange	Yes	No	Yes

[a] The leading terms of the asymptotic long-range potential are proportional to $(-1/R^4)$ for $e^\pm$ and $(-A/R^6 + B/R^8)$ for Ps. In the case of Ps, R is the coordinate between its center of mass and that of the other atom.

asymmetries in the annihilation modes of Ps (e.g. Vetter and Freedman (2003)) through the characterization of open volume defects in thin films (e.g. Petkov et al. (1999) and Dull et al. (2001)) to engineering diagnostics (Parker et al., 1997), aiding the analysis of energetic events in the galactic center (where it is estimated that approximately 93 per cent of all annihilation events occur through the decay of Ps) (e.g. Belanger et al. (2006)) and of medical applications (e.g. Champion and Le Loirec (2006)).

The dominant interactions arising from the scattering of the three light purely-leptonic projectiles (e^-, e^+, Ps) are indicated in Table 2 where fundamental differences may be discerned. For positrons, the opposite signs of the static and polarization interactions give rise, at low energies where target polarization is efficiently induced, to a reduced scattering probability in comparison with electrons, as manifested in Figure 1 (*rhs*) for $e^\pm$–H scattering. The lack of exchange and the possibility of electron capture – the latter responsible for the rapid rise of $Q_t(e^+)$ above ~ 7 eV – further distinguish e^+ from e^- scattering. Positronium formation, the inclusion of which as an open (real) or closed (virtual) channel is crucial for an accurate description of positron–atom interactions, is a tangible manifestation of the strong correlations between the electron and positron arising from their mutual Coulomb attraction. Annihilation is the eventual culmination.

Ps is structurally equivalent to a hydrogen atom with half the reduced mass and thus Bohr energy levels. However, the large magnetic moment of the positron (658 times that of the proton) and QED effects (such as virtual annihilation) give rise to considerable differences between the fine and hyperfine structure of the two atoms (e.g. Berko and Pendleton (1980)). The mass equality of the constituent particles in Ps makes the centers of charge and mass coincident, yielding a zero static interaction. The neutrality of Ps results in a zero first-order polarization and, as a consequence, short-range distortions due to exchange are comparatively more important than in electron scattering. Because of the light mass of Ps and its relatively large dipole polarizability ($72a_o^3$), the long-range Van der Waals interaction of Ps with another atom may also be significant.

Recently, Surko et al. (2005) have reviewed results for low-energy positron interactions with atoms and molecules, including elastic scattering, electronic and vibrational excitation. Here, we shall focus primarily on experimental studies of ionizing reactions induced by both positrons and positronium impact on the noble gases which, due to their atomic form at standard temperature and pressure, have been more extensively investigated and for which convergence is beginning to emerge among the results of different experimental methods and between these and theories. An appraisal of available data will be made where appropriate, results will be compared with theory where possible and outstanding questions highlighted. Reviews of results on positron scattering from hydrogen, alkali-metal and other atoms may be found in e.g. Stein et al. (1998) and Walters et al. (2005). Additional perspective on the subject may be acquired from a recent book on positron physics (Charlton and Humberston, 2001).

2. ION PRODUCTION BY POSITRON IMPACT

At low energies, the main processes by which an atom (A) can be ionized by a positron are annihilation, Ps formation and direct single ionization as expressed, respectively, by the following reactions:

$$e^+ + A \rightarrow A^+ + 2\gamma \qquad (1)$$
$$e^+ + A \rightarrow A^+ + Ps \qquad (2)$$
$$e^+ + A \rightarrow A^+ + e^+ + e^-. \qquad (3)$$

Whilst annihilation is always exothermic, Ps formation has a threshold energy given by $E_{Ps} = (E_i - 6.8 \text{ eV}/n^2)$ where E_i is the threshold energy for target ionization, n the principal quantum number of Ps and 6.8 eV its binding energy in the ground state. Thus Ps formation is an endothermic reaction for all the inert atoms.

As discussed in the following sections, there is very good agreement among experiments on the values for the total ionization cross-section by positron impact; its partitioning among the various channels is less certain. $Q_i^t(e^+)$ is defined as

$$Q_i^t(e^+) = Q_{Ps} + Q_i^+(e^+) + \sum Q_{HO}(e^+) \qquad (4)$$

where Q_{Ps} is the cross-section for Ps formation as per reaction (2), Q_i^+ that for direct single ionization and $\sum Q_{HO}$ is the sum over those for higher-order processes. The latter includes Q_i^{z+}, the cross-section for multiple direct ionization

$$e^+ + A \rightarrow A^{z+} + e^+ + ze^-, \qquad (5)$$

where $z = 1, 2, \ldots$ denotes the degree of ionization, and, the cross-section for transfer ionization

$$e^+ + A \rightarrow A^{z+} + Ps + (z-1)e^- \qquad (6)$$

which for $z = 1$ corresponds to Q_{Ps}, as per reaction (2).

$$Q_i^{z+,t} = Q_{Ps}^{z+} + Q_i^{z+} \qquad (7)$$

defines the total ionization cross-section for the zth ionic charge state. The total ionization cross-section for electron impact is analogously defined as

$$Q_i^t(e^-) = Q_i^+(e^-) + \sum Q_{HO}(e^-) \qquad (8)$$

where $\sum Q_{HO}(e^-)$ is the sum over the cross-sections for higher-order processes such as multiple ionization and excitation ionization.

Due primarily to Ps formation, $Q_i^t(e^+)$ exceeds $Q_i^t(e^-)$ at low and intermediate energies, as illustrated in Figure 2. The double-peaked structure, which develops with decreasing target ionization energy in $Q_i^t(e^+)$, arises from the increasing separation between the maxima of the leading constituent cross-sections, namely Q_{Ps} and Q_i^+, with the former peaking at $\approx 2E_{Ps}$ and the latter at $\approx 5 - 7E_i$ (Szluinska et al., 2002).

As remarked by Kauppila and Stein (1990), it has been 'known (Massey, 1976) that annihilation is not a significant effect except at energies which are well below those (>0.2 eV)' used in e^+ scattering experiments. Although, as discussed by Van Reeth et al. (2005), there are now expectations that non-trivial enhancements in the annihilation probability may also occur in the vicinity of various inelastic thresholds, thus far this has only been experimentally observed for molecules (e.g. Surko et al. (2005)). A rich structure in the energy dependence of the

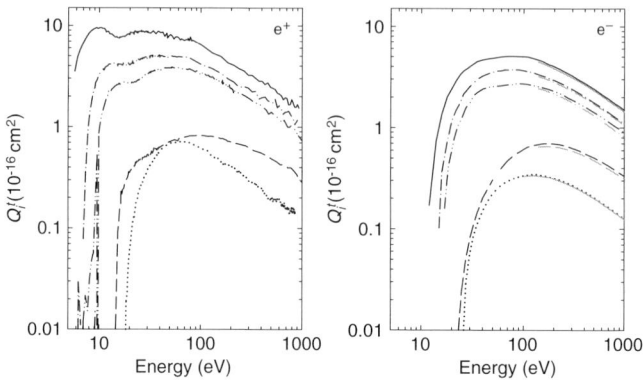

FIGURE 2 Total ionization cross-sections Q_i^t of the inert atoms for positrons (Laricchia et al., 2002; Murtagh et al., 2005) and electrons (Rejoub et al., 2002; Sorokin et al., 2000, 2004). Lines: dotted, He; dashed, Ne; dash-dot-dotted, Ar; dash dotted, Kr; solid, Xe

annihilation probability of positrons from complex hydrocarbons has been observed in the vicinity of the thresholds for vibrational excitation (Gilbert et al., 2002; Barnes et al., 2003, 2006) using the high-energy resolution ($\Delta E_+ \approx (25 \pm 10)$ meV) beam developed at the University of California at San Diego (UCSD) and discussed in the next section. A downward shift of the structures from the thresholds themselves has been identified with the binding energy of the positron to the molecule via vibrational Feshbach resonances (Gribakin and Gill, 2004), and linked with the magnitude of the annihilation rates (Young and Surko, 2007). Analogous investigations on atoms (i.e. in the vicinity of, and above, the thresholds for Ps formation, electronic excitation, etc.) have been frustrated by the large background of γ-rays arising from Ps formation (Marler et al., 2004) or impurities with low ionization thresholds (Szluinska and Laricchia, 2004; Szluinska et al., 2005). However, measurements for argon and xenon up to a couple of eV below E_{Ps} have been performed at UCSD by Marler et al. (2004) who have observed Q_{ann} to be several orders of magnitude smaller than Q_t. Their results are displayed in Figure 3 in terms of the dimensionless parameter $Z_{eff} = Q_{ann}v/(\pi r_o^2 c)$ where Q_{ann} is the annihilation cross-section and v the relative velocity between the electron and the positron involved in the annihilation event, r_o the classical electron radius and c the speed of light (e.g. Van Reeth et al. (2005)).

2.1 Experimental Methods for Integral Cross-Sections

With reference to reactions ((2) and (3)), the main methods employed for measuring a particular e^+ induced ionization cross-section involve counting the number of:

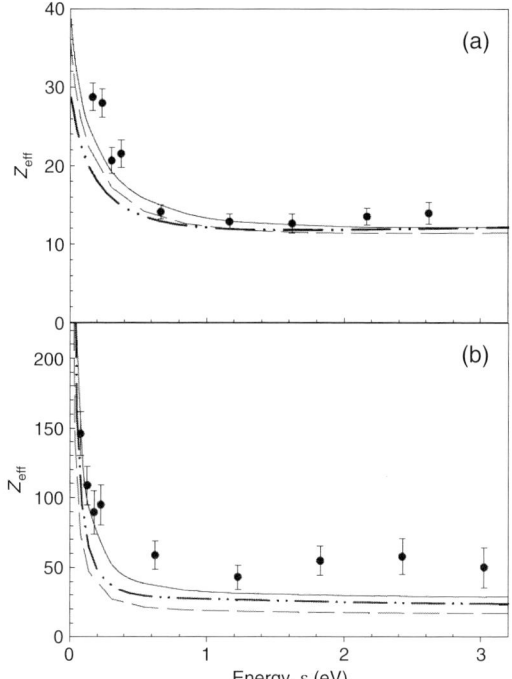

FIGURE 3 The annihilation parameter $Z_{eff} \propto Q_{ann}v$ as a function of the incident positron energy measured for (a) argon and (b) xenon (Marler et al., 2004). The thresholds for Ps formation are 7.2 eV for argon and 5.33 eV for xenon. Also shown are results from the polarized-orbital theory, (– –) (McEachran et al., 1979), many-body theory, (–.. –) (Ludlow, 2003; Ludlow J. and Gribakin G.F., (2004) (unpublished)), and a model-potential calculation (—) (Mitroy and Ivanov, 2002a), tuned to reproduce the polarized-orbital scattering results and experimental room temperature Z_{eff} values. Reprinted with permission from Surko et al. (2005) © 2005, by IOP Publishing Ltd

(i) ions in the final state: a quantity directly proportional to Q_i^t (e.g. Fromme et al. (1986), Laricchia et al. (2002) and Murtagh et al. (2005)) or $Q_i^{z+,t}$, if charge resolved (Bluhme et al., 1999a,b);

(ii) the correlated coincidences between ions and scattered positrons with an energy, $E_+^S \leq (E_+ - E_i)$, where E_+ is the positron incident energy. This determines Q_i^{z+} (e.g. Jacobsen et al. (1995), Moxom et al. (1996), Ashley et al. (1996) and Kara et al. (1997));

(iii) positrons scattered as in (ii) above, but without the coincident ion. This yields strictly $\sum_{z=1}^{\infty} Q_i^{z+}$, although Q_i^+ dominates (Mori and Sueoka, 1994; Marler et al., 2005);

(iv) all final-state positrons (i.e. scattered, irrespectively of their energy, plus unscattered); then, by assuming Q_{ann} to be negligible, any loss with respect to the incident flux corresponds to Ps formation,

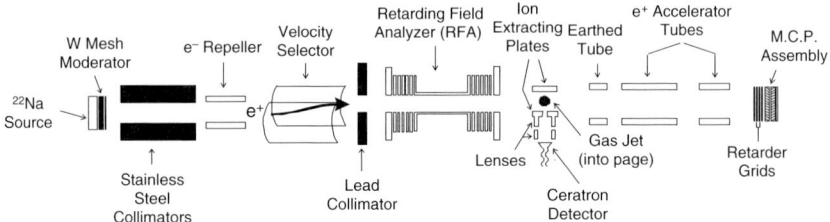

FIGURE 4 A schematic diagram of the positron beam apparatus employed by Ashley et al. (1996), Kara et al. (1997) to measure integrated ionization cross-sections

precisely to $\sum_{z=1}^{\infty} Q_{Ps}^{z+}$ with Q_{Ps} being the major contribution (e.g. Fornari et al. (1983) and Marler et al. (2005)).

Most of the above measurements have been performed with 'conventional' beams which are characterized by energy resolutions $\Delta E_+ \sim$ (0.3 − 3) eV, the lower limit usually achieved at a severe reduction (∼80%) of the beam intensity. An example of this type of beam is given in Figure 4. The slow positrons are obtained from a β^+-emitter (typically Na-22 or Co-58) in conjunction with a moderator (usually, tungsten) and are radially confined by a magnetic field of ∼0.01 T. A cylindrical electrode held at −1 kV removes secondary e^- prior to the interaction region whilst the slow positrons are separated from residual fast particles by a velocity selector. A retarding-field analyzer prior to the interaction region can be employed to bias off a portion of the beam to reduce its energy spread and/or reflect backscattered positrons and/or aid the extraction of final-state positrons with close-to-zero kinetic energy. Collisions occur at the intersection with a gas jet and, upon detection of a positron at the end of the flight path, an electric field is pulsed across the interaction region to extract the ions whose charge-to-mass ratios are identified by their flight times.

The high-energy resolution ($\Delta E_+ \geqslant (0.025)$ eV) of the positron beam at UCSD has been achieved by allowing the positron to thermalize in a trap (Gilbert et al., 1997; Marler et al., 2005). The slow positrons from a frozen neon moderator are guided to a three-stage buffer-gas Penning–Malmberg trap where cooling occurs by inelastic collisions with N_2 and CF_4 in a confining field of 0.15 T. Then, with reference to Figure 5, the electric potential in the accumulator is raised to accelerate positrons out of the trap, so that $E_+ = e(V - V_c)$. Positrons that have not annihilated or formed positronium in the scattering cell are guided by the magnetic field through a retarding-potential analyzer (RPA) to an annihilation plate viewed by a γ-ray detector. The magnetic field in the gas cell is 0.09 T whilst, in the RPA region, it is adjustable from zero to 0.09 T. Cross-section measurements are executed by reducing the magnetic field in the RPA region (B_{RPA}) with respect to that in the cell (B_C)

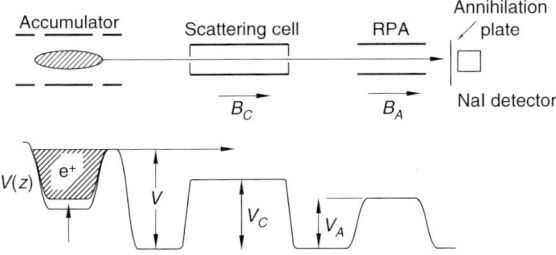

FIGURE 5 Schematic diagram of the electrode structure (above) and the electric potentials (below) used to study scattering with the trap-based positron beam at UCSD. Reprinted with permission from Marler et al. (2005) © 2005, by The American Physical Society

thus parallelizing the scattered beam, according to $(B_{RPA}/B_C) \propto (\sin^2\theta_{RPA}/\sin^2\theta_c)$ where θ_{RPA}, θ_C are the corresponding pitch angles, and enabling an analysis of the total energy loss.

2.2 Total Ionization Cross-Sections

Typically, the absolute scale of $Q_i^t(e^+)$ determined by counting ions (Fromme et al., 1986; Laricchia et al., 2002; Murtagh et al., 2005) is set by normalization to $Q_i^t(e^-)$ at energies (~600–1000 eV) where positronium formation is negligible, Q_i^+ dominates and $Q_i^t(e^+)$ exhibits (to within a few per cent) the same energy dependence as $Q_i^t(e^-)$ (Van Reeth et al., 2002). This procedure assumes independence of the cross-section upon the sign of the projectile charge (q), in accordance with the First Born Approximation (FBA) which predicts $Q_i^+ \propto q^2$. The results of these studies are displayed in Figure 6 together with those obtained by Marler et al. (2005) who summed their values for $\sum_{z=1}^{\infty} Q_i^{z+}$ and $\sum_{z=1}^{\infty} Q_{Ps}^{z+}$. Bluhme et al. (1999a,b) employed a gas mixture method, using the single-ionization cross-section for e^+–helium as the standard, in order to determine the absolute scale of $Q_i^{+,t}$ and $Q_i^{2+,t}$.

For helium, a fair agreement may be discerned at low and high energies between the UCL data (Murtagh et al., 2005) and those of Fromme et al. (1986) with discrepancies of ~15 per cent around and above the peak. In the figure, the UCL data are compared also with the close-coupling results of Wu et al. (2004) and those of Campbell et al. (1998), the former not resolving positronium formation from single ionization. For neon, the divergence between the UCL data (Laricchia et al., 2002) and those from UCSD (Marler et al., 2005) augments as the peak is approached whilst the agreement for argon, krypton and xenon is very good. The results for argon are overlapped by the total single-ionization cross-sections $Q_i^{+,t}(e^+)$ of Bluhme et al. (1999b) whilst in the case of krypton and xenon considerable discrepancies arise. Included in Figure 6 are the

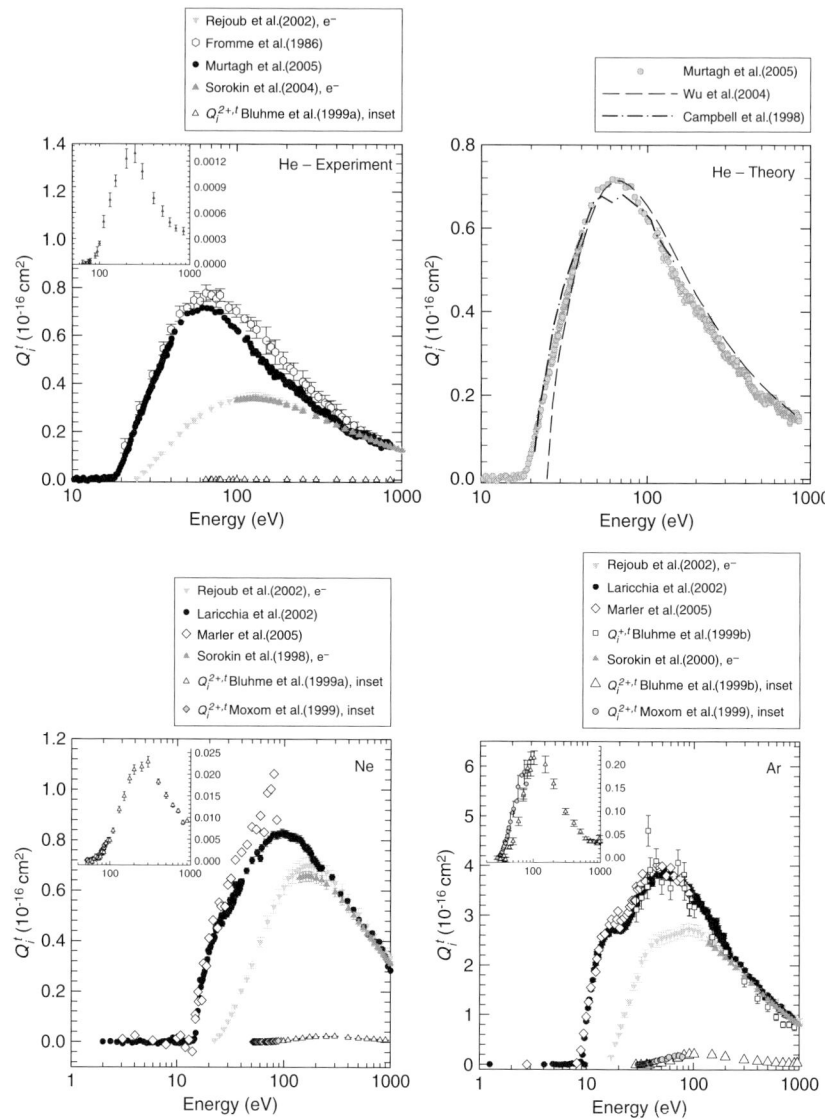

FIGURE 6 Total ionization cross-sections $Q_i^t(e^+)$ compared with corresponding values for electrons. Insets highlight available $Q_i^{z+,t}$

total multiple ionization data of Moxom et al. (1999), Moxom (2000) for $z = 2, 3, 4$ and the total double-ionization cross-sections $Q_i^{2+,t}$ of Bluhme et al. (1999a,b) which for Xe amount up to 8 per cent of $Q_i^{+,t}(e^+)$ in the range 100–250 eV.

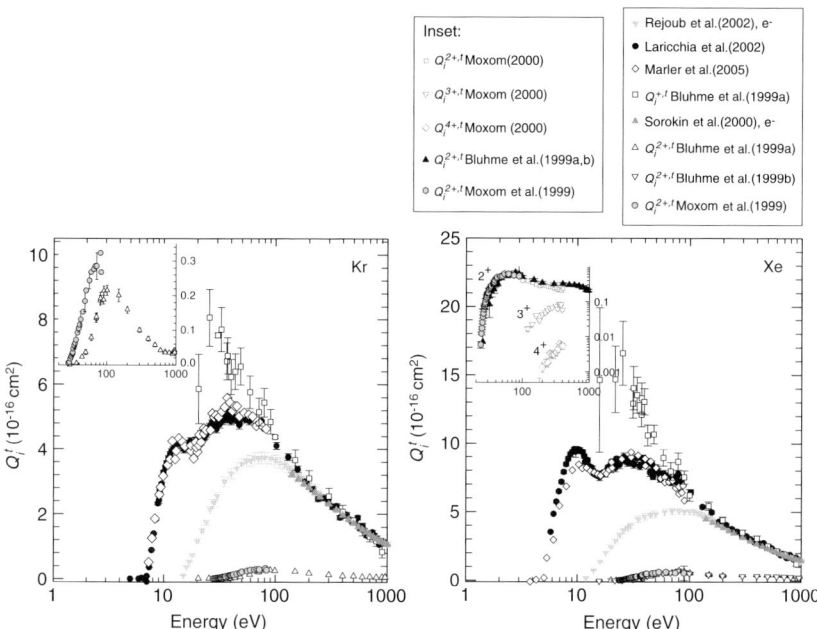

FIGURE 6 (*continued*)

2.3 Direct Ionization

2.3.1 Integral Cross-Sections

The cross-sections for direct ionization are shown in Figure 7. For helium, the UCL data for $Q_i^+(e^+)$ (Moxom et al., 1996; Ashley et al., 1996) are presented as renormalized by Murtagh et al. (2005) who used for this purpose the highly-concordant electron data of Sorokin et al. (2004) and Rejoub et al. (2002). In general, good agreement exists with the results of other experiments as well as of the coupled-pseudostate method of Campbell et al. (1998) and of the distorted-Coulomb and plane-wave approach of Campeanu et al. (1996).

Near threshold, the most precise experimental data are those of Ashley et al. (1996), shown in the inset. In the energy range (1–3) eV above threshold, these data are proportional to $E_{\text{ex}}^{1.99\pm0.19}$ (with $E_{\text{ex}} = E_+ - E_i$, the excess energy above threshold), the exponent being greater than for electrons – namely 1.127, as predicted by Wannier (1953) – but smaller than envisaged by extensions of this theory to positrons ($E_{\text{ex}}^{2.65}$) (Klar, 1981; Rost and Heller, 1994). By employing a hidden-crossing method, Ihra et al. (1997) found good agreement with experiment by including anharmonic terms in the three-particle potential, thus modifying the

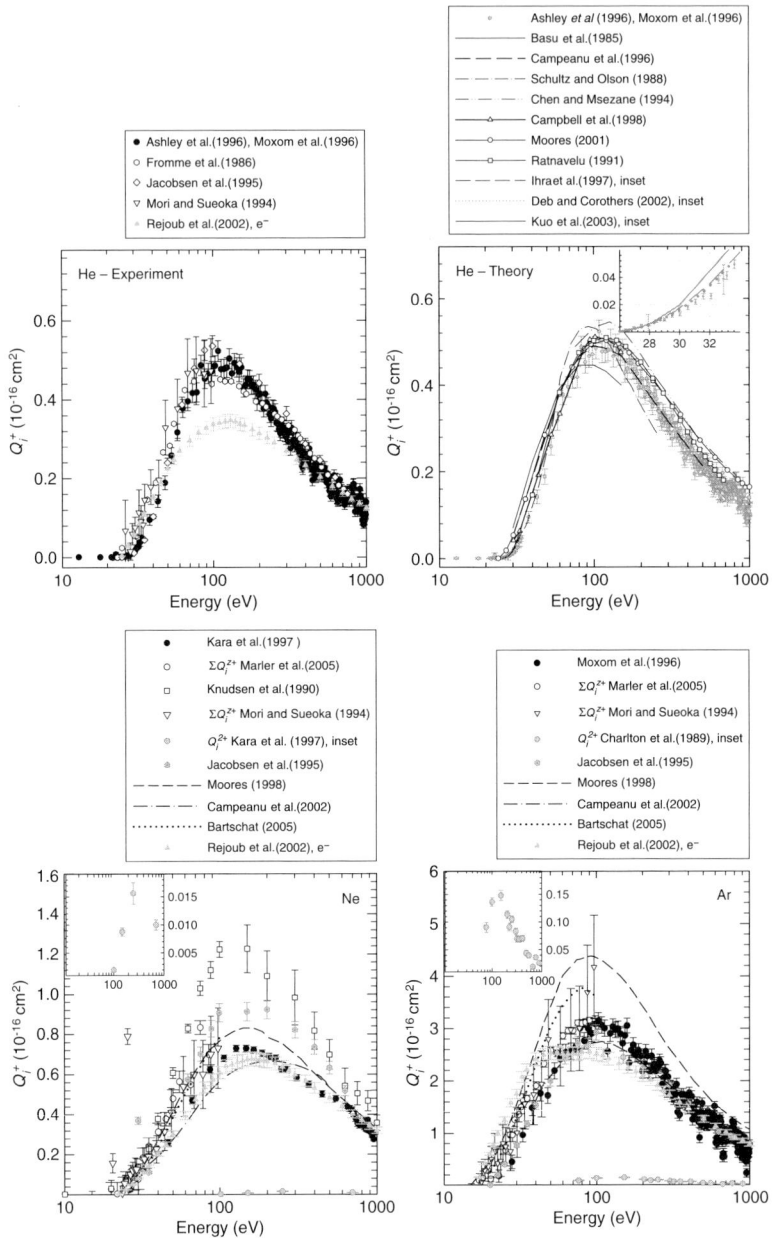

FIGURE 7 Direct ionization cross-sections of the inert atoms for single-(Q_i^+) and double-(Q_i^{2+}) ionization as well as the sum over all charge states $\sum_{z=1}^{\infty} Q_i^{z+}$. The inset for He zooms on the first few eV above $E_i = 24.58$ eV

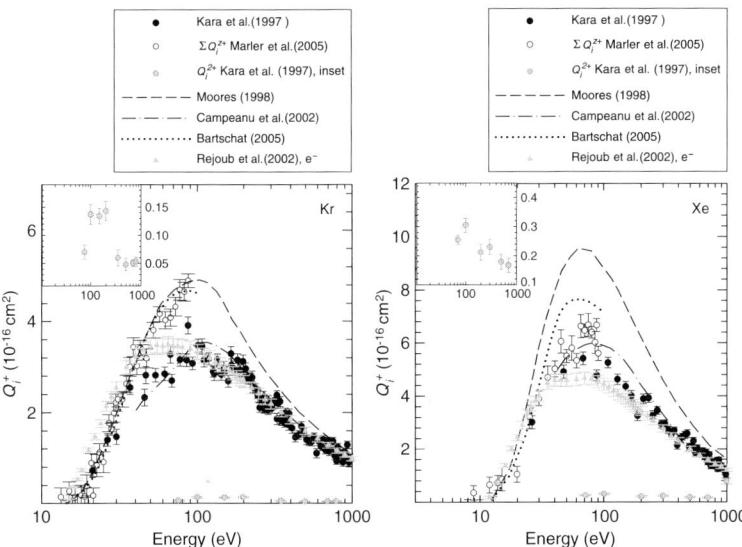

FIGURE 7 (*continued*)

Wannier law for positrons according to $Q_i^+(E_{ex}) \propto E_{ex}^{2.640} e^{-0.73\sqrt{E_{ex}}}$. This energy dependence has recently been confirmed by Kadyrov et al. (2007) for the e$^+$–H system. The absolute quantal-semi-classical calculation of Deb and Crothers (2002), which comprises values of up to 11 for the total angular momentum azimuthal quantum number, agrees quantitatively with the experimental data for helium of Ashley et al. (1996). Kuo et al. (2003), who employ models with various asymptotic charges in a two-potential distorted-wave approximation, also find a reasonable agreement with experiment from ~3 eV above threshold. Experimentally, the energy region within 1 eV above the threshold remains essentially unexplored.

Concerning neon, argon, krypton and xenon, the $Q_i^+(e^+)$ data of Kara et al. (1997) and Moxom et al. (1996) are plotted in Figure 7 as normalized by Van Reeth et al. (2002) to the accurate electron-impact ionization results of Sorokin et al. (1998, 2000). For both neon and argon, there is good agreement in shape from threshold to ~1 keV between the experimental $Q_i^+(e^+)$ from UCL and those of Jacobsen et al. (1995); the difference in magnitude for neon stems mainly from corresponding discrepancies in the electron data used for the normalization (Van Reeth et al., 2002). The values for $\sum_{z=1}^{\infty} Q_i^{z+}$ of Mori and Sueoka (1994), whose normalization method uses instead positron total scattering cross-sections, are also consistent. The more precise absolute $\sum_{z=1}^{\infty} Q_i^{z+}$ values from UCSD (Marler et al., 2005) are significantly higher than the $Q_i^+(e^+)$ results

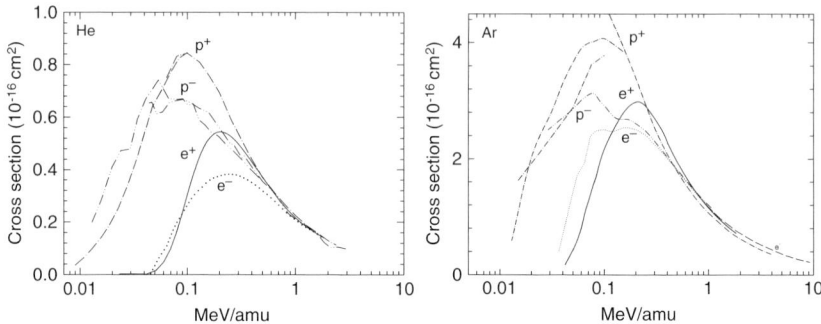

FIGURE 8 Comparison of the direct single-ionization cross-sections Q_i^+ for helium and argon by equivelocity e^+, e^-, p^+, p^- impact

from UCL, particularly for krypton and xenon. For the latter targets, however, the differences are much greater than the contributions from multiple ionization measured by Kara et al. (1997), Charlton et al. (1989) and illustrated in the insets. Also included in the figure are the $Q_i^+(e^+)$ results of the distorted-wave calculations of Moores (1998), Campeanu et al. (2002) and Bartschat (2005), the latter describing the initial bound state and the ejected-electron–residual-ion interaction by an R-matrix (close-coupling) expansion. The cross-sections of Moores (2001) reach a maximum at an energy close to the experimental $Q_i^+(e^+)$ but are generally larger; in neon and krypton good correspondence is observed with the theory of Bartschat (2005). The magnitudes of the cross-sections of Campeanu et al. (2002) are in fair agreement with the experimental data across the common energy range, and especially at high energies.

Comparative studies of atomic and molecular ionization by the impact of particle/antiparticle pairs have been carried out in order to differentiate effects due to the charge and mass of the projectile (e.g. Schultz et al. (1991), Knudsen and Reading (1992) and Paludan et al. (1997)). Examples are given in Figure 8 where the single direct ionization cross-sections for equivelocity $e^\pm$, $p^\pm$ scattering from helium and argon are seen to merge at high velocities whilst, at lower values, kinetic energy effects result in the heavier pair having the larger cross-sections, i.e. $Q_i^+(p^+) > Q_i^+(e^+)$ and $Q_i^+(p^-) > Q_i^+(e^-)$. Generally, the cross-sections for the positively-charged projectiles are smaller near the threshold than for their antimatter counterparts due to competition from electron capture and binding/antibinding effects experienced by the target electrons in close collisions with a positive/negative projectile. At intermediate velocities, positrons and protons have higher cross-sections due to polarization drawing target electrons closer to the incoming positively-charged projectiles. As may be noted in Figure 8, however, for the light projectiles this effect appears to diminish with Z, possibly because of the larger (smaller) impact parameters and deceleration (acceleration) of a positron (electron) in the growing Coulomb field of the nucleus.

Whilst the single-ionization cross-sections for equivelocity $e^\pm$, $p^\pm$ are the same at high energies, in double-ionization differences arise which have been linked to the dependence upon the sign of charge of the quantal interference between possible ionization mechanisms, e.g. between the sequential interaction of the projectile with two target electrons and that yielding the ejection of a second electron upon ionic rearrangement which results in the cross-section comprising a term proportional to q^3 (McGuire, 1982; McGuire and Deb, 1987; Knudsen and Reading, 1992).

In Figure 9, the ratios of the experimental cross-sections $R_2 = Q_i^{2+}/Q_i^+$ for electron (Charlton et al., 1988; Krishnakumar and Srivastava, 1988) and positron impact (Charlton et al., 1988, 1989; Kruse et al., 1991; Kara et al., 1997) are plotted together with $R_2^t = Q_i^{2+,t}/Q_i^{+,t}$ (Helms et al., 1994a,b, 1995; Bluhme et al., 1999a,b; Moxom et al., 1999; Moxom, 2000), $R_3 = Q_i^{3+}/Q_i^+$ (Kara et al., 1998; Kruse et al., 1991; Helms et al., 1994a,b), $R_3^t = Q_i^{3+,t}/Q_i^{+,t}$ (Helms et al., 1995; Moxom et al., 1999) and $R_4^t = Q_i^{4+,t}/Q_i^{+,t}$ (Moxom, 2000). The near-threshold behavior of double-to-single-total ionization cross-sections has been investigated in detail by Bluhme et al. (1998) and Moxom et al. (1999).

At low energies the data of Kara et al. (1997) are smaller than most, arguably because of efficient discrimination against ions due to Ps formation whilst the convergence of R_2 and R_2^t at high energies in argon, krypton and xenon infers negligible Ps formation in double ionization above a few hundreds eV. Also shown in the figure are the contributions from inner-shell ionization of the heavier targets estimated by Helms et al. (1995) who adapted the semi-empirical formula of Lotz (1967, 1968) to positrons. The results, which reproduce the high-energy dependence of the experimental data, suggest a strong influence of M- and N-shell ionization followed by Auger decay in krypton and xenon, respectively.

The predictions obtained for positrons from the corresponding ratios for the other projectiles, namely $R_2(e^+) = R_2(e^-)R_2(p^+)/R_2(p^-)$, are seen to be in remarkably good agreement for all five targets. This infers that trajectory effects cancel in the ratio and are thus the same in single and double ionization.

2.3.2 Differential Ionization Cross-Sections

Studies in which the angular and/or energy distributions of the final-state particles are determined offer greater insights into the dynamics of the collision and more stringent tests of the physical accuracy of theoretical descriptions. Experimental values of differential cross-sections for positron-impact ionization are relatively scarce but they

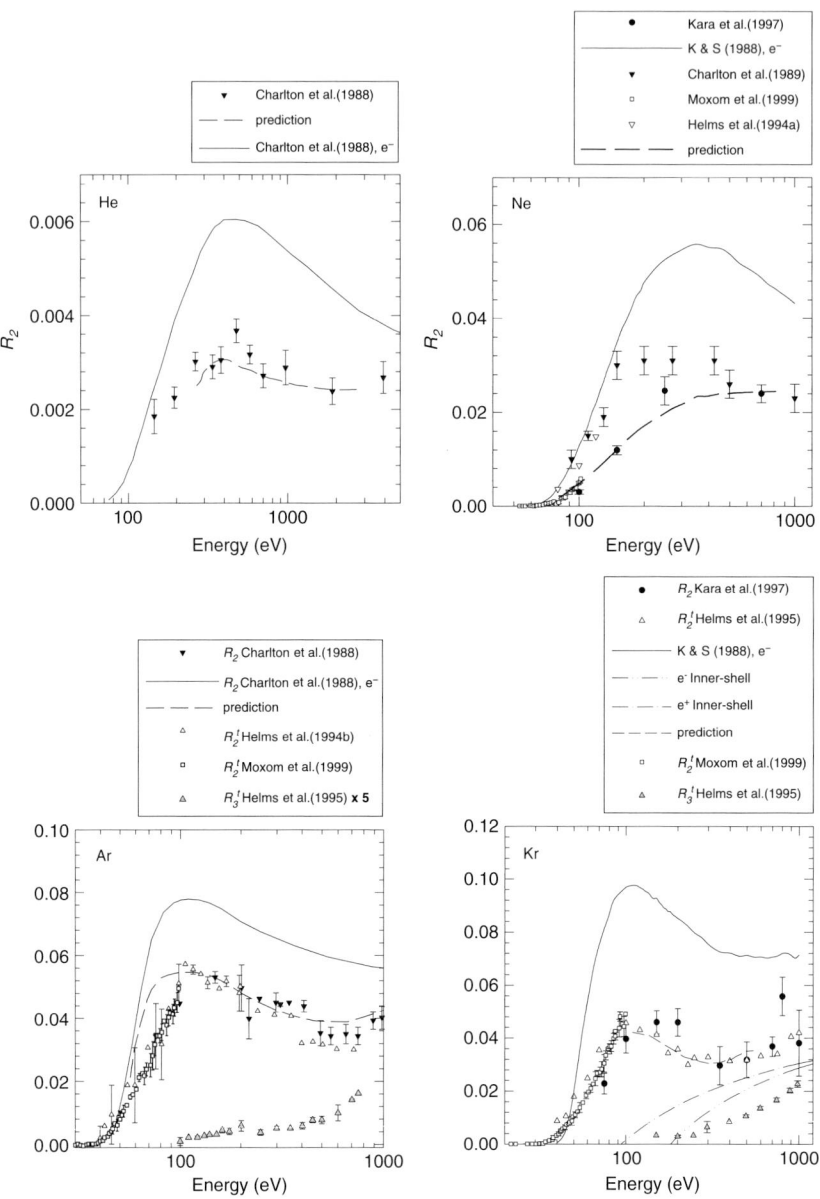

FIGURE 9 Ratios of double-to-single-ionization cross-section for He, Ne, Ar, Kr, Xe, triple-to-single-ionization cross-section ratios for Ar, Kr and Xe, and quadruple-to-single-ionization cross-section for Xe. Filled symbols refer to direct ionization measurement (R_z), hollow symbols to total ionization (R_z^t). The 'prediction' for e^+ is extracted from corresponding ratios for e^-, p^+, p^- (see text). 'K&S' refers to Krishnakumar and Srivastava. Inner-shell contributions estimated by Helms et al. (1995)

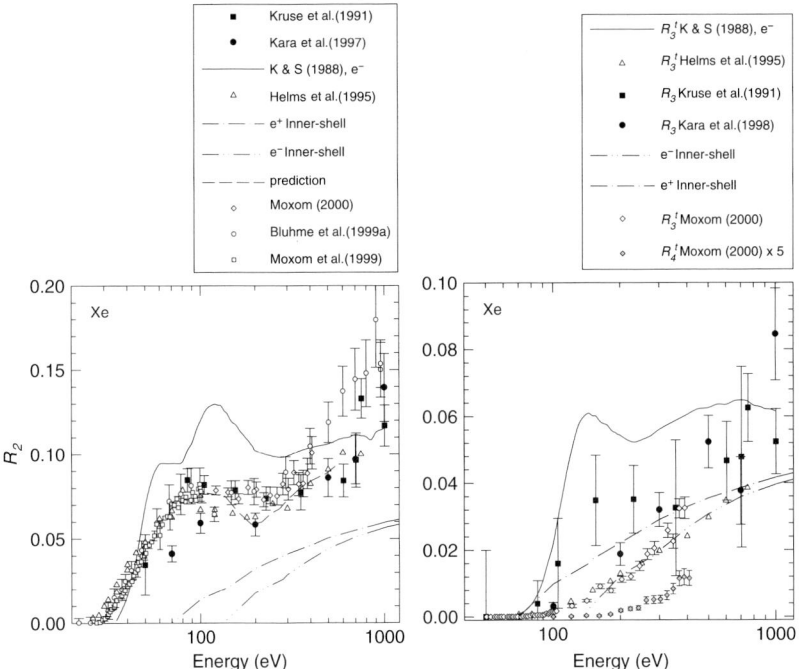

FIGURE 9 (*continued*)

comprise singly-, doubly- and even triply-differential measurements, although over restricted ranges of energies, angles and targets. Among the determinations are the singly-differential ionization cross-section ($dQ_i^{z+}/d\Omega$ with $z = 1, 2$) by Falke et al. (1997) for argon and krypton at 75, 90 and 120 eV incident positron energies. Doubly-differential cross-sections ($d^2Q_i^+/dEd\Omega$) have also been measured close to 0° for $e^\pm$ impact on argon at 100, 150 and 250 eV by Kövér et al. (1993). These authors, in accordance with the investigation of Moxom et al. (1992), located no structure arising from electron capture to a low-lying continuum state of positronium (ECC). This phenomenon, whose counterpart is well documented in positive-ion–atom collision (e.g. Rødbro and Andersen (1979), Schultz et al. (1991)), corresponds to events in which scattered positron and ejected electron emerge from the collision unbound but with similar velocities. The attractive Coulomb correlations between them produces a cusp-like peak in the energy spectra of the ejected electron (Brauner and Briggs, 1986; Brauner et al., 1989) whilst an 'anticusp' (or dip) results in the case of negatively-charged projectiles (see e.g. Guang-yan et al. (1993) and Golden et al. (1996)). For heavy projectiles, these

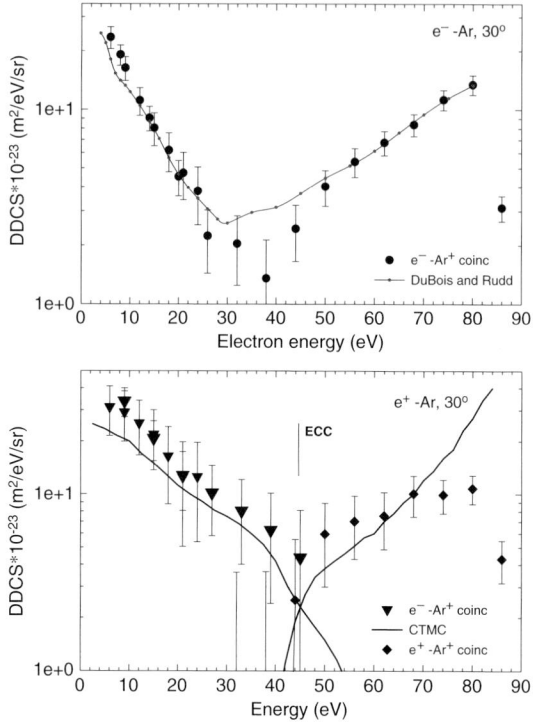

FIGURE 10 Comparison of e^+ and e^- scattering from argon at 100 eV incident energy (Kövér et al., 1994; DuBois and Rudd, 1978). The solid lines correspond to a CTMC calculation (Sparrow and Olson, 1994)

features are prominent in the doubly-differential cross-sections measured at small angles around the incident beam direction.

Measurements of $d^2 Q_i^+/dEd\Omega$ for e^+ scattering from argon at angles other than zero have been reported at 100 eV (Kövér et al., 1994, 1997; Schmitt et al., 1994) and at 60 eV (Kövér et al., 1997). The results of Kövér et al. (1994) at a scattering angle of 30° are reproduced in Figure 10. Unlike the case of electron impact where exchange prevents a clear-cut distinction, for positron impact the scattered projectile and the ejected electron are unambiguously identifiable: the electron lying mainly in the lower half of the energy range and the scattered positron in the higher. The CTMC calculation of Sparrow and Olson (1994) agrees well for the ejected electron spectra but somewhat less for positrons. Energy-loss measurements for single, double and triple ionization of argon (DuBois et al., 2001) have been carried out at 750 eV incident energy for positrons scattered into a forward cone ($\sim\pm 17°$). Within this angular range, the ratio of double-to-single ionization has been found to rise rapidly and approach the ratios of the total cross-sections for photoionization of the M-shell.

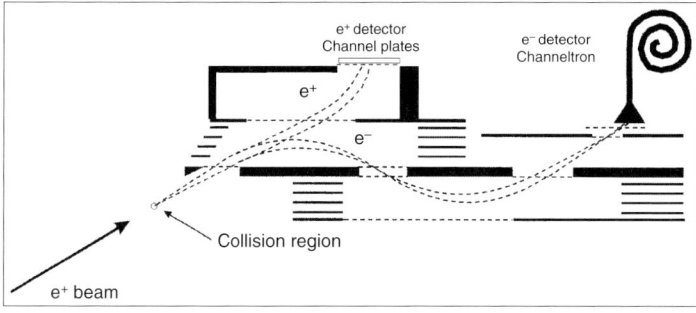

FIGURE 11 Schematic diagram of the experimental setup of Kövér and Laricchia (1998) for measuring $d^3 Q_i^+/dE_- d\Omega_- d\Omega_+$

Santos et al. (2004) have extended these measurements to krypton, finding that single and multiple ionization of the outermost N-shell is dominant.

The observation of ECC by positrons was eventually achieved at the triply-differential level $d^3 Q_i^+/dE_- d\Omega_- d\Omega_+$, where the subscripts $-/+$ refer to electrons and positrons, respectively (Kövér and Laricchia, 1998). Using the parallel-plate analyzer shown in Figure 11, the scattered positron and the ejected electron were both detected in coincidence around 0°, and the electron was also energy analyzed. Its spectrum, displayed in Figure 12, contains a small broad peak at half of the residual energy, consistently with the expectations of ECC and well described theoretically by Berakdar (1998) who employed a three-body Coulomb wavefunction (or BBK, after Brauner et al. (1989)) to describe the final state of the system. The calculation of Fiol et al. (2001) is also based on a BBK approach, their treatment enabling them to deal consistently with either heavy or light projectiles. It is thought that the failure to observe the phenomenon in $d^2 Q_i^+/dE d\Omega$ of e^+ scattering from argon (a relatively high-Z target) is due to the influence of the static interaction which, in contrast to heavier ion impact, diffuses the effect over a wide angular spread of the scattered projectile and, correspondingly, of the ejected electron (Sparrow and Olson, 1994).

Studies were extended to 50 eV by Kövér et al. (2001), as well as to helium and to the determination of the scattered projectile spectrum (Arcidiacono et al., 2005). As depicted in Figure 13, these authors registered a shift in the electron spectrum to lower (and the scattered positron to correspondingly higher) energies than expected from the calculation of Fiol et al. (2001). Fiol and Olson (2002) examined the problem further and observed discrepancies between the results obtained using the perturbative BBK approach and the Classical Trajectory Monte Carlo (CTMC) method, with the former giving better agreement at 100 eV

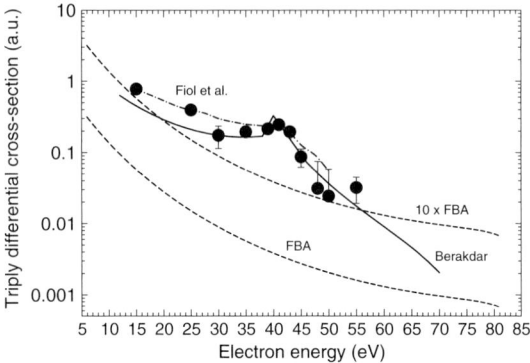

FIGURE 12 The triply-differential electron spectrum in $e^+(100\text{ eV}) + H_2 \rightarrow e^+(0°) + e^-(0°, E_-)$. circles, experiment (Kövér and Laricchia, 1998). curves, theories: solid, Berakdar (1998); dash dotted, Fiol et al. (2001). The curve labeled "FBA" corresponds to the first Born approximation result – which neglects electron capture – of Berakdar (1998). These are also shown multiplied by 10 for a comparison with the energy dependence of the different determinations

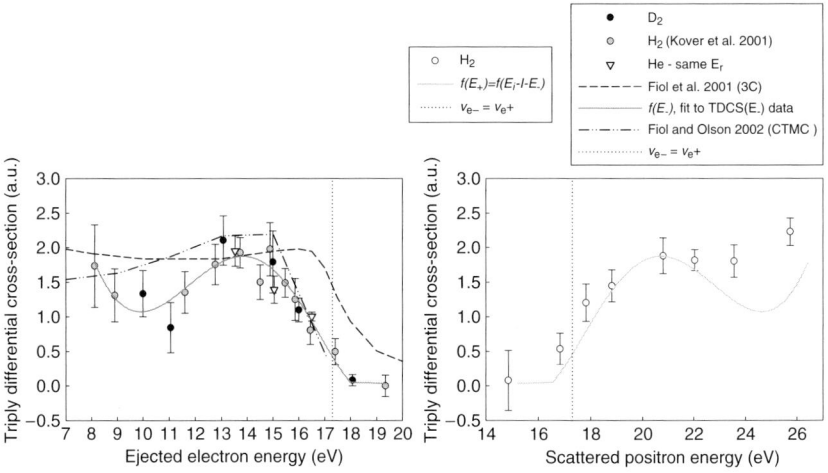

FIGURE 13 Experimental and theoretical results for the triply-differential ionization cross-sections for ejected electrons (lhs) and scattered positrons (rhs) in 50 eV positron collision with molecular hydrogen. Data for ejected electrons from helium at the same residual energy are included as triangles. The gray lines are guides to the eye only (Arcidiacono et al., 2005)

and the latter at 50 eV. At this lower energy, a strong correlation between the momenta of the positron and the recoil ion was also noticed.

The physical origin of the observed experimental asymmetry is not yet understood but it has been conjectured that it might be due to interference with formation of Ps in bound states, a significant channel at 50 eV but much less so at 100 eV. This hypothesis has gained fresh support from Benedek and Campeanu (2007) who have applied a molecular BBK model to represent more realistically the H_2 target. Clearly, further work is needed in order to understand the observations.

A recent development has been the determination of triply-differential cross-sections for the single ionization of argon by 500 eV positron impact over a large range of electron-emission angles (de Lucio et al., 2006). A schematic diagram of their apparatus is shown in Figure 14. Forward scattered positrons are energy analyzed and imaged onto a position-sensitive detector (PSD). Ejected electrons are detected by a second PSD. Target ions are extracted by means of a weak uniform electric field. An example of the detailed information obtainable from this experiment is illustrated by the polar plots which allow binary and recoil interactions to be distinguished. These types of studies, which have provided in the case of electron impact (e.g. Lahmam-Bennani (1991) and Ullrich et al. (2003)) important insights into the physics of many-body correlated systems, are in their infancy for positrons and hold much promise for the future.

2.4 Positronium Formation

2.4.1 Integral Cross-Sections

As discussed in Section 2.1, one method for determining Q_{Ps} requires that all scattered positrons are collected and ascribes the loss of positrons with respect to the incident flux to Ps formation (e.g. Fornari et al. (1983) and Overton et al. (1993); upper limits (UL) of Stein et al. (1998) and Marler et al. (2005)). Additional techniques include:

(a) 3γ-ray coincidences (Charlton et al., 1983): a clear signature of Ps formation although an energy-dependent loss of signal is difficult to avoid, e.g. due to ortho-Ps escaping the detection region or quenching on the cell walls (Charlton and Laricchia, 1990).
(b) 2γ-ray coincidences (e.g. Stein et al. (1998)): this has been interpreted as setting a lower limit (LL) on Q_{Ps} corresponding to para-Ps plus any quenched ortho-Ps.
(c) measurement of all the ions and of the (e^+–ion) coincidences. Ions not accompanied by a positron in the final state are attributed to Ps formation (e.g. Fromme et al. (1986), Laricchia et al. (2002) and Murtagh et al. (2005)).

All these measurements assume negligible annihilation (except (a)) and comprise transfer ionization (reaction 6) which, by inspection of Figures 6 and 15, may be surmised to amount at most to a few per cent of Q_{Ps}. However, in extracting Q_{Ps} according to Equation (4),

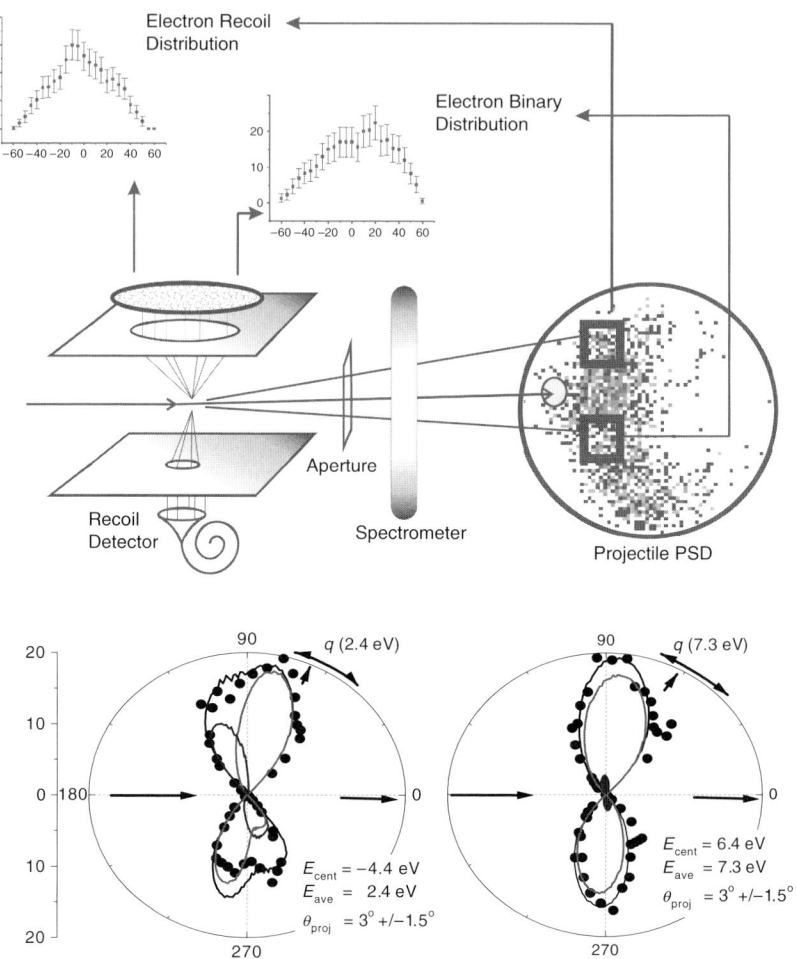

FIGURE 14 Top: schematic view of experimental apparatus for measuring electron emission and positron scattering and energy loss for single ionization. Bottom: triply-differential electron emission yields for single ionization of argon by 500 eV positrons. experimental data: 'horizontal' arrows represent the initial and scattered positron directions; radial and curved arrows denote momentum transfer direction and range. the solid and dashed lines and the filled areas are simulated curves for the binary and recoil emission convoluted over experimental conditions. Reprinted with permission from de Lucio et al. (2006) © 2006, by The American Physical Society

Laricchia et al. (2002) subtracted from their measured Q_i^t, as well as Q_i^+ (Kara et al., 1997; Moxom et al., 1995), the total double-ionization cross-sections $Q_i^{2+,t}$ of Bluhme et al. (1999b) which for Xe amount up to 8 per cent of $Q_i^{+,t}(e^+)$.

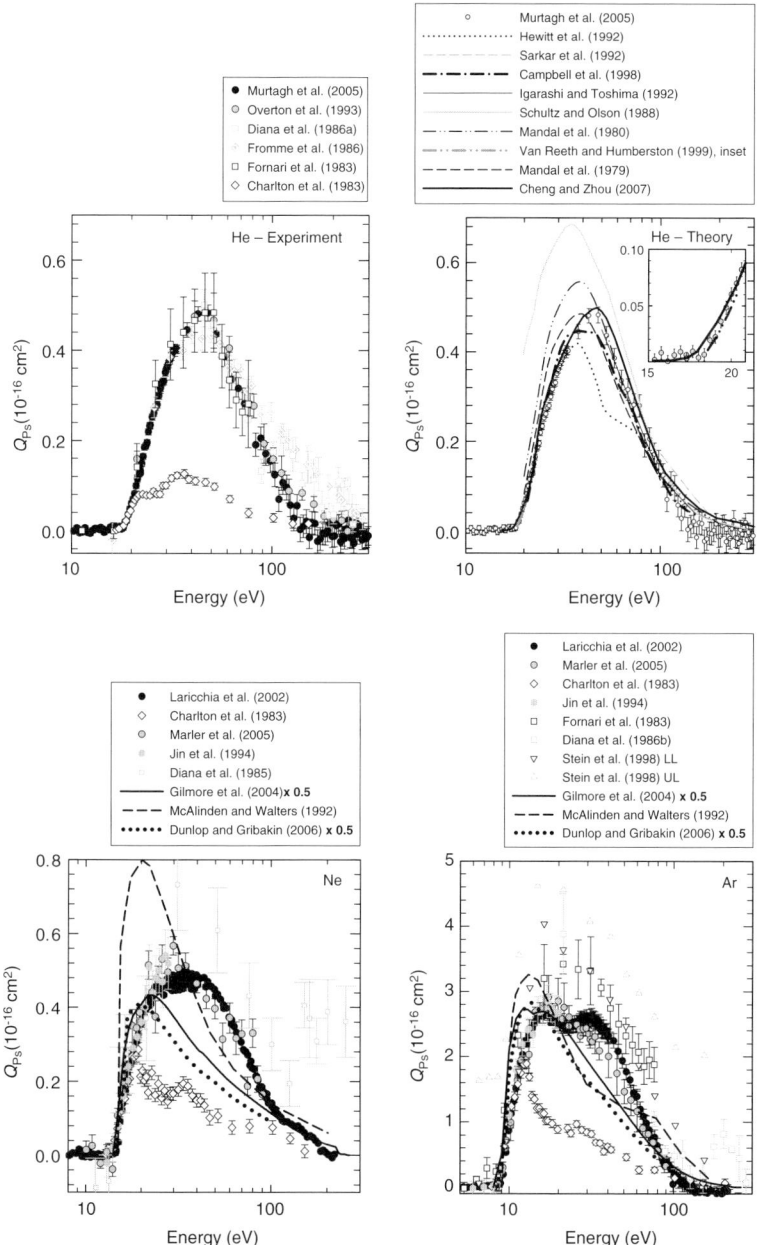

FIGURE 15 Positronium formation cross-sections. For He, available theories are compared with the data of Murtagh et al. (2005), the inset zooming on the first few eV above $E_{Ps} = 17.78$ eV

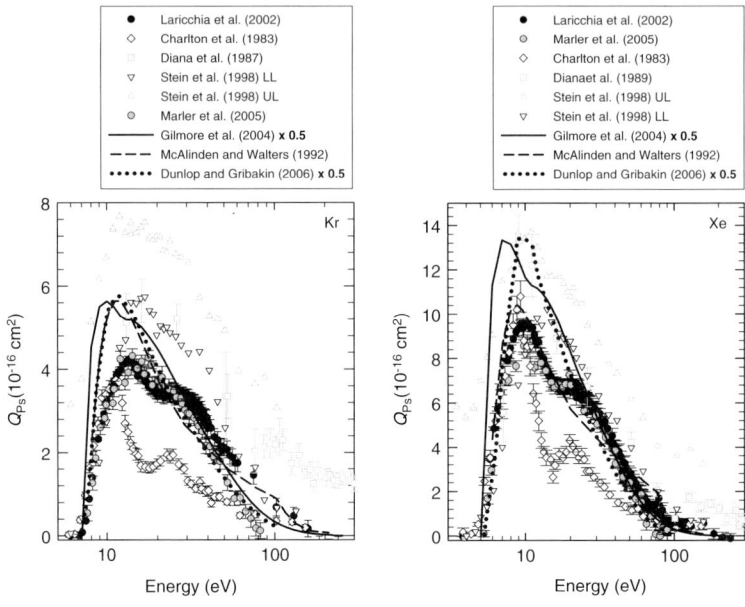

FIGURE 15 (continued)

In Figure 15, the results of various determinations of Q_{Ps} are illustrated. In helium, there is a very good accord, except at the highest energies, between the most recent measurements (Overton et al., 1993; Murtagh et al., 2005). Whilst the agreement between these and the elaborate pseudostate calculation of Campbell et al. (1998) is good, the position of the peak is better reproduced by the recent momentum-space coupled-channel optical method of Cheng and Zhou (2007). In the inset, Q_{Ps} of Murtagh et al. (2005) is compared with the results of Van Reeth and Humberston (1999), who state a $\sim$10 per cent uncertainty arising from the convergence of their variational calculations, and those of Cheng and Zhou (2007) whose magnitude is close to the experimental values.

For neon and argon, there is good agreement between the results of Jin et al. (1994) and those of Laricchia et al. (2002) over the common energy range. For these targets and krypton, agreement also exists over a wider energy range between the studies of Laricchia et al. (2002) and Marler et al. (2005) except on the size of the structures observed above the first maximum. In contrast, the accord in xenon is poor around the first maximum but it improves above.

Also depicted in the figure are the results from: the truncated coupled-static calculation of McAlinden and Walters (1992) for Ps(1s) formation, including capture from the 3s-shell in the case of argon; the distorted-

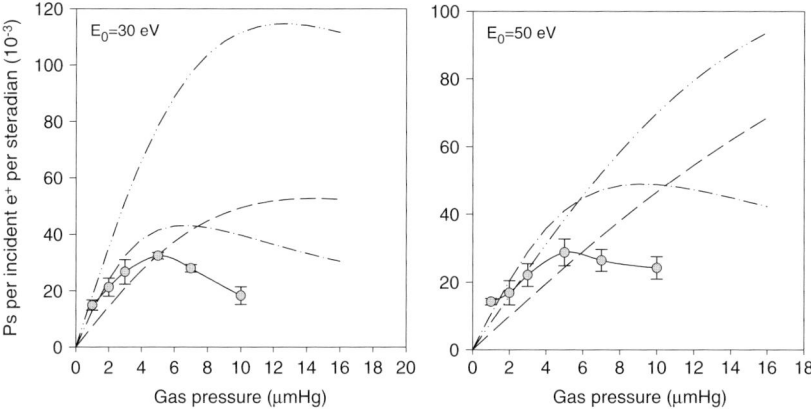

FIGURE 16 Variation of the beam production efficiency with gas pressure at the Ps energies shown on the plots: Xe (circles); H_2 (dash-double-dotted line); helium (dashed line) and argon (dash dotted) (Laricchia et al., 2004)

wave Born approximation calculations of Gilmore et al. (2004) which allow for excited-state Ps formation and for capture from the first inner ns-shells, the latter process also incorporated in the many-body calculation of Dunlop and Gribakin (2006). The treatment of McAlinden and Walters (1992) yields the best agreement in magnitude with the experimental cross-sections, while their energy dependences are best described by the results of Gilmore et al. (2004), rescaled in the figure by 0.5 for ease of comparison. The same factor is applied to the values of Dunlop and Gribakin (2006) which are in good agreement with experiment at high energies but overestimate the cross-sections by a factor of ~2 around the peak. The precise origin of the observed structures at intermediate energies remains unclear but they appear consistent with higher thresholds for Ps formation, e.g. capture of inner ns-electrons (Stein et al., 1998; Dunlop and Gribakin, 2006) or into an excited state (Laricchia et al., 2002). The latter authors employed the empirical formula of Szluinska et al. (2002) to deduce a surprisingly large contribution from Ps ($n > 1$) formation. Clearly, explicit experimental investigation of both these processes would be helpful.

2.4.2 Differential Positronium Formation Cross-Sections

The theoretical prediction of a forward-peaked differential Ps formation cross-section ($dQ_{Ps}/d\Omega$) from few eV above threshold (e.g. Mandal et al. (1979); Brown and Humberston (1985)) encouraged experimental investigations aimed at the production of positronium beams by neutralizing a positron beam by charge exchange in a gas (see Section 3 below). The Ps beam production efficiency ε_{Ps} is defined as the number of Ps atoms (N_{Ps}) produced per incident positron (N_+) per steradian according to

$$\varepsilon_{Ps} = \frac{N_{Ps}}{\Omega N_+} D \qquad (9)$$

where Ω corresponds to the detection solid angle and D corrects for in-flight annihilation of Ps. In terms of $dQ_{Ps}/d\Omega$, ε_{Ps} may be expressed as

$$\varepsilon_{Ps} \propto \{1 - \exp(-\rho \ell Q_{t+})\} \left\{ \frac{2\pi}{Q_{t+}} \int_0^{\theta'} \frac{dQ_{Ps}}{d\Omega} \sin\theta d\theta \right\} \exp(-\rho \ell_0 Q_{t0}) \qquad (10)$$

where the first term corresponds to the fraction of scattered positrons, the second to the probability of forming Ps within a small forward cone of half-angle θ' and the third to the transmission probability of Ps through the cell (ρ is the gas density, ℓ the effective cell length, Q_t the total cross-sections, the subscripts + and 0 referring to the positron and Ps, respectively). Examples of the target and pressure dependence of ε_{Ps} at two incident positronium energies for a variety of targets are given in Figure 16, where the rise and fall with pressure is due to the competing effects between the first and last terms in Equation (10). Both at 30 and 50 eV Ps energy, H_2 may be seen to yield the highest efficiency whilst xenon appears somewhat inefficient, possibly due to the considerably larger static repulsion resulting in broader $dQ_{Ps}/d\Omega$.

If the third term in Equation (10) is negligible, ε_{Ps} is directly proportional to $dQ_{Ps}/d\Omega$. In Figure 17, the energy dependence of the low density values of ε_{Ps}, normalized for pressure and energy dependence of the positron-to-positronium detection efficiency ratio $\varepsilon_+/\varepsilon_0$ (Armitage, 2002), are compared with theoretical results at $\theta = 0^0$ (Laricchia et al., 2004). These comprise the truncated-coupled-static interaction calculation of McAlinden and Walters (1994), the distorted-wave approximation of Mandal et al. (1979), the close-coupling approach of Chaudhuri and Adhikari (1998) and the FBA of Biswas et al. (1991). The agreement with experiment is in general poor although, in the case of H_2, it is improved upon retention of the Ps scattering term in Equation (10) (Armitage et al., 2005). Evidently, further work is necessary both experimentally and theoretically.

At non-zero angles, relative values of $dQ_{Ps}/d\Omega$ with positron incident energy have been measured by Finch et al. (1996) in argon at 60° and observed to have a different dependence from that predicted by McAlinden and Walters (1994). Falke et al. (1995, 1997) have also investigated $dQ_{Ps}/d\Omega$ over the angular range 0°–120° for e^+ scattering from argon and krypton at incident energies 30–120 eV. As displayed

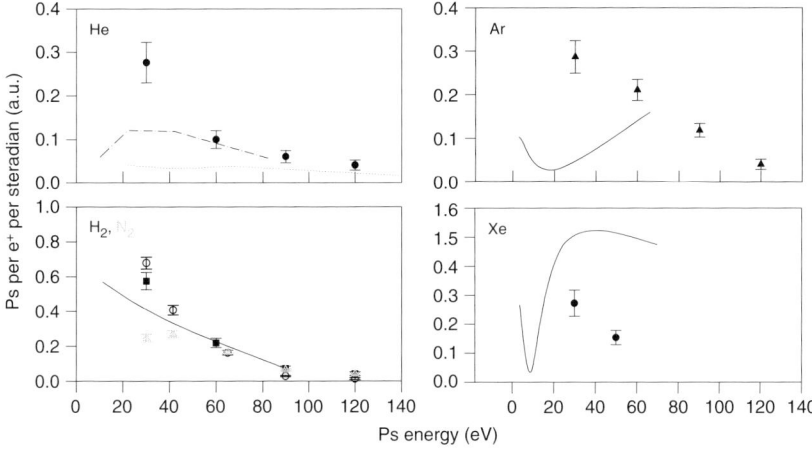

FIGURE 17 Comparison of the energy dependence of ε_{Ps} per unit pressure with those of theoretical differential Ps formation cross-sections at $0°$. In all cases, the absolute magnitude of the theoretical data (in a_0^2) may be regained by dividing by 0.019. He: experimental data (Garner et al., 1996); dot-dashed line (Mandal et al., 1979); dotted line (Chaudhuri and Adhikari, 1998); Ar: experimental data (Garner et al., 1996); solid line (McAlinden and Walters, 1994), H_2: experimental data (black squares) (Garner et al., 1996), hollow circles (Leslie et al., 2002); theory (Biswas et al., 1991), N_2: experimental data (hollow triangles) (Leslie et al., 2002), Xe: solid circles (Leslie et al., in preparation); theory (McAlinden and Walters, 1994)

in Figure 18, good agreement is found with the result of the calculation of McAlinden and Walters (1994) in argon at 75 eV but not at 90 eV. For krypton at 75 eV, the minimum in the theory around $85°$ is not confirmed by experiment. The physical origin of the structure apparent in the experimental data is not understood.

3. POSITRONIUM IONIZING COLLISIONS

3.1 Background

Although, as discussed in the preceding sections, Ps is readily formed in collisions of energetic positrons with atoms and molecules, it is not easy to perform experimentally-controlled studies of its interactions with matter. The difficulties arise because it is unstable against annihilation, it cannot be accelerated for efficient detection, and its weak beam intensities impose severe demands on permissible background levels as well as instrumental stability.

Fast positronium can be efficiently produced from solid targets by epithermal positrons capturing an electron upon re-emerging, or scattering at glancing angles, from a metal surface. However, the resulting

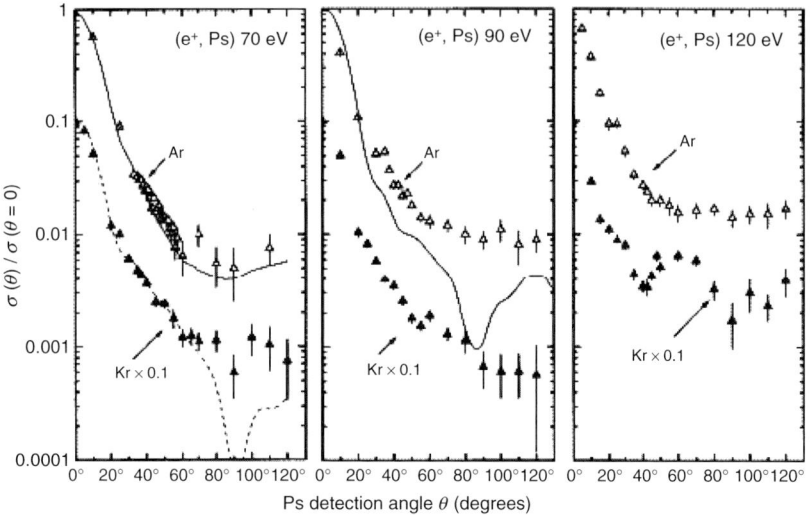

FIGURE 18 Angular dependence of Ps formation normalized to the 0° value: Ar (open symbols, full curve) and Kr (full symbols, broken curve). The curves represent the theoretical results of McAlinden and Walters (1994) folded with the angular resolution of the apparatus). Reprinted with permission from Falke et al. (1997) © 1997, by IOP Publishing Ltd

fluxes have wide distributions in energy and quantum states (e.g. Laricchia (1995)). Brown (1985) was the first to identify energy-tunable Ps produced by charge exchange of positrons in a gas. He observed the γ-rays produced from the in-flight annihilation of energetic para-Ps to be Doppler-shifted, consistently with changes in E_+. The short lifetime of para-Ps (see Table 1) make it unsuitable for beam production at typical atomic velocities; however, a useful beam of ortho-Ps may be generated (Laricchia et al., 1987). This remains the basic principle of production for the UCL beam (Zafar et al., 1996; Garner et al., 1996) which has been used to measure total cross-sections of Ps scattering from simple atoms (He, Ne, Ar and Xe) and molecules (H_2, N_2, O_2 and H_2O) (Zafar et al., 1996; Garner et al., 1996, 1998, 2000; Leslie et al., 2002; Beale et al., 2006) in the energy range $E_0 \sim$ 10–150 eV. More recently, ionizing collisions by Ps impacting on helium and xenon have been investigated (Armitage et al., 2002, 2006; Brawley et al., 2008). The lower limit of $E_0 \sim$ 10 eV for the Ps beam energy originates from the decline of Q_{Ps} and broadening of $dQ_{Ps}/d\Omega$ as $E_0 \to 0$ (see Section 2.4.2), as well as a reduced probability for Ps survival and detection for a given flight path. However, valuable information on Ps–matter interactions at lower energies may be gained using 'swarm' type techniques such as lifetime spectroscopy (e.g. Canter et al. (1975)), Doppler broadening of the energy spectrum of the annihilation quanta (e.g. Skalsey et al. (1998, 2003)) or by measuring the deviation from collinearity of two gamma rays (Angular Correlation of the Annihilation Radiation, ACAR) (e.g. Coleman et al. (1994), Nagashima et al. (1998), Saito et al. (2003) and Saito and Hyodo (2006)). Discussions on the convergence of the momentum transfer cross-sections may be found in Ghosh et al. (2001) and Walters et al. (2004).

Theoretically, Ps–atom scattering is taxing because of exchange being a two-center problem and because of the need to include in its description also excitation and ionization of both projectile and target, even as closed channels (e.g. Ghosh et al. (1998)). In fact, it has been established that the addition of virtual target excitations substantially reduces the elastic scattering cross-sections at low energies (Ghosh et al., 2001; Basu et al., 2001a,b; Blackwood et al., 2002a). The effect is believed to stem from a partial cancellation of short-range exchange repulsions and attractive long-range interactions, although the explicit 'switching off' of the van der Waals term in the Ps–H potential only produced a small modification (Blackwood et al., 2002a), unlike the virtual formation of H^- which plays a major role (Biswas, 2001; Blackwood et al., 2002b). Calculations have addressed primarily atomic hydrogen and helium targets, however theoretical results are also available for the alkali-metal and the heavier inert atoms (e.g. Walters et al. (2004), Blackwood et al. (2002c) and Chakraborty et al. (2004)). The latter authors have found the van der Waals and higher-order long-range interactions to play a crucial role in Ps scattering from the highly polarizable alkali-metal atoms.

The main point of contact between theory and experiment is the determination of the total cross-section for the Ps–He system depicted in Figure 19 where, for clarity, only some of the available theories have been included. The most extensive calculation is the frozen-core coupled-state approach of Blackwood et al. (1999) who incorporated up to 22 Ps states but only the ground state for the target, and used pseudostates to represent the Ps continuum. The results of an earlier target-elastic (TE) calculation by McAlinden et al. (1996) are also shown. In this work, a simpler coupled-positronium-pseudostate approximation was employed with no exchange but taking into account excitation and ionization of the target via the First Born approximation. The target-inelastic (TI) contributions were found to be negligible below 50 eV and to account for more than 50 per cent of the total cross-section above 100 eV. In comparison with the results of Sarkar et al. (1999), the coupled-eigenstate calculation of Basu et al. (2001b) reveals the effect of including target excitation in the expansion scheme. The reduction of the elastic scattering cross-section, close to zero energy, upon the inclusion of target excitation channels has also been rigorously confirmed for the Ps–H system (Blackwood et al., 2002a). Included in the figure are the zero energy elastic scattering cross-sections of Mitroy and Ivanov (2002b), the error bar

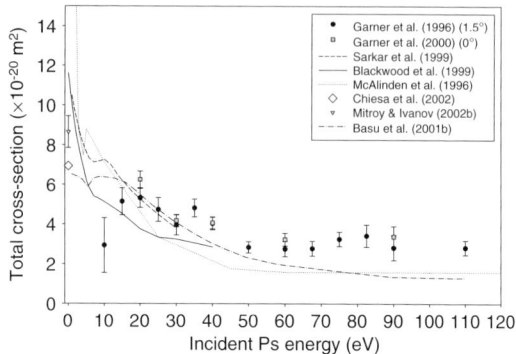

FIGURE 19 Total cross-sections for Ps–He scattering. The angles noted for Garner et al. (1996, 2000). Refer to the angular acceptance for forward scattered projectiles

reflecting the range of values obtained with different values of parameters in their model potential, and of Chiesa et al. (2002) who used a diffusion Monte Carlo approach.

The main ionizing reactions induced by Ps at low energies are listed below together with the thresholds (in eV) for helium and xenon, the only two targets which have been experimentally investigated thus far:

$$\text{Ps} + A \rightarrow e^+ + e^- + A \quad \text{Ps fragmentation (6.8)} \tag{11a}$$

$$\text{Ps} + A \rightarrow \text{Ps} + e^- + A^+ \quad \text{target ionization (24.6, 12.1)} \tag{11b}$$

$$\text{Ps} + A \rightarrow e^+ + e^- + A^*$$

$$\quad \text{Ps fragmentation with target excitation (27.8, 15.1)} \tag{11c}$$

$$\text{Ps} + A \rightarrow \text{Ps}^* + e^- + A^+$$

$$\quad \text{projectile excitation with target ionization (29.7, 17.2)} \tag{11d}$$

$$\text{Ps} + A \rightarrow e^+ + 2e^- + A^+$$

$$\quad \text{projectile and target ionization (31.4, 18.9)} \tag{11e}$$

Collision (11a) is the target-elastic reaction and (11b) its projectile-elastic equivalent; (11c), (11d) and (11e) relate to doubly-inelastic processes where both target and projectile are ionized, or one is ionized and the other is excited. Experimentally, these may be probed by detecting the positron or the electron in the final state. The former defines:

$$Q_f^+ = Q_a + Q_c + Q_e \tag{12}$$

which comprises contributions involving the fragmentation of the projectile (i.e. reactions (11a), (11c) and (11e), if energetically allowed). Detection of final-state electrons, on the other hand, encompasses all of

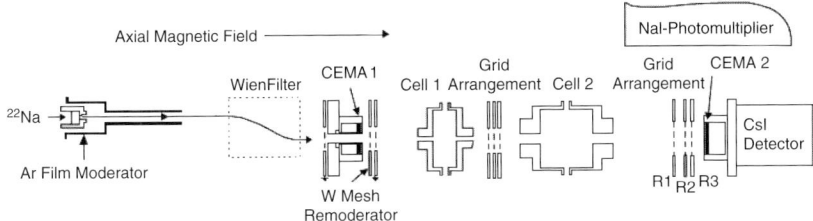

FIGURE 20 Schematic diagram of the positronium beam at UCL

the above reactions defining:

$$Q_f^- = Q_a + Q_b + Q_c + Q_d + Q_e. \quad (13)$$

The difference between Q_f^+ and Q_f^- thus corresponds to the cross-section for target ionization without fragmentation of the projectile:

$$Q_f^T = Q_b + Q_d. \quad (14)$$

Because Ps fragmentation is expected to dominate over target-inelastic processes at low energies (Blackwood et al., 1999; Basu et al., 2001b) and because the positron is an unambiguous signature of the process, as well as easily separable from the ubiquitous background of secondary electrons, experimental investigations thus far (Armitage et al., 2002, 2006; Brawley et al., 2008) have focused primarily on the break-up of positronium.

3.2 Experimental Methods

The energy-tunable positronium beam in operation at UCL is illustrated in Figure 20 (Garner et al., 2000; Leslie et al., 2002; Laricchia et al., 2004). The slow e^+ from the source-moderator arrangement are magnetically confined and separated from the flux of high-energy particles by a Wien filter. The beam is then converted into Ps in the first cell and a grid arrangement between the two gas cells is used to prevent transmitted e^+ from entering the second cell which contains the gas under investigation for Ps interactions. Time-of-flight measurements are achieved via coincidences between the 'tagger' (Laricchia et al., 1988), consisting of a channel-electron-multiplier-array (CEMA1) and a remoderator, and the detector at the end of the beamline. The latter detector (CEMA2) may also be set in coincidence with a number of γ-ray counters, if required. The timing signal at the tagger is provided by detecting the secondary electrons produced at the remoderator. From the flight times, the energy of Ps in the beam and, to some extent, its quantum state may be determined (e.g. Laricchia et al. (2004)). The kinetic energy of the Ps atoms, $E_0 = E_+ - E_i + 6.8$ eV/n^2, may be varied in the range ~10–250 eV by tuning E_+. The angular and energy distributions, as well as the beam intensity, depend on the differential positronium formation cross-section $dQ_{Ps}(E_+)/d\Omega$ for the neutralizing gas (as discussed in Section 2.4.2) and on the characteristics of the e$^+$ beam itself (angular divergence, energy spread, etc.).

Ionization by positronium impact has been investigated using two different experimental techniques:

(i) time-of-flight (TOF): used at first to identify the positrons released through Ps fragmentation (Armitage et al., 2002);
(ii) retarding field analysis (RFA): in order to circumvent the restrictions on the energy resolution and signal-to-background ratios imposed by the TOF at high and low energies, respectively, the grid arrangement in front of CEMA was used to obtain the longitudinal energy (E_ℓ) distribution of the e$^+$ and e$^-$ resulting from Ps impact on helium and xenon (Armitage et al., 2006).

3.3 Results

3.3.1 Integral Cross-Sections

In both experimental techniques (Armitage et al 2002, 2006), the absolute integrated ionization cross-sections for the ejected e$^-$, Q_f^-, and e$^+$, Q_f^+, have been determined using

$$Q_f^{+/-}(E_0) = \frac{N_{+/-}}{(N_{Ps})_{scatt}} Q_t(E_0) \, SG\left(\frac{\varepsilon_0}{\varepsilon_{+/-}}\right) \quad (15)$$

where $N_{+/-}$ is the net number of final-state e$^+$/e$^-$, $(N_{Ps})_{scatt}$ is the scattered Ps flux given by the difference between the incident- and transmitted-Ps fluxes, measured in vacuum and gas respectively; Q_t is the corresponding total cross-section of Ps of incident energy E_0; S accounts for in-flight annihilation and corresponds to the ratio between the number of Ps atoms that traverse the scattering cell and the corresponding number that reach the detector; G accounts for the differing solid angles for e$^+$ and Ps atoms as the e$^+$ are confined by the magnetic field. ε_0 and $\varepsilon_{+/-}$ are the detection efficiencies of Ps atoms and e$^+$/e$^-$ respectively (Armitage, 2002).

In Figure 21, the absolute integrated ionization cross-sections ($Q_f^\pm$) for Ps impact on helium are presented where consistency may be noted between the data obtained for Q_f^+ using the two experimental methods. Also at 30 eV, the experimental $Q_f^\pm$ are within errors, indicating negligible contribution from reaction (11b), target ionization (comparison with the FBA calculation of Walters et al. (2006) is not meaningful in this case

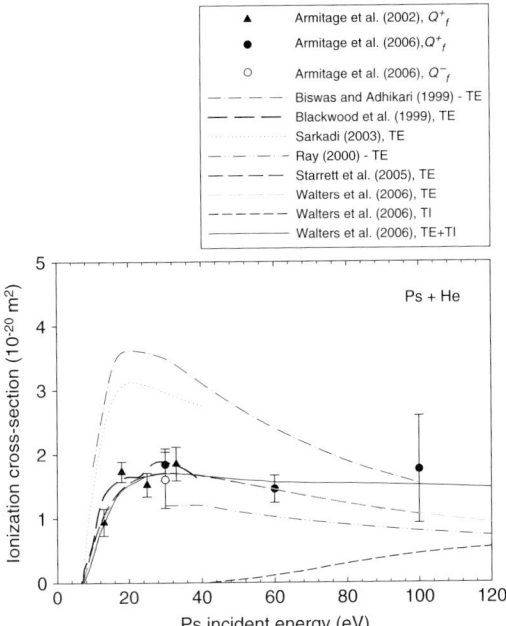

FIGURE 21 The ionization cross-sections for Ps impact on helium

because, as discussed by these authors, the FBA gives zero cross-section for transitions in which the Ps state does not change parity). Comprised in Figure 21 are other theoretical determinations of Q_f^+ for target-elastic collisions. Both the Born approximation calculation of Biswas and Adhikari (1999) and the Classical Trajectory Monte Carlo (CTMC) results of Sarkadi (2003) exceed the measured values by a factor of ≈ 2, whilst the Coulomb–Born approximation calculation of Ray (2002) falls below the experimental data across the entire energy range. A good agreement is observed between the experimental data and the coupled-state theoretical determination of Blackwood et al. (1999) below 40 eV. A recent calculation using the impulse approximation (Starrett et al., 2005), which treats the scattering of Ps as a coherent sum of the individual scattering of its constituents, agrees well with the measurements across the entire energy range. The sum of the target-elastic (Starrett et al., 2005) and target-inelastic (Walters et al., 2006) cross-sections is also depicted and found to be in good agreement with the experimental data, although the experimental uncertainty of the datum at 100 eV is too large to distinguish between the TE and the TE + TI results.

In Figure 22, the magnitude of various channels contributing to the scattering of Ps from helium may be gleaned. The partitioning of the total cross-section into elastic scattering, projectile excitation and

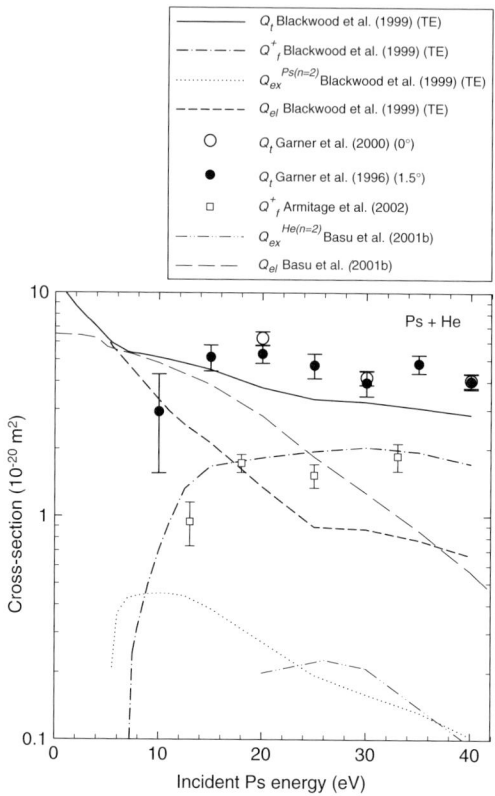

FIGURE 22 Cross-sections for Ps + He scattering: Q_t, total; Q_{el}, elastic; Q_{ex}^{Ps}, projectile excitation to $n = 2$; Q_{ex}^{He} target excitation to $n = 2$. TE denotes 'target elastic'. The angles in Garner et al. (1996, 2000). Refer to angular acceptance of forward scattered projectiles

fragmentation cross-sections computed by Blackwood et al. (1999) is displayed together with the results of Basu et al. (2001b) for target excitation and elastic scattering cross-sections, the latter allowing for virtual excitation of the target, and the experimental results of Garner et al. (1996, 2000) and Armitage et al. (2002). The comparison illustrates the dominance of projectile fragmentation among inelastic processes over this energy range.

Initial results for $Q_f^{\pm}$ for Ps scattering from xenon are shown in Figure 23 (Armitage et al., 2006; Brawley et al., 2008). The data at 30 eV for both the ejected e^+ and e^- were obtained using the RFA method. As can be seen from the figure, Q_f^+ is in good agreement with the TE Impulse Approximation calculation of Starrett et al. (2005) whilst the excess of Q_f^- suggests that target ionization may be significant (albeit the precision of

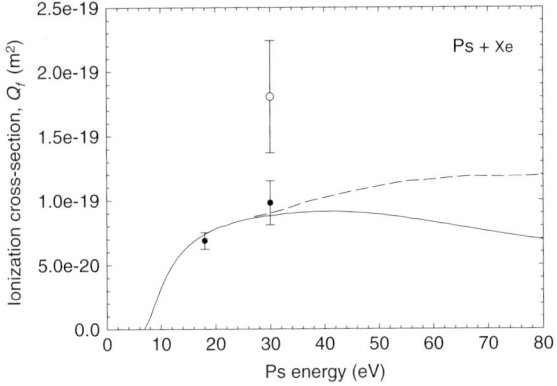

FIGURE 23 The ionization cross-sections for Ps impact on xenon: positron detection, Q_f^+ (filled circles) and electron detection, Q_f^- (hollow) (Armitage et al., 2006; Brawley et al., 2008). Theory: TE, solid curve (Starrett et al., 2005); TE + TI, dashed line, (Starrett and Walters, 2007)

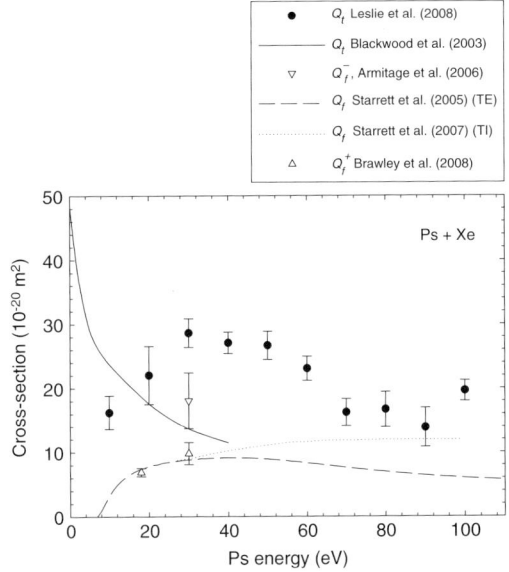

FIGURE 24 Cross-sections for Ps + Xe scattering: Q_{el}, elastic; $Q_f^{+/-}$, fragmentation from e^+/e^- detection. TE denotes 'target elastic'

this datum is low). This finding is in contrast with TI results of the FBA calculation of Starrett and Walters (2007).

In Figure 24, the experimental cross-section for ionization is plotted together with the total cross-section of Ps scattering from xenon. Also included are the theoretical results for ionization (Starrett et al., 2005; Starrett and Walters, 2007) and for elastic scattering calculated in the static-exchange approximation (Blackwood et al., 2002a,b,c, 2003), below 5.1 eV the latter representing the total cross-section. Once again the low-energy experimental datum falls below theory whilst at high energies, the TE + TI results of Starrett and Walters (2007) account for ~70 per cent of the experimental total cross-section.

3.3.2 Differential Cross-Sections

The absolute differential cross-section for the fragmentation of Ps in collision with helium atoms, dQ_f^+/dE_ℓ, determined for the ejected e^+ using both the TOF and RFA is shown in Figure 25. A peak, just below half of the excess energy (indicated by the arrows in the figure), may be seen to become more pronounced as the impact energy is augmented. This peak, qualitatively reproduced by both theories, is suggestive of the final-state particles (i.e. e^+ and e^-) traveling with similar velocities along the incident beam direction, the corresponding phenomenon in ion–atom physics being known as electron-*loss*-to-the-continuum (Crooks and Rudd, 1970) and related to the electron-capture-to-the-continuum process discussed in Section 2.3.2. The CTMC calculation of Sarkadi (2003) in the figure has been multiplied by a factor of 0.5 for shape comparison. The Impulse approximation results of Starrett et al. (2005) are in fair agreement with experiment both in magnitude and shape.

In Figure 26, a target comparison is made between dQ_f^+/dE_ℓ measured from helium and xenon at $E_0 = 18$ eV (for ease of comparison the He spectrum has been multiplied by 4). At this energy, the shapes of the positron spectra for the two targets are very similar, displaying a peak just below $E_r/2$, in accordance with the expectations of the ELC process. No evidence of target excitation (reaction (11c)) is discernable in the xenon spectrum at this energy.

4. CONCLUSIONS AND OUTLOOK

A detailed understanding of positron induced ionization phenomena is emerging, aided by the increasing convergence of experimental results and between these and theories. One onus on experimentalists is to resolve observational parameters (i.e. final-state angles and/or energy) in order to explore in-depth collision dynamics, reveal interference effects and refine comparison with theoretical descriptions. High-level information in this regard is beginning to come forward (e.g. Arcidiacono et al. (2005), de Lucio et al. (2006)) and the pace of this research is expected to quicken with the application of reaction microscope techniques (e.g. Ullrich et al. (2003)) to positrons.

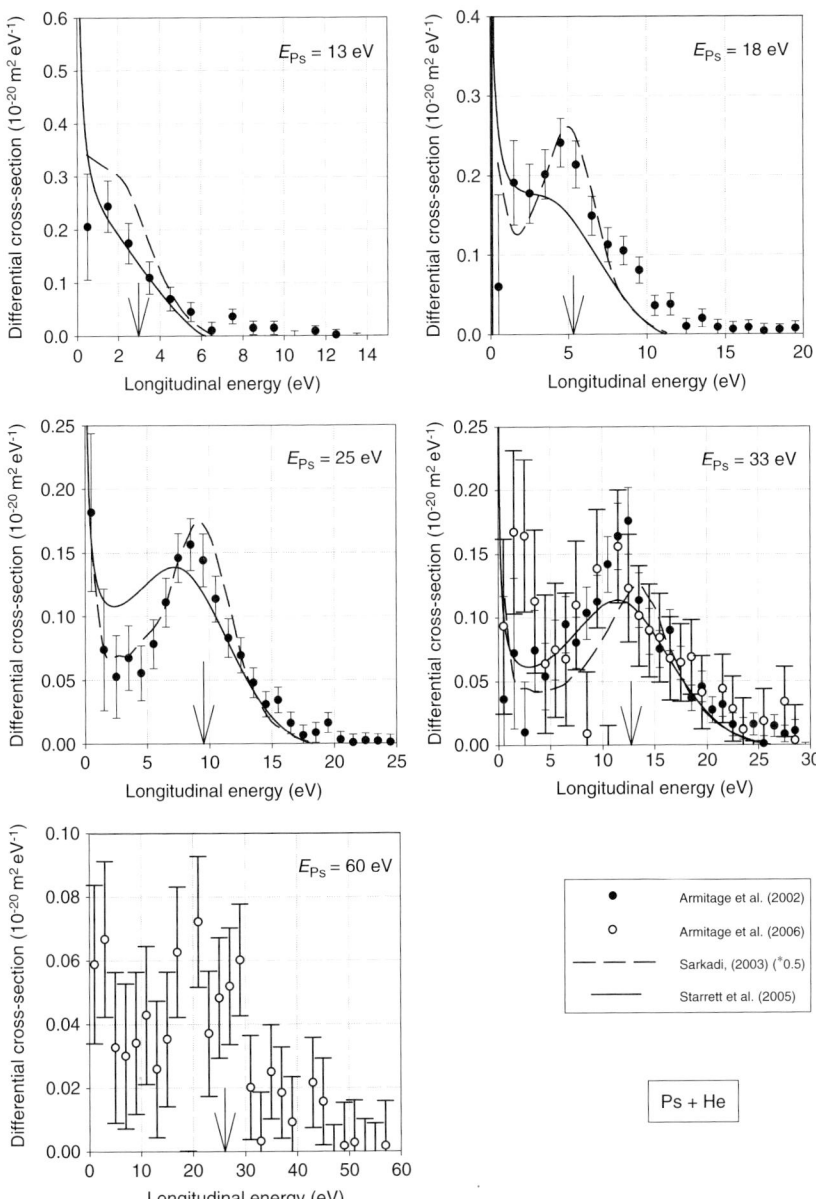

FIGURE 25 The absolute single differential cross-section dQ_f^+/dE_ℓ for the fragmentation of Ps in collision with helium atoms. In this figure, longitudinal energy refers to that of the ejected positron and E_{Ps} to the incident energy of the Ps projectile

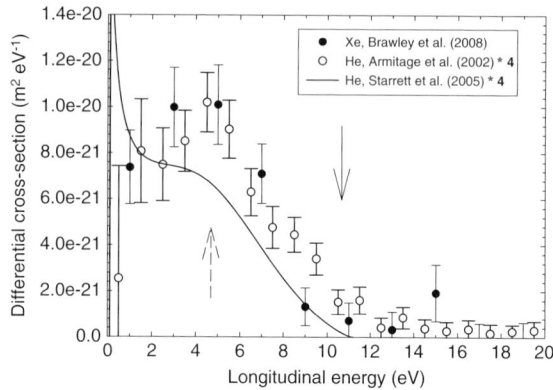

FIGURE 26 The longitudinal energy distributions of the ejected positrons from Ps collisions with xenon at 18 eV (•). The corresponding data for helium (Armitage et al) are also shown multiplied by 4 for shape comparison (○). The solid and dashed arrows show $E_{\max}$ and $E_r/2$ for TE fragmentation. The theoretical results of Starrett et al. (2005) multiplied by 4 are also shown (Brawley et al., 2008)

The production of Ps in excited states, via inner-shell electron capture and accompanied by the production of multiply-charged ions remains largely unexplored, as does its angular distribution. The trap-based positron beams (Gilbert et al., 1997), instrumental also in achieving the synthesis of antihydrogen (Amoretti et al., 2002) and molecular positronium (Ps$_2$) (Cassidy and Mills, 2007), offer the prospect of being able to probe the near-threshold energy dependence of Q_{Ps} and Q_i^+ with an energy resolution an order of magnitude better than achieved thus far (Moxom et al., 1994; Ashley et al., 1996) as well as the behavior of Q_{ann} in the vicinity of various positron–atom inelastic thresholds, an experimentally difficult task but an important fundamental problem (e.g. Van Reeth et al. (2005)).

Concerning Ps scattering, studies of target-inelastic and doubly-inelastic processes have commenced; energy and angle resolved investigations are also envisaged. Alternative methods to positron neutralization in a gas may enable the production of positronium with energies well beyond those presently achievable. Such methods might include electron photodetachment after accelerating Ps$^-$ formed by bombarding surfaces (Mills, 1981; Nagashima and Sakai, 2006), possibly with a cesium coverage (Laricchia, 1995).

ACKNOWLEDGMENTS

We wish to thank our co-workers, past and present, at UCL and elsewhere, for their contribution to many of the results presented in this article and express our gratitude to EPSRC, The European Union and The Royal Society for supporting positron and positronium research at UCL.

REFERENCES

Amoretti, M., Amsler, C., Bonomi, G., Bouchta, A., Bowe, P., Carraro, C., Cesar, C.L., Charlton, M., Collier, M.J.T., Doser, M., Filippini, V., Fine, K.S., Fontana, A., Fujiwara, M.C., Funakoshi, R., Genova, P., Hangst, J.S., Hayano, R.S., Holzscheiter, M.H., Jørgensen, L.V., Lagomarsino, V., Landua, R., Lindelöf, D., Lodi Rizzini, E., Macrì, M., Madsen, N., Manuzio, G., Marchesotti, M., Montagna, P., Pruys, H., Regenfus, C., Riedler, P., Rochet, J., Rotondi, A., Rouleau, G., Testera, G., Variola, A., Watson, T.L. and van der Werf, D.P. (2002). Production and Detection of Cold Antihydrogen Atoms. *Nature*, **419**, 456–459.

Arcidiacono, C., Kövér, A. and Laricchia, G. (2005). Energy-Sharing Asymmetries in Ionization by Positron Impact. *Phys. Rev. Lett.*, **95**, 223202-4.

Armitage, S. (2002). Positronium Ionization in Collision with He Atoms. PhD Thesis, University College London.

Armitage, S., Leslie, D.E., Garner, A.J. and Laricchia, G. (2002). Fragmentation of Positronium in Collision with He Atoms. *Phys. Rev. Lett.*, **89**, 173402-4.

Armitage, S., Beale, J., Leslie, D. and Laricchia, G. (2005). Production and Scattering of a Positronium Beam. *Nucl. Instrum. Methods B*, **233**, 87–94.

Armitage, S., Leslie, D., Beale, J. and Laricchia, G. (2006). Collisions Involving Positronium. *Nucl. Instrum. Methods B*, **247**, 98–104.

Ashley, P., Moxom, J. and Laricchia, G. (1996). Near-Threshold Ionization of He and H_2 by Positron Impact. *Phys. Rev. Lett.*, **77**, 1250-3.

Barnes, L.D., Gilbert, S.J. and Surko, C.M. (2003). Energy-resolved Positron Annihilation for Molecules. *Phys. Rev. A*, **67**, 032706-11.

Barnes, L.D., Young, J.A. and Surko, C.M. (2006). Energy-Resolved Positron Annihilation Rates for Molecules. *Phys. Rev. A*, **74**, 012706-10.

Bartschat, K. (2005). Direct Ionization of Heavy Noble Gases by Positron Impact. *Phys. Rev. A*, **71**, 032718-5.

Basu, M., Mazumdar, P.S. and Ghosh, A.S. (1985). Ionisation Cross Sections in Positron-Helium Scattering. *J. Phys. B: At., Mol. Opt. Phys*, **18**, 369–377.

Basu, A., Sinha, P.K. and Ghosh, A.S. (2001a). Effect of Target Inelastic Channels in Positronium-Hydrogen Scattering. *Phys. Rev. A*, **63**, 012502-10.

Basu, A., Sinha, P.K. and Ghosh, A.S. (2001b). Scattering of Orthopositronium off a Helium Atom. *Phys. Rev. A*, **63**, 052503-11.

Beale, J., Armitage, S. and Laricchia, G. (2006). Positronium- and Positron- H_2O Total Cross Sections. *J. Phys. B: At., Mol. Opt. Phys.*, **39**, 1337–1344.

Belanger, G., Goldwurm, A., Renaud, M., Terrier, R., Melia, F., Lund, N., Paul, J., Skinner, G. and Zadeh, Y.F. (2006). A Persistent High Energy Flux from the Heart of the Milky Way: INTEGRAL's View of the Galactic Center. *Astrophys. J.*, **636**, 275–289.

Benedek, A. and Campeanu, R.I. (2007). Molecular 3C Approximation for the Ionization of H_2 by Positron Impact. *J. Phys. B: At., Mol. Opt. Phys.*, **40**, 1589–1596.

Berakdar, J. (1998). Asymmetric Formation of Positronium Continuum States Following Positron-Impact Ionization of H_2. *Phys. Rev. Lett.*, **81**, 1393–1396.

Berko, S. and Pendleton, H.N. (1980). Positronium. *Annu. Rev. Nucl. Particle Sci.*, **30**, 543–581.

Biswas, P.K., Mukherjee, T. and Ghosh, A.S. (1991). Ground-State Ps Formation in e^+-H_2 Scattering using the FBA. *J. Phys. B: At., Mol. Opt. Phys.*, **24**, 2601–2607.

Biswas, P.K. and Adhikari, S.K. (1999). Electron Exchange Model Potential: Application to Positronium-Helium Scattering. *Phys. Rev. A*, **59**, 363–370.

Biswas, P.K. (2001). Effect of H^- ion Formation on Positronium-Hydrogen Elastic Scattering. *J. Phys. B: At., Mol. Opt. Phys.*, **34**, 4831–4844.

Blackwood, J.E., Campbell, C.P., McAlinden, M.T. and Walters, H.R.J. (1999). Positronium Scattering by Helium. *Phys. Rev. A*, **60**, 4454–4460.

Blackwood, J.E., McAlinden, M.T. and Walters, H.R.J. (2002a). Positronium Scattering by Atomic Hydrogen with Inclusion of Target Excitation Channels. *Phys. Rev. A*, **65**, 032517-10.

Blackwood, J.E., McAlinden, M.T. and Walters, H.R.J. (2002b). Importance of the H⁻ Channel in Ps-H Scattering. *Phys. Rev. A*, **65**, 030502-4.

Blackwood, J.E., McAlinden, M.T. and Walters, H.R.J. (2002c). Positronium Scattering by Ne, Ar, Kr and Xe in the Frozen Target Approximation. *J. Phys. B: At., Mol. Opt. Phys.*, **35**, 2661–2682.

Blackwood, J.E., McAlinden, M.T. and Walters, H.R.J. (2003). Reply to Comment on Positronium Scattering by Ne, Ar, Kr and Xe in the Frozen Target Approximation. *J. Phys. B: At., Mol. Opt. Phys*, **36**, 797–798.

Bluhme, H., Knudsen, H., Merrison, J.P. and Poulsen, M.R. (1998). Strong Suppression of the Positronium Channel in Double Ionization of Noble Gases by Positron Impact. *Phys. Rev. Lett.*, **81**, 73–76.

Bluhme, H., Knudsen, H., Merrison, J.P. and Nielsen, K.A. (1999a). Ionization of Helium, Neon and Xenon by Positron Impact. *J. Phys. B: At., Mol. Opt. Phys.*, **32**, 5237–5245.

Bluhme, H., Knudsen, H., Merrison, J.P. and Nielsen, K.A. (1999b). Ionization of Argon and Krypton by Positron Impact. *J. Phys. B: At., Mol. Opt. Phys.*, **32**, 5835–5842.

Brauner, M. and Briggs, J.S. (1986). Ionisation to the Projectile Continuum by Positron and Electron Collisions with Neutral Atoms. *J. Phys. B: At., Mol. Opt. Phys.*, **19**, L325–L330.

Brauner, M., Briggs, J.S. and Klar, H. (1989). Triply-Differential Cross Sections for Ionisation of Hydrogen Atoms by Electrons and Positrons. *J. Phys. B: At., Mol. Opt. Phys.*, **22**, 2265–2287.

Brawley, S., Beale, J., Armitage, S., Leslie, D.E., Kövér, A. and Laricchia, G. (2008). Fragmentation of Positronium in Collision with Xenon. *Nucl. Instrum. Methods B* (in press).

Brown, B.L. (1985). Creation of Monoenergetic Positronium (Ps) in a Gas. *Bull. Am. Phys. Soc.*, **30**, 614.

Brown, C.J. and Humberston, J.W. (1985). Positronium Formation in Positron-Hydrogen Scattering. *J. Phys. B: At., Mol. Opt. Phys*, **18**, L401–L406.

Campbell, C.P., McAlinden, M., Kernoghan, A.A. and Walters, H.R.J. (1998). Positron Collisions with One- and Two- Electron Atoms. *Nucl. Instrum. Methods B*, **143**, 41–56.

Campeanu, R.I., McEachran, R.P. and Stauffer, A.D. (1996). Positron Impact Ionization of He, Ne, and Ar. *Can. J. Phys.*, **74**, 544–547.

Campeanu, R.I., McEachran, R.P. and Stauffer, A.D. (2002). Distorted-Wave Models in Positron Impact Ionization of Atoms. *Nucl. Instrum. Methods B*, **192**, 146–149.

Canter, K.F., Mills, A.P. and Berko, S. (1975). Observations of Positronium Lyman-Alpha Radiation. *Phys. Rev. Lett.*, **34**, 177–180.

Cassidy, D.B. and Mills, A.P. (2007). The Production of Molecular Positronium. *Nature*, **449**, 195–197.

Chakraborty, S., Sinha, P.K. and Ghosh, A.S. (2004). S-wave Ps-Li and Ps-Na Scattering. *Phys. Rev. A*, **69**, 052506-7.

Champion, C. and Le Loirec, C.L. (2006). Positron Follow-Up in Liquid Water: I. A New Monte Carlo Track-Structure Code. *Phys. Medicine Biol.*, **51**, 1707–1723.

Charlton, M., Clark, G., Griffith, T.C. and Heyland, G.R. (1983). Positronium Formation Cross Sections in the Inert Gases. *J. Phys. B: At., Mol. Opt. Phys.*, **16**, L465–L470.

Charlton, M., Andersen, L.H., Brun-Nielsen, L., Deutch, B.I., Hvelplund, P., Jacobsen, F.M., Knudsen, H., Laricchia, G., Poulsen, M.R. and Pedersen, J.O. (1988). Positron and Electron Impact Double Ionisation of Helium. *J. Phys. B: At., Mol. Opt. Phys.*, **21**, L545–L549.

Charlton, M., Brun-Nielsen, L., Deutch, B.I., Hvelplund, P., Jacobsen, F.M., Knudsen, H., Laricchia, G. and Poulsen, M.R. (1989). Double Ionisation of Noble Gases by Positron and Electron Impact. *J. Phys. B: At., Mol. Opt. Phys.*, **22**, 2779–2788.

Charlton, M. and Laricchia, G. (1990). Positron Impact Phenomena. *J. Phys. B: At., Mol. Opt. Phys.*, **23**, 1045–1078.

Charlton, M. and Humberston, J.W. (2001). *Positron Physics*. Cambridge University Press.

Chaudhuri, P. and Adhikari, S.K. (1998). Close-Coupling Calculations of Positronium Formation in Positron-Helium Scattering. *J. Phys. B: At., Mol. Opt. Phys.*, **31**, 3057–3063.

Chen, Z. and Msezane, A.Z. (1994). Calculation of the Cross Sections for Positron- and Proton-Impact Ionization of Helium. *Phys. Rev. A*, **49**, 1752–1756.

Cheng, Y. and Zhou, Y. (2007). Momentum-Space Coupled-Channel Calculation for Positron-Helium Scattering. *Phys. Rev. A*, **76**, 012704-6.

Chiesa, S., Mella, M. and Morosi, G. (2002). Orthopositronium Scattering off H and He. *Phys. Rev. A*, **66**, 042502-8.

Coleman, P.G., Rayner, S., Jacobsen, F.M., Charlton, M. and West, R.N. (1994). Angular Correlation Studies of Positron Annihilation in the Noble Gases. *J. Phys. B: At., Mol. Opt. Phys.*, **27**, 981–991.

Crooks, G.B. and Rudd, M.E. (1970). Experimental Evidence for the Mechanism of Charge Transfer into Continuum States. *Phys. Rev. Lett.*, **25**, 1599–1601.

Deb, N.C. and Crothers, D.S.F. (2002). Near-Threshold Ionization of Helium by Positron Impact. *J. Phys. B: At., Mol. Opt. Phys.*, **35**, L85–L89.

de Lucio, O.G., Gavin, J. and DuBois, R.D. (2006). Differential Electron Emission for Single and Multiple Ionization of Argon by 500 eV Positrons. *Phys. Rev. Lett.*, **97**, 243201-4.

Diana, L.M., Fornari, L.S., Sharma, S.C., Pendleton, P.K. and Coleman, P.G. (1985). Measurements of Total Ionization Cross Sections for Positrons. in: Jain, P.C., Singru, R.M. and Gopinathan, K.P. (Eds.), *Positron Annihilation*. World Scientific, pp. 342–343.

Diana, L.M., Coleman, P.G., Brooks, D.L., Pendleton, P.K. and Norman, M., D. (1986a). Positronium Formation Cross Sections in He and H_2 at Intermediate Energies. *Phys. Rev. A*, **34**, 2731–2737.

Diana, L.M., Coleman, P.G., Brooks, D.L., Pendleton, P.K., Norman, D.M., Seay, B.E. and Sharma, S.C. (1986b). Measurement of Total Positronium Formation Cross-Section in Argon to 441.3 eV. in: Kauppila, W.E., Stein, T.S. and Wadehra, J.M. (Eds.), *Positron (Electron) - Gas Scattering*. World Scientific, Singapore, pp. 296–298.

Diana, L.M., Coleman, P.G., Brooks, D.L. and Chaplin, R.L. (1987). Studies of Inelastic Positron Scattering using 2.3 and 3 m Spectrometers. in: Humberston, J.W. and Armour, E.A.G. (Eds.), *Atomic Physics with Positrons*. Plenum, pp. 55–69.

Diana, L.M., Brooks, D.L., Coleman, P.G., Chaplin, R.L. and Howell, J.P. (1989). Total Cross-Sections for Positronium Formation in Xenon. in: Dorikens-Vanpraet, L., Dorikens, M. and Segers, D. (Eds.), *Positron Annihilation*. World Scientific, pp. 311–313.

DuBois, R.D. and Rudd, M.E. (1978). Absolute Doubly Differential Cross Sections for Ejection of Secondary Electrons from Gases by Electron Impact. II. 100-500-eV Electrons on Neon, Argon, Molecular Hydrogen, and Molecular Nitrogen. *Phys. Rev. A*, **17**, 843–848.

DuBois, R.D., Doudna, C., Lloyd, C., Kahveci, M., Khayyat, K., Zhou, Y. and Madison, D.H. (2001). Energy-Loss Measurements for Single and Multiple Ionization of Argon by Positron Impact. *J. Phys. B: At., Mol. Opt. Phys.*, **34**, L783–L789.

Dull, T., Frieze, W., Gidley, D., Sun, J. and Yee, A. (2001). Determination of Pore Size in Mesoporous thin Films from the Annihilation Lifetime of Positronium. *J. Phys. Chem. B*, **105**, 4657–4662.

Dunlop, L. and Gribakin, G. (2006). Positronium Formation from Valence and inner Shells in Noble Gas Atoms. *Nucl. Instrum. Methods B*, **247**, 61–67.

Falke, T., Raith, W., Weber, M. and Wesskamp, U. (1995). Differential Positronium Formation in Positron-Argon Collisions. *J. Phys. B: At., Mol. Opt. Phys.*, **28**, L505–L509.

Falke, T., Brandt, T., Kuhl, O., Raith, W. and Weber, M. (1997). Differential Ps-Formation and Impact-Ionization Cross Sections For Positron Scattering on Ar and Kr Atoms. *J. Phys. B: At., Mol. Opt. Phys.*, **30**, 3247–3256.

Finch, R.M., Kövér, A., Charlton, M. and Laricchia, G. (1996). Positron-Argon Elastic, Positronium Formation and Ionizing Collisions at. *J. Phys. B: At., Mol. Opt. Phys.*, **29**, L667–L672.

Fiol, J., Rodriguez, V.D. and Barrachina, R.O. (2001). Electron Capture to the Continuum by Proton and Positron Impact. *J. Phys. B: At., Mol. Opt. Phys.*, **34**, 933–944.

Fiol, J. and Olson, R.E. (2002). Three-Body Dynamics in the Ionization of Hydrogen by Positron Impact. *J. Phys. B: At., Mol. Opt. Phys.*, **35**, 1173–1184.
Fornari, L.S., Diana, L.M. and Coleman, P.G. (1983). Positronium Formation in Collisions of Positrons with He, Ar, and H_2. *Phys. Rev. Lett.*, **51**, 2276–2279.
Fromme, D., Kruse, G., Raith, W. and Sinapius, G. (1986). Partial-Cross-Section Measurements for Ionization of Helium by Positron Impact. *Phys. Rev. Lett.*, **57**, 3031–3034.
Garner, A.J., Laricchia, G. and Ozen, A. (1996). Ps Beam Production and Scattering from Gaseous Targets. *J. Phys. B: At., Mol. Opt. Phys.*, **29**, 5961–5968.
Garner, A.J., Ozen, A. and Laricchia, G. (1998). Positronium Beam Scattering from Atoms and Molecules. *Nucl. Instrum. Methods B*, **143**, 155–161.
Garner, A.J., Ozen, A. and Laricchia, G. (2000). The Effect of Forward-Angle Scattering on Positronium-Gas Total Cross Sections. *J. Phys. B: At., Mol. Opt. Phys.*, **33**, 1149–1157.
Ghosh, A.S., Sinha, P.K. and Ray, H. (1998). Theory of Positronium-Atom Scattering. *Nucl. Instrum. Methods B*, **143**, 162–169.
Ghosh, A.S., Basu, A., Mukherjee, T. and Sinha, P.K. (2001). Ps–He Scattering below the First Target Excitation Threshold. *Phys. Rev. A*, **63**, 042706-5.
Gilbert, S.J., Kurz, C., Greaves, R.G. and Surko, C.M. (1997). Creation of a Monoenergetic Pulsed Positron Beam. *Appl. Phys. Lett.*, **70**, 1944–1946.
Gilbert, S.J., Barnes, L.D., Sullivan, J.P. and Surko, C.M. (2002). Vibrational-Resonance Enhancement of Positron Annihilation in Molecules. *Phys. Rev. Lett.*, **88**, 043201-4.
Gilmore, S., Blackwood, J.E. and Walters, H.R.J. (2004). Positronium Formation in Positron-Noble Gas Collisions. *Nucl. Instrum. Methods B*, **221**, 129–133.
Golden, D.E., Xu, Z., Bernhard, J. and Mueller, D.W. (1996). Electron Anticapture to the Continuum in Double- Binary Electron–Atom Collisions. *J. Phys. B: At., Mol. Opt. Phys.*, **29**, 3741–3746.
Gribakin, G.F. and Gill, P.M.W. (2004). The Role of Vibrational Doorway States in Positron Annihilation with Large Molecules. *Nucl. Instrum. Methods B*, **221**, 30–35.
Guang-yan, P., Hvelplund, P., Knudsen, H., Yamazaki, Y., Brauner, M. and Briggs, J.S. (1993). Electron Correlation in the Continuum: The e-2e Process for Small Relative Momenta of the Outgoing Electrons. *Phys. Rev. A*, **47**, 1531–1534.
Helms, S., Brinkmann, U., Deiwiks, J., Schneider, H. and Hippler, R. (1994a). Double Ionisation of Rare Gas Atoms by Positron Impact. *Hyperfine Interactions*, **89**, 395–400.
Helms, S., Brinkmann, U., Deiwiks, J., Hippler, R., Schneider, H., Segers, D. and Paridaens, J. (1994b). Inner Shell Contributions to Multiple Ionization of Argon by Positron Impact. *J. Phys. B: At., Mol. Opt. Phys.*, **27**, L557–L562.
Helms, S., Brinkmann, U., Deiwiks, J., Hippler, R., Schneider, H., Segers, D. and Paridaens, J. (1995). Multiple Ionization of Argon, Krypton and Xenon Atoms by Positron Impact. *J. Phys. B: At., Mol. Opt. Phys.*, **28**, 1095–1103.
Hewitt, R.N., Noble, C.J. and Bransden, B.H. (1992). He(2^1S, 2^1P) Excitation and Positronium Formation in Positron-Helium Collisions at Intermediate Energies. *J. Phys. B: At., Mol. Opt. Phys.*, **25**, 557–570.
Higgins, K., Burke, P.G. and Walters, H.R.J. (1990). Positron Scattering by Atomic-Hydrogen at Intermediate Energies. *J. Phys. B: At., Mol. Opt. Phys.*, **23**, 1345–1357.
Igarashi, A. and Toshima, N. (1992). Positronium Formation in Positron-Helium Collisions at Intermediate Energies. *Phys. Lett. A*, **164**, 70–72.
Ihra, W., Macek, J.H., Mota-Furtado, F. and O'Mahony, P.F. (1997). Threshold Law for Positron Impact Ionization of Atoms. *Phys. Rev. Lett.*, **78**, 4027–4030.
Jacobsen, F.M., Frandsen, N.P., Knudsen, H., Mikkelsen, U. and Schrader, D.M. (1995). Single Ionization of He, Ne and Ar by Positron Impact. *J. Phys. B: At., Mol. Opt. Phys.*, **28**, 4691–4695.
Jin, B., Miyamoto, S., Sueoka, O. and Hamada, A. (1994). *At. Collision Res. Japan*, **20**, 9.
Jones, G.O., Charlton, M., Slevin, J., Laricchia, G., Kövér, A., Poulsen, M.R. and Chormaic, S.N. (1993). Positron Impact Ionization of Atomic Hydrogen. *J. Phys. B: At., Mol. Opt. Phys.*, **26**, L483–L488.

Kadyrov, A.S. and Bray, I. (2002). Two-Center Convergent Close-Coupling Approach to Positron-Hydrogen Collisions. *Phys. Rev. A*, **66**, 012710-13.

Kadyrov, A.S., Bray, I. and Stelbovics, A.T. (2007). Near-Threshold Positron-Impact Ionization of Atomic Hydrogen. *Phys. Rev. Lett.*, **98**, 263202-4.

Kamali, M.Z.M. and Ratnavelu, K. (2001). Positron-Hydrogen Scattering at Low Intermediate Energies. *Phys. Rev. A*, **65**, 014702-4.

Kara, V., Paludan, K., Moxom, J., Ashley, P. and Laricchia, G. (1997). Single and Double Ionization of Neon, Krypton and Xenon by Positron Impact. *J. Phys. B: At., Mol. Opt. Phys.*, **30**, 3933–3949.

Kara, V., Paludan, K., Moxom, J., Ashley, P. and Laricchia, G. (1998). Positron Impact Ionisation of Atoms. *Nucl. Instrum. Methods B*, **143**, 94–99.

Kauppila, W.E. and Stein, T.S. (1990). Comparisons of Positron and Electron Scattering by Gases. *Adv. in At. Mol. Opt. Phys.*, **26**, 1–50.

Kernoghan, A.A., Robinson, D.J.R., McAlinden, M.T. and Walters, H.R.J. (1996). Positron Scattering by Atomic Hydrogen. *J. Phys. B: At., Mol. Opt. Phys.*, **29**, 2089–2102.

Klar, H. (1981). Threshold Ionisation of Atoms by Positrons. *J. Phys. B: At., Mol. Opt. Phys.*, **14**, 4165–4170.

Knudsen, H., Brun-Nielsen, L., Charlton, M. and Poulsen, M.R. (1990). Single Ionization of H_2, He, Ne and Ar by Positron Impact. *J. Phys. B: At., Mol. Opt. Phys.*, **23**, 3955–3976.

Knudsen, H. and Reading, J.F. (1992). Ionization of Atoms by Particle and Antiparticle Impact. *Phys. Rep.*, **212**, 107–222.

Kövér, A., Laricchia, G. and Charlton, M. (1993). Ionization by Positrons and Electrons of Ar at Zero Degrees. *J. Phys. B: At., Mol. Opt. Phys.*, **26**, L575–L580.

Kövér, A., Laricchia, G. and Charlton, M. (1994). Doubly Differential Cross Sections for Collisions of 100 eV Positrons and Electrons with Argon Atoms. *J. Phys. B: At., Mol. Opt. Phys.*, **27**, 2409–2416.

Kövér, A., Finch, R.M., Charlton, M. and Laricchia, G. (1997). Double-Differential Cross Sections for Collisions of Positrons with Argon Atoms. *J. Phys. B: At., Mol. Opt. Phys.*, **30**, L507–L512.

Kövér, A. and Laricchia, G. (1998). Triply Differential Study of Positron Impact Ionization of H_2. *Phys. Rev. Lett.*, **80**, 5309–5312.

Kövér, A., Paludan, K. and Laricchia, G. (2001). Triply Differential Ionization Cross-Section of H_2 by 50 eV Impact-Energy Positrons. *J. Phys. B: At., Mol. Opt. Phys.*, **34**, L219–L222.

Krishnakumar, E. and Srivastava, S.K. (1988). Ionisation Cross Sections of Rare-Gas Atoms by Electron Impact. *J. Phys. B: At., Mol. Opt. Phys.*, **21**, 1055–1082.

Kruse, G., Quermann, A., Raith, W., Sinapius, G. and Weber, M. (1991). Multiple Ionization of Xenon by Positron Impact. *J. Phys. B: At., Mol. Opt. Phys.*, **24**, L33–L37.

Kuo, T., Sun, H. and Huang, K. (2003). Theoretical Threshold Law of Positron-Impact Ionization of He. *Phys. Rev. A*, **67**, 012705-6.

Lahmam-Bennani, A. (1991). Recent Developments and New Trends in (e, 2e) and (e, 3e) Studies. *J. Phys. B: At., Mol. Opt. Phys.*, **24**, 2401–2442.

Laricchia, G., Charlton, M., Davies, S.A., Beling, C.D. and Griffith, T.C. (1987). The Production of Collimated Beams of o-Ps Atoms using Charge Exchange in Positron-Gas Collisions. *J. Phys. B: At., Mol. Opt. Phys.*, **20**, L99–L105.

Laricchia, G., Davies, S.A., Charlton, M. and Griffith, T.C. (1988). The Production of a Timed Tunable Beam of Positronium Atoms. *J. Phys. E: Scientific Instruments*, **21**, 886–888.

Laricchia, G. (1995). Positronium Beams and Surfaces. in: Dupasquier, A. and Mills, A.P. (Eds.), *Proceedings of the International School of Physics Enrico Fermi*. In: Positron Spectroscopy of Solids, IOS, Amsterdam, pp. 401–418.

Laricchia, G., Reeth, P.V., Szluinska, M. and Moxom, J. (2002). Total Positron-Impact Ionization and Positronium Formation from the Noble Gases. *J. Phys. B: At., Mol. Opt. Phys.*, **35**, 2525–2540.

Laricchia, G., Armitage, S. and Leslie, D.E. (2004). Positronium Induced Collisions. *Nucl. Instrum. Methods B*, **221**, 60–68.

Leslie, D.E., Armitage, S. and Laricchia, G. (2002). Production of Collimated Positronium from Molecular Nitrogen. *J. Phys. B: At., Mol. Opt. Phys.*, **35**, 4819–4827.
Leslie, D.E., Armitage, S. and Laricchia, G. (2008). (in preparation).
Lotz, W. (1967). Electron-Impact Ionization Cross-Sections and Ionization Rate Coefficients for Atoms and Ions (Electron Impact Ionization Cross Sections and Rate Coefficients for Atoms and Ions of Hg, Rare Gas and Alkali Metal Groups). *Astrophys. J. Suppl. Ser.*, **14**, 207–238.
Lotz, W. (1968). Electron-Impact Ionization Cross-Sections and Ionization Rate Coefficients for Atoms and Ions from Hydrogen to Calcium. *Zeitschrift fur Physik*, **216**, 241.
Ludlow, J. (2003), PhD Thesis, Queens University Belfast.
Mandal, P., Guha, S. and Sil, N.C. (1979). Positronium Formation in Positron Scattering from Hydrogen and Helium Atoms: The Distorted-Wave Approximation. *J. Phys. B: At., Mol. Opt. Phys.*, **12**, 2913–2924.
Mandal, P., Guha, S. and Sil, N.C. (1980). Positron-Helium Collisions: Positronium Formation into an Arbitrary Excited S State. *Phys. Rev. A*, **22**, 2623–2629.
Marler, J.P., Barnes, L.D., Gilbert, S.J., Sullivan, J.P., Young, J.A. and Surko, C.M. (2004). Experimental Studies of the Interaction of Low Energy Positrons with Atoms and Molecules. *Nucl. Instrum. Methods B*, **221**, 84–92.
Marler, J.P., Sullivan, J.P. and Surko, C.M. (2005). Ionization and Positronium Formation in Noble Gases. *Phys. Rev. A*, **71**, 022701-10.
Massey, H. (1976). Slow Positrons in Gases. *Physics Today*, **29**, 42.
McAlinden, M. and Walters, H. (1992). Positron Scattering by the Noble Gases. *Hyperfine Interactions*, **73**, 65–83.
McAlinden, M.T. and Walters, H.R.J. (1994). Differential Cross Sections for Elastic Scattering and Positronium Formation for Positron Collisions with Ne, Ar, Kr and Xe. *Hyperfine Interactions*, **89**, 407–418.
McAlinden, M.T., MacDonald, F.G.R.S. and Walters, H.R.J. (1996). Positronium-Atom Scattering. *Can. J. Phys.*, **74**, 434–444.
McEachran, R.P., Ryman, A.G. and Stauffer, A.D. (1979). Positron Scattering from Argon. *J. Phys. B: At., Mol. Opt. Phys.*, **12**, 1031-1031.
McGuire, J.H. (1982). Double Ionization of Helium by Protons and Electrons at High Velocities. *Phys. Rev. Lett.*, **49**, 1153–1157.
McGuire, J.H. and Deb, N.C. (1987). New Understanding of Atomic Physics using High Velocity Positrons. in: Humberston, J.W. and Armour, E.A.G. (Eds.), *Atomic Physics with Positrons*. Plenum Press, pp. 83–94.
Mills, A.P. (1981). Observation of the Positronium Negative Ion. *Phys. Rev. Lett.*, **46**, 717–720.
Mitroy, J. (1996). An L(2) Calculation of Positron-Hydrogen Scattering at Intermediate Energies. *J. Phys. B: At., Mol. Opt. Phys.*, **29**, L263–L269.
Mitroy, J. and Ivanov, I.A. (2002a). Semiempirical Model of Positron Scattering and Annihilation. *Phys. Rev. A*, **65**, 042705-15.
Mitroy, J. and Ivanov, I.A. (2002b). Positronium Scattering from Closed-Shell Atoms and Ions. *Phys. Rev. A*, **65**, 012509-12.
Moores, D.L. (1998). Positron Impact Ionisation of Rare Gas Atoms by a Distorted Wave Method with Close Coupled Target States. *Nucl. Instrum. Methods B*, **143**, 105–111.
Moores, D.L. (2001). Calculations of Integral Cross-Sections for Electron and Positron Impact of Rare Gas Atoms. *Nucl. Instrum. Methods B*, **179**, 316–324.
Mori, S. and Sueoka, O. (1994). Excitation and Ionization Cross Sections of He, Ne and Ar by Positron Impact. *J. Phys. B: At., Mol. Opt. Phys.*, **27**, 4349–4364.
Moxom, J., Laricchia, G., Charlton, M., Jones, G.O. and Kövér, A. (1992). Ejected-Electron Energy Spectra in low Energy Positron-Argon Collisions. *J. Phys. B: At., Mol. Opt. Phys.*, **25**, L613–L619.
Moxom, J., Laricchia, G., Charlton, M., Kövér, A. and Meyerhof, W.E. (1994). Threshold Effects in Positron Scattering on Noble Gases. *Phys. Rev. A*, **50**, 3129–3133.
Moxom, J., Laricchia, G. and Charlton, M. (1995). Ionization of He, Ar and H_2 by Positron-Impact at Intermediate Energies. *J. Phys. B: At., Mol. Opt. Phys.*, **28**, 1331–1347.

Moxom, J., Ashley, P. and Laricchia, G. (1996). Single Ionization by Positron Impact. *Can. J. Phys.*, **74**, 367–372.

Moxom, J., Schrader, D.M., Laricchia, G., Xu, J. and Hulett, L.D. (1999). Double Ionization of Noble Gases by Positron Impact. *Phys. Rev. A*, **60**, 2940–2943.

Moxom, J. (2000). Multiple Ionization of Xenon by Positron Impact. *J. Phys. B: At., Mol. Opt. Phys.*, **33**, L481–L485.

Murtagh, D.J., Szluinska, M., Moxom, J., Reeth, P.V. and Laricchia, G. (2005). Positron-Impact Ionization and Positronium Formation from Helium. *J. Phys. B: At., Mol. Opt. Phys.*, **38**, 3857–3866.

Nagashima, Y., Hyodo, T., Fujiwara, K. and Ichimura, A. (1998). Momentum-Transfer Cross Section for Slow Positronium-He Scattering. *J. Phys. B: At., Mol. Opt. Phys.*, **31**, 329–339.

Nagashima, Y. and Sakai, T. (2006). First Observation of Positronium Negative Ions Emitted from Tungsten Surfaces. *New J. Phys.*, **8**, 319-7.

Overton, N., Mills, R.J. and Coleman, P.G. (1993). The Energy Dependence of the Positronium Formation Cross Section in Helium. *J. Phys. B: At., Mol. Opt. Phys.*, **26**, 3951–3957.

Paludan, K., Laricchia, G., Ashley, P., Kara, V., Moxom, J., Bluhme, H., Knudsen, H., Mikkelsen, U., Moller, S.P., Uggerhoj, E. and Morenzoni, E. (1997). Ionization of Rare Gases by Particle — Antiparticle Impact. *J. Phys. B: At., Mol. Opt. Phys.*, **30**, L581–L587.

Parker, D.J., Dijkstra, A.E., Martin, T.W. and Seville, J.P.K. (1997). Positron Emission Particle Tracking Studies of Spherical Particle Motion in Rotating Drums. *Chemical Engineering Science*, **52**, 2011–2022.

Petkov, M.P., Weber, M.H., Lynn, K.G., Rodbell, K.P. and Cohen, S.A. (1999). Open Volume Defects (Measured by Positron Annihilation Spectroscopy) in Thin Film Hydrogen-Silsesquioxane Spin-on-Glass, Correlation with Dielectric Constant. *J. Appl. Phys.*, **86**, 3104–3109.

Ratnavelu, K. (1991). Positron Impact Ionization of H and He Atoms—The Continuum Model. *Aust. J. Phys.*, **44**, 265–270.

Ray, H. (2002). Comparative Studies on Ionizations in Ps-Atom Scattering Using the Coulomb–Born Approximation. *J. Phys. B: At., Mol. Opt. Phys.*, **35**, 3365–3376.

Rejoub, R., Lindsay, B.G. and Stebbings, R.F. (2002). Determination of the Absolute Partial and Total Cross Sections for Electron-Impact Ionization of the Rare Gases. *Phys. Rev. A*, **65**, 042713-8.

Rødbro, M. and Andersen, F.D. (1979). Charge Transfer to the Continuum for 15 to 1500 keV H^+ in He, Ne, Ar and H_2 Gases under Single-Collision Conditions. *J. Phys. B: At., Mol. Opt. Phys.*, **35**, 2883–2903.

Rost, J.M. and Heller, E.J. (1994). Ionization of Hydrogen by Positron Impact Near the Fragmentation Threshold. *Phys. Rev. A*, **49**, R4289–R4292.

Saito, F., Nagashima, Y. and Hyodo, T. (2003). Momentum-Transfer Cross Sections for Slow Positronium-Gas Collisions. *J. Phys. B: At., Mol. Opt. Phys.*, **36**, 4191–4199.

Saito, H. and Hyodo, T. (2006). Experimental Evidence for Spin-Orbit Interactions in Positronium-Xe Collisions. *Phys. Rev. Lett.*, **97**, 253402-4.

Santos, A.C.F., Hasan, A. and DuBois, R.D. (2004). Doubly Differential Multiple Ionization of Krypton by Positron and Electron Impact. *Phys. Rev. A*, **69**, 032706-5.

Sarkadi, L. (2003). Fragmentation of Positronium in Collision with He Atoms: A Classical Theoretical Approach. *Phys. Rev. A*, **68**, 032706-8.

Sarkar, N.K., Basu, M. and Ghosh, A.S. (1992). Positronium Formation in the $n = 1$ and $n = 2$ States in e^+–He Scattering. *Phys. Rev. A*, **45**, 6887-93.

Sarkar, N.K., Chaudhury, P. and Ghosh, A.S. (1999). Ps–He Scattering using the Three-State Positronium Close-Coupling Approximation. *J. Phys. B: At., Mol. Opt. Phys.*, **32**, 1657–1667.

Schmitt, A., Cerny, U., Möller, H., Raith, W. and Weber, M. (1994). Positron-Atom Doubly Differential Ionization Cross Sections. *Phys. Rev. A*, **49**, R5–R7.

Schultz, D.R. and Olson, R.E. (1988). Single-Electron Removal Processes in Collisions of Positrons And Protons with Helium at Intermediate Velocities. *Phys. Rev. A*, **38**, 1866–1876.

Schultz, D.R., Olson, R.E. and Reinhold, C.O. (1991). Recent Advances in the Comparison of Matter- and Antimatter-Atom Collisions. *J. Phys. B: At., Mol. Opt. Phys.*, **24**, 521–558.
Skalsey, M., Engbrecht, J.J., Bithell, R.K., Vallery, R.S. and Gidley, D.W. (1998). Thermalization of Positronium in Gases. *Phys. Rev. Lett.*, **80**, 3727–3730.
Skalsey, M., Engbrecht, J.J., Nakamura, C.M., Vallery, R.S. and Gidley, D.W. (2003). Doppler-Broadening Measurements of Positronium Thermalization in Gases. *Phys. Rev. A*, **67**, 022504-15.
Sorokin, A.A., Shmaenok, L.A., Bobashev, S.V., Möbus, B. and Ulm, G. (1998). Measurements of Electron-Impact Ionization Cross Sections of Neon by Comparison with Photoionization. *Phys. Rev. A*, **58**, 2900–2910.
Sorokin, A.A., Shmaenok, L.A., Bobashev, S.V., Möbus, B., Richter, M. and Ulm, G. (2000). Measurements of Electron-Impact Ionization Cross Sections of Argon, Krypton, and Xenon by Comparison with Photoionization. *Phys. Rev. A*, **61**, 022723-11.
Sorokin, A.A., Beigman, I.L., Bobashev, S.V., Richter, M. and Vainshtein, L.A. (2004). Total Electron-Impact Ionization Cross Sections of Helium. *J. Phys. B: At., Mol. Opt. Phys.*, **37**, 3215–3226.
Sparrow, R.A. and Olson, R.E. (1994). Projectile and Electron Spectra Resulting from Positron-Argon Collisions. *J. Phys. B: At., Mol. Opt. Phys.*, **27**, 2647–2655.
Starrett, C., McAlinden, M.T. and Walters, H.R.J. (2005). Fragmentation of Positronium. *Phys. Rev. A*, **72**, 012508-14.
Starrett, C. and Walters, H. (2007). Ionization in Positronium Collisions with Noble Gases. *J. El. Spec. and Rel. Phenom.*, **161**, 194–198.
Stein, T.S., Harte, M., Jiang, J., Kauppila, W.E., Kwan, C.K., Li, H. and Zhou, S. (1998). Measurements of Positron Scattering by Hydrogen, Alkali Metal, and Other Atoms. *Nucl. Instrum. Methods B*, **143**, 68–80.
Surko, C.M., Gribakin, G.F. and Buckman, S.J. (2005). Low-Energy Positron Interactions with Atoms and Molecules. *J. Phys. B: At., Mol. Opt. Phys.*, **38**, R57–R126.
Szluinska, M., Reeth, P.V. and Laricchia, G. (2002). Empirical Scaling of Positron- and Electron-Impact Ionization Cross Sections. *J. Phys. B: At., Mol. Opt. Phys.*, **35**, 4059–4067.
Szluinska, M. and Laricchia, G. (2004). Investigation into the Energy Dependence of the Annihilation Cross-Sections for Ne and Xe below the Positronium Formation Threshold. *Nucl. Instrum. Methods B*, **221**, 100–106.
Szluinska, M., Murtagh, D. and Laricchia, G. (2005). Low Energy Positron-Induced Ionization of He and Ne. *Nucl. Instrum. Methods B*, **229**, 51–54.
Ullrich, J., Moshammer, R., Dorn, A., Dorner, R., Schmidt, L.P.H. and Schmidt-Bocking, H. (2003). Recoil-Ion and Electron Momentum Spectroscopy: Reaction-Microscopes. *Reports on Progress in Physics*, **66**, 1463–1545.
Van Reeth, P. and Humberston, J.W. (1999). Elastic Scattering and Positronium Formation in Low-Energy Positron-Helium Collisions. *J. Phys. B: At., Mol. Opt. Phys.*, **32**, 3651–3667.
Van Reeth, P., Szluinska, M. and Laricchia, G. (2002). On the Normalization of the Positron-Impact Direct Ionization Cross-Section in the Noble Gases. *Nucl. Instrum. Methods B*, **192**, 220–224.
Van Reeth, P., Laricchia, G. and Humberston, J.W. (2005). On the Energy Dependence of Annihilation Cross-Sections for Positrons in Collision with Atoms and Molecules. *Physica Scripta*, **71**, C9–C13.
Vetter, P.A. and Freedman, S.J. (2003). Search for CPT-odd Decays of Positronium. *Phys. Rev. Lett.*, **91**, 263401-4.
Walters, H.R.J. (1988). Positron Scattering by Atomic Hydrogen at Intermediate Energies: 1s to 1s, 1s to 2s and 1s to 2p Transitions. *J. Phys. B: At., Mol. Opt. Phys.*, **21**, 1893–1906.
Walters, H.R.J., Yu, A.C.H., Sahoo, S. and Gilmore, S. (2004). Positronium-Atom Collisions. *Nucl. Instrum. Methods B*, **221**, 149–159.
Walters, H.R.J., Sahoo, S. and Gilmore, S. (2005). Atomic Collisions Involving Positrons. *Nucl. Instrum. Methods B*, **233**, 78–87.
Walters, H.R.J., Starrett, C. and McAlinden, M.T. (2006). Differential Cross Sections for Fragmentation of Positronium. *Nucl. Instrum. Methods B*, **247**, 111–116.

Wannier, G.H. (1953). The Threshold Law for Single Ionization of Atoms or Ions by Electrons. *Phys. Rev.*, **90**, 817–825.

Winick, J.R. and Reinhardt, W.P. (1978). Moment T-Matrix Approach to e^+-H Scattering. I. Angular Distribution and Total Cross Section for Energies Below the Pickup Threshold. *Phys. Rev A*, **18**, 910–914; Moment T-matrix approach to e^+-H scattering. II. Elastic scattering and total cross section at intermediate energies. *Phys. Rev A*, 18, 925-934.

Wu, H., Bray, I., Fursa, D.V. and Stelbovics, A.T. (2004). Convergent Close-Coupling Calculations of Positron-Helium Scattering at Intermediate to High Energies. *J. Phys. B: At., Mol. Opt. Phys.*, **37**, 1165–1172.

Yang, C.N. (1950). Selection Rules for the Dematerialization of a Particle into Two Photons. *Phys. Rev.*, **77**, 242–245.

Young, J.A. and Surko, C.M. (2007). Role of Binding Energy in Feshbach-Resonant Positron-Molecule Annihilation. *Phys. Rev. Lett.*, **99**, 133201-4.

Zafar, N., Laricchia, G., Charlton, M. and Garner, A. (1996). Positronium-Argon Scattering. *Phys. Rev. Lett.*, **76**, 1595–1598.

Zhou, S., Li, H., Kauppila, W.E., Kwan, C.K. and Stein, T.S. (1997). Measurements of Total and Positronium Formation Cross Sections for Positrons and Electrons Scattered by Hydrogen Atoms and Molecules. *Phys. Rev. A*, **55**, 361–368.

CHAPTER 2

Interactions Between Thermal Ground or Excited Atoms in the Vapor Phase: Many-Body Dipole–Dipole Effects, Molecular Dissociation, and Photoassociation Probed By Laser Spectroscopy

J.G. Eden, B.J. Ricconi, Y. Xiao[1], F. Shen and A.A. Senin[2]

Laboratory for Optical Physics and Engineering, Department of Electrical and Computer Engineering, University of Illinois, Urbana, IL 61801, United States

Contents			
	1.	Introduction	51
	2.	Detection of Quantum Beating in Atoms and Molecules by Wavepacket Interferometry and a Coherent Nonlinear Optical Process	54
		2.1 Historical Background, Wavepacket Detection	54
		2.2 Brief Review of Theory	57
		2.3 Experimental Arrangement and Data Acquisition	59
		2.4 Rb Wavepackets: Quantum Beating at 2–18.2 THz	64

---

[1] Present address: Alfalight Inc., 1832 Wright St., Madison, WI 53705, United States.
[2] Present address: VLOC, Subsidiary of II-VI Inc., YAG Business Unit, 6716 Industrial Ave., Port Richey, FL 34668, United States.

Advances in Atomic, Molecular, and Optical Physics, Volume 56 © 2008 Elsevier Inc.
ISSN 1049-250X, DOI 10.1016/S1049-250X(08)00011-6 All rights reserved.

  2.5 Na 5s–4d$_{5/2}$ Quantum Beating at 1348 cm^{-1} (40.41 THz)   69
  2.6 PFWM and PSWM in Other Atoms and Molecules   69
 3. Many-Body, Dipole–Dipole Interactions Among Excited Alkali Atoms   70
  3.1 Introductory Comments   70
  3.2 Estimates of Many-Body Interaction Energies   71
  3.3 Experimental Results: Sideband Splittings Observed in the Fourier and Temporal Domains   75
  3.4 Impact of the Pair Distribution Function   81
 4. Observation of Molecular Dissociation and Nascent Product State Distributions by the Dipole–Dipole Interaction   82
  4.1 Introduction   82
  4.2 Electronic Structure and Predissociation of Rb$_2$: Generation of Excited Atomic Fragments   84
  4.3 Molecular Dissociation Transients, Atomic Product State Distributions   85
  4.4 Diffusion Description of Amplitude Transients   90
  4.5 Dominant Rb$_2$ Predissociation Channels   93
 5. Coherent Control of Rb$_2$ Predissociation   95
 6. Photoassociation of Rare-Gas–Halogen Atomic Pairs at Ambient Temperature   100
  6.1 Introduction, Stationary Phase Approximation   100
  6.2 Theoretical Considerations   103
  6.3 Simulations and Derived Xenon Monoiodide Spectroscopic Constants   106
 7. Application of Photoassociation to High Intensity Discharge Lighting   110
 8. Summary and Conclusions   112
 Acknowledgments   114
 References   114

Abstract Atom–atom and atom–molecule interactions in the absence or presence of an optical radiation field are among the fundamental processes in atomic, molecular, and optical physics. This review addresses recent advances that provide insight into the dynamics of many-body fipole–dipole interactions between excited atoms at long range, and the production of a transient molecule from a pair of thermal, ground state atoms by photoassociation at small atomic separations (<5 Å). Ultrafast pump–probe experiments have probed the dipole–dipole interactions within ensembles of excited alkali atoms by combining wavepacket (Ramsey) interferometry with a coherent nonlinear optical process. Monitoring the temporal behavior of quantum beating in the atomic species with parametric four (or six) wave mixing, for example, provides an *in situ* probe of the net interaction between two or more excited atoms, and the coherence of the nonlinear process preserves phase information. In the Fourier domain, the dipole–dipole interaction is manifested by the appearance of sidebands associated with the quantum beating frequency. Analysis of the sideband splittings observed as the mean atom–atom separation is varied suggests that n-atom ensembles ($2 \leq n \leq 5$) are detected. Quantum beating in the atomic species also serves as a sensitive detector of molecular dissociation by monitoring the approach of atomic fragments through the dipole–dipole interaction. With this new spectroscopic tool, the predissociation of electronic excited states of Rb_2 and Na_2 have been observed and nascent excited state distributions determined. Control of the Rb_2 dissociation process so as to yield Rb fragments in a specific atomic excited state has been accomplished through varying the chirp of the pump and probe pulses. Complementary experiments on the nanosecond time scale have recorded the excitation spectra for the photoassociation of thermal pairs of ground state rare-gas–halogen atoms. Analysis of the experimental Kr–F and Xe–I photoassociative excitation spectra, acquired at ambient temperature, by coupled vibration–rotation (VR) calculations yields molecular excited state ($B^2\Sigma^+$) constants more precise than those available from perturbative theoretical models or bound $\rightarrow$ free emission spectroscopy. An application of the photoassociation of thermal atomic pairs to the operation and efficiency of high temperature, metal-halide arc lamps is presented.

1. INTRODUCTION

Interactions among ground and excited atoms or molecules undergird much of atomic, molecular and optical (AMO) physics. Since the nature of such interactions is governed primarily by the electronic structure of the participants and, hence, the characteristic potential for the system of interest, experimental and theoretical efforts continue to pursue tools capable of precisely determining potential surfaces or extracting interaction parameters from measurements. Advances of the past decade in this field have, to a significant degree, been propelled by the rapid development of techniques for generating and trapping cold atoms and molecules. The demonstration of the magneto-optical trap, in particular, and the realization of Bose–Einstein condensates provide new avenues for the study of atom–atom and molecule–molecule interactions at long range and in the virtual absence of thermal motion. Sample densities in the trap of nominally $\lesssim 10^{10}$ cm^{-3} correspond to mean particle separations of at least a few μm and μs interaction times, and analysis of diatomic interaction potentials acquired by photoassociation spectroscopy [1,2] yields the dispersion coefficients associated with electric multipole interactions (C_6, C_8, ...) [3]. Near the separated atom limit, the behavior of atom–atom excited state potentials is dictated by the two-body dipole–dipole interaction, and precise measurements of the asymptotic coefficient C_3 in this region of internuclear separation yield values for alkali atomic excited state lifetimes, s-wave scattering lengths, and retardation corrections that pose rigorous tests for theory [4,5]. Ensembles of cold Rydberg atoms generated in traps have also opened windows onto long-range interactions culminating in Rydberg blockades [6,7], cold plasma formation, [8,9] one-dimensional Rydberg gases with permanent dipole moments, [10] and the existence of Rydberg molecules at ultralong range–bound states of the alkali dimers having equilibrium internuclear separations of 10^2–10^5 a.u. (Refs. [11–13]). Because the dipole moment of a Rydberg atom scales quadratically with the principal quantum number n, most experiments and theory have focused on states with $n > 20$, and Boisseau et al. [13] have succinctly summarized the scientific landscape with respect to cold Rydberg atom studies by noting that 'The exaggerated properties of Rydberg atoms provide a fertile ground for new physics.' For a thorough review of recent developments, the reader is encouraged to consult Ref. [14].

Although providing a unique environment for the study of many-body systems comprising Rydberg or ground state atoms or molecules, [15, 16] ultracold gases are restricted at present in several respects, including the available range in mean interatomic separation, complications in data interpretation introduced by Rydberg state emission, and the limited number of atoms and molecules that have been trapped successfully to date. A wealth of structural and dynamic information concerning atom–atom interactions is also available from laser spectroscopic experiments in which the atoms are thermalized at ambient or higher temperatures. This chapter is devoted to a review of several sets of such experiments, conducted at the University of Illinois, and their theoretical descriptions. Specifically, Sections 2–5 describe an ultrafast laser pump–probe approach to detecting long-range interactions within ensembles of excited atoms, based upon *in situ* monitoring of the frequency characteristics of a wavepacket by Ramsey (wavepacket) interferometry in combination with a coherent nonlinear optical process.

Coupling a wavepacket to a coherent optical process such as parametric four or six wave mixing (denoted PFWM and PSWM, respectively) provides for monitoring the temporal history of the wavepacket while preserving both amplitude and phase information. Quantum beating is a sensitive probe of the net dipole–dipole interaction of an atom with its electronically-excited neighbors, and absolute measurements of many-body, dipole–dipole interaction energies are now accessible because of the standard provided by atomic quantum beating frequencies. In the Fourier domain, dipole–dipole interactions are evident by the appearance of sidebands associated with an $ns - (n-2)d$ quantum beating frequency in an alkali atom. Measurements of the sideband splitting in both Rb and Na indicate that the dipole–dipole interactions within n-atom ensembles ($n \leq 5$) are being observed.

A brief review of the theoretical foundation for the detection of an atomic wavepacket with a coherent nonlinear optical process is presented in Section 2, along with exemplary Fourier spectra for Rb and Na illustrating quantum beating at several frequencies but primarily 18.2 THz and 40.4 THz, respectively. Many-body dipole–dipole interactions are treated in Section 3. Sections 3.3 and 3.4 summarize recent experimental results and discuss the implications of the data for extracting atomic pair spatial distribution functions. It should be emphasized that the ability of wavepacket interferometry and PFWM (for example) to monitor slight shifts in the frequency components of a wavepacket, despite the temperature of the thermalized velocity distribution of the background atoms, is a result of freezing the atom–atom interaction with $\sim$120 fs laser pulses and a $\sim$100 ps measurement interval. This experimental capability offers a new dimension to the study of atomic and molecular dynamics over an extended range in internuclear separation.

Once the ability to detect wavepacket–atom (or molecule) interactions has been established, a door appears through which one is able to explore a broad range of fundamental processes, and to do so with the temporal resolution afforded by ultrafast spectroscopy. As one example, quantum beating in atoms and molecules serves as a sensitive detector of molecular dissociation by enabling the observation of approaching atomic fragments through the wavepacket–atom linkage provided by the dipole–dipole interaction. That is, the temporal histories of the amplitude and phase of the Fourier (frequency) components comprising the wavepacket are perturbed by the proximity of newly-produced excited atoms. Fourier analysis of the variation of the PFWM signal wave intensity with the pump–probe delay time reveals clear molecular dissociation transients that are attributable to specific dissociation channels (atomic fragment states). Section 4 describes experiments in which the predissociation of Rb_2 electronic states in the 24 000–28 000 cm^{-1} region has been observed and the nascent atomic product state distributions have been determined.

Armed with the detection of molecular dissociation fragments by wavepacket interferometry (also known as Ramsey interferometry), the control of the atomic products of Rb_2 dissociation can now be pursued, and Section 5 discusses the selectivity that has been realized with chirped pump and probe pulses. The theme, therefore, of Sections 2–5 is that wavepacket interferometry, in concert with a coherent nonlinear optical process, offers a new and versatile *in situ* probe with which atomic and molecular dynamics can be observed to a level of detail not available previously.

A second atom–atom process pursued in our laboratory is the photoassociation of thermal pairs of ground state atoms at small values of internuclear separation ($R \lesssim 5$ Å). As described in Section 6, excitation spectroscopy of the photoassociation of rare-gas–halogen atomic pairs from the thermal and vibrational continua of the ground electronic state reveals deeply-modulated Franck–Condon structure that provides a demanding test for theoretical excited state potentials. In particular, representation of the Xe–I photoassociative excitation spectrum in the ultraviolet in terms of a coupled vibration–rotation (VR) description yields excited and ground state structural constants that are more precise than those available from *ab initio* calculations or the analysis of conventional bound → free emission spectra. This chapter concludes with an application of photoassociation spectra to high intensity discharge (HID) lighting. Interaction of the intense, broadband optical field generated by an HID lamp with the high background Hg and I number densities ($\gtrsim 10^{19}$ cm^{-3}) characteristic of metal-halide arc lamps has a deleterious effect on the efficiency and emission spectrum of the lamp. Calculations of the photoassociation spectrum of Hg–I collision pairs indicates that the impact of this bound ← free absorption process is significant, altering the blue-green portion of the spectral profile and absorbing at least several per cent of the energy radiated by the core of the arc.

The examples selected for discussion here vividly illustrate that the simplest of AMO processes – the interaction of two or more atoms in the presence of an optical radiation field – are, indeed, fundamental to a host of physical and chemical phenomena and yet are exquisitely elegant.

2. DETECTION OF QUANTUM BEATING IN ATOMS AND MOLECULES BY WAVEPACKET INTERFEROMETRY AND A COHERENT NONLINEAR OPTICAL PROCESS

2.1 Historical Background, Wavepacket Detection

Reported independently in 1964 by Alexandrov [17] and Dodd and co-workers, [18] quantum beat spectroscopy offers a time domain approach to characterizing non-stationary states of an atom or molecule with exceptional resolution. A coherent, time-dependent superposition of eigenstates is produced upon populating two or more atomic (or molecular) excited states with a laser pulse having sufficient bandwidth. Measurements of the radiative decay of these states to a common lower level exhibit oscillations in the fluorescence intensity as a result of interference, known as quantum beating, between the wave functions of the coherently-driven states. The periodic variation of the emission from the coupled states has a frequency directly proportional to the energy defect between the photoexcited levels. Prominent early experiments demonstrating the capability of this powerful spectroscopic tool include the observation of hyperfine structure in Cs by Haroche et al. [19] in 1973, and the measurements by Andrä [20] of the Lamb-shift in atomic excited states. Salour [21] noted in 1976 that the fine structure intervals in the 7^2D and 8^2D states of Na had not yet been measured by quantum beat spectroscopy because the detection system bandwidths were insufficient at the time to accommodate fine structure splittings of 173 and 273 MHz, respectively. Indeed, the observation of quantum beating requires that the bandwidth of the optical pulses photoexciting an atom or molecule be sufficient to encompass the states of interest. However, Beach and Hartmann [22] and Morita and Yajima [23] demonstrated the feasibility of exciting coherent transient processes such as photon echoes without the need for ultrashort laser pulses or temporal coherence. Quantum beating between hyperfine levels of Na was observed in Ref. [22] with incoherent optical pulses having a bandwidth of ~10 GHz. Subsequently, Golub and Mossberg [24] combined the degenerate four wave mixing process proposed in Ref. [23] with broadband, 5 ns pulses from an optically-pumped dye cell to yield quantum beating at 7.5 THz between the $J = \frac{1}{2}$ and $\frac{3}{2}$ fine structure levels of the Rb($5p^2P_J$) state. Brief but well-written reviews of the seminal experiments in the quantum beat spectroscopy of both atoms and molecules can be found in Refs. [25] and [26].

Coherent superpositions of atomic or molecular states, given by the expression:

$$|\psi(t)\rangle = \sum_k c_k |k\rangle \exp(-iE_k t/\hbar), \qquad (1)$$

where $|k\rangle$ are stationary states of an atom or molecule having energies E_k, are known as wavepackets. Consequently, as Alber et al. [27] have noted in reference to wavepackets constructed from Rydberg states of principal quantum number n, '...Rydberg wave packets can be interpreted as quantum beats between different n states'. The general characteristics of atomic wavepackets were predicted in 1986 (Refs. [27,28]) and experimental observations of the dynamics of radially-localized wavepackets were reported two years later [29]. In experimental

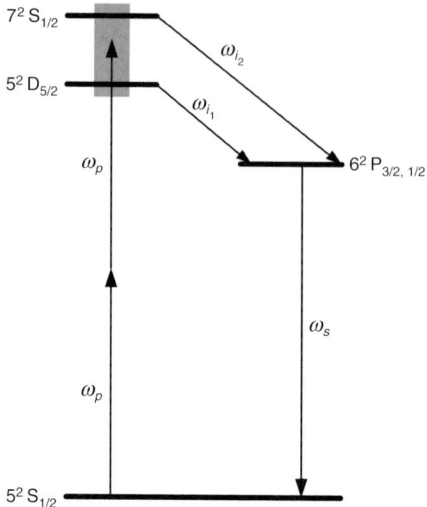

FIGURE 1 Partial energy level diagram for atomic Rb illustrating the difference frequency PFWM process $\omega_s = 2\omega_p - \omega_{i_{1,2}}$, driven by two-photon excitation of the 7s and $5d_{5/2}$ states with ultrafast (~120 fs) laser pulses having a center wavelength of $\lambda_p \sim 770$ nm. The shaded box qualitatively represents twice the bandwidth of the laser, and $\omega_{i_{1,2}}$ denotes the two idler frequencies generated. For this axially-phase-matched process, the signal wave is produced at $\lambda_s \sim 420$ nm

investigations of Kepler orbital periods as well as the dephasing and revival of Rydberg wavepackets in the alkalis [30–33] and Ca [34] (and virtually all studies since), photoionization served exclusively as the detection mechanism. Nonlinear coherent optical processes were introduced approximately a decade ago as an alternative approach to detecting and analyzing both molecular [35–41] and atomic [36, 42] wavepackets. Investigating the vibrational and rotational structure of several diatomics, including I_2 and NaI, was accomplished by coupling the ground and excited state potentials by coherent anti-Stokes Raman scattering (CARS), coherent Stokes Raman scattering (CSRS), or degenerate four wave mixing (DFWM) which require spatially separating the four wave mixing signal from the three laser pulses driving the third-order polarization of the system. Siebert et al. [40] showed that adopting a magic angle geometry for the polarization of the beams in a CARS experiment eliminates the contribution of rotation to the FWM signal, thereby permitting the vibrational dynamics of the molecule to be observed selectively.

In 1998, Tran et al. [42] demonstrated that atomic wavepackets could be detected, and their temporal dynamics monitored, by parametric four wave mixing (PFWM) in ultrafast pump–probe experiments. Figure 1 is a partial energy level diagram of the Rb atom, illustrating the PFWM difference frequency process: $\omega_s = 2\omega_p - \omega_{i_{1,2}}$, where ω_p, $\omega_{i_{1,2}}$, and ω_s are the frequencies of the pump, idlers and signal, respectively. Wavepackets comprising the $7^2S_{1/2}$ and $5^2D_{5/2}$ states (hereafter denoted as 7s and $5d_{5/2}$, respectively) are generated and detected by exciting the resonantly-enhanced, two-photon 7s, $5d_{5/2}$ ←← 5s transitions with two identical 100–150 fs optical pulses, one of which (probe) is delayed in time with respect to the other (pump) by Δt. The shaded box in Figure 1 indicates that the bandwidth of the pump and probe pulses is sufficient to encompass the 7s ←← 5s and $5d_{5/2}$ ←← 5s transitions. Consequently, both pulses generate quantum beating at 607.94 cm^{-1} (7s–$5d_{5/2}$ energy defect) and the interference between the pump and probe-produced 7s–5d coherent superpositions provides the means by which the wavepacket is monitored. The composition of the wavepacket, reflected by the phase and amplitude of the quantum beating, is imprinted onto the intensity of the macroscopic signal wave produced by FWM near the $6^2P_J \rightarrow 5^2S_{1/2}$ transitions ($\lambda \sim 420$ nm), and is recovered by Fourier analysis of the dependence of the modulated signal wave intensity on Δt. Parametric six wave mixing (PSWM), a $\chi^{(5)}$ process, was demonstrated recently by Zhu et al. [43] to also be suitable for detecting and monitoring atomic Rb wavepackets. Although these results suggest that the concept of monitoring atomic or molecular wavepackets by a coherent nonlinear optical process is broad and it is likely that a variety of other $\chi^{(n)}$ processes, such as third harmonic radiation, will also prove to be viable detection mechanisms, the discussion to follow will be restricted to the detection of wavepackets by PFWM.

2.2 Brief Review of Theory

Modulation of the PFWM signal wave intensity upon scanning the pump–probe time delay Δt reflects the linkage between wavepacket interferometry and the nonlinear optical process. Since this coupling between the microscopic (wavepacket) and the macroscopic (signal wave intensity) undergirds all of the experiments described in Sections 2–6, we summarize here one theoretical approach but only the salient aspects are presented. A density matrix formalism describing the interaction of the pump and probe optical fields with the nonlinear polarization of the medium was presented in Ref. [44] but in the discussion to follow we follow Reintjes [45] in expressing the third-order nonlinear polarization of the medium, $P^{(3)}(\omega)$, in terms of the complex field amplitudes. For the sake of simplicity, only one idler wave is considered initially, the probe pulse is ignored, and the quasi-static and slowly-varying envelope approximations are assumed to be valid. Under these conditions, the total electric field can be expressed by the sum:

$$E_t(z,t) = \frac{1}{2}\left[A_p(z)e^{-j(\omega_p t - k_p z)} + A_i(z)e^{-j(\omega_i t - k_i z)} + A_s(z)e^{-j(\omega_s t - k_s z)} + \text{c.c.}\right], \tag{2}$$

where the pump, idler, and signal are taken to be plane waves denoted by the subscripts p, i, and s, respectively. Thus, the set of coupled equations obeyed by the field amplitudes (A) when only the pump optical field is present can be written:

$$2ik_p \frac{dA_p}{dz} = -\frac{4\pi \omega_p^2}{c^2} P^{(3)}(\omega_p) e^{-i\Delta k z} \tag{3}$$

$$2ik_i \frac{dA_i}{dz} = -\frac{4\pi \omega_i^2}{c^2} P^{(3)}(\omega_i) e^{-i\Delta k z} \tag{4}$$

$$2ik_s \frac{dA_s}{dz} = -\frac{4\pi \omega_s^2}{c^2} P^{(3)}(\omega_s) e^{-i\Delta k z} \tag{5}$$

in which the group velocities at ω_p, ω_i, and ω_s are assumed to be equal, the wave vector mismatch Δk is given by $\Delta k = 2\omega_p n_p - \omega_i n_i - \omega_s n_s$, and the third-order nonlinear polarizations induced by the pump field at ω_p, ω_i, and ω_s are given by

$$P^{(3)}(\omega_p) = \tfrac{3}{2}\chi^{(3)}(-\omega_p, \omega_s, -\omega_p, \omega_i) A_s A_p^* A_i, \tag{6}$$

$$P^{(3)}(\omega_i) = \tfrac{3}{4}\chi^{(3)}(-\omega_i, -\omega_s, \omega_p, \omega_p) A_s^* A_p^2, \tag{7}$$

and

$$P^{(3)}(\omega_s) = \tfrac{3}{4}\chi^{(3)}(-\omega_s, \omega_p, \omega_p, -\omega_i) A_p^2 A_i^*. \tag{8}$$

In this situation, the intensity of the signal wave is expressed as:

$$I_s \propto \left|P^{(3)}(\omega_s)\right|^2 \propto \left|\chi^{(3)} A_p^2 A_i^*\right|^2. \tag{9}$$

If the pump and probe pulses are identical, Equations (3)–(8) can be modified to incorporate the probe optical field by replacing the field amplitudes $A_p(z)$ and $A_i(z)$ with the respective expressions: $A'_p(z) = \left(1 + e^{-i\omega_p \Delta t}\right) A_p(z)$ and $A'_i(z) = \left(1 + e^{-i\omega_i \Delta t}\right) A_i(z)$. Substituting the revised pump and idler amplitudes into Equation (9) yields

$$I'_s(\Delta t) \propto \left|\chi^{(3)} A'^2_p A'^*_i\right|^2 \propto \left|\left(1 + e^{-i\omega_p \Delta t}\right)^2 \left(1 + e^{i\omega_i \Delta t}\right) \chi^{(3)} A^2_p A^*_i\right|^2. \quad (10)$$

The dependence of the signal wave intensity on the pump–probe delay time is now explicit but the expression of Equation (10) does not yet accurately reflect the interference between two PFWM pathways (5s →→ 7s → 6p → 5s and 5s →→ 5d → 6p → 5s), both of which contribute to the nonlinear polarization induced at the signal frequency. Assuming the two pathways to be independent of one another, Equation (10) can be modified by the factor

$$\left|\left(1 + e^{-i\omega_{p_1}\Delta t}\right)^2 \left(1 + e^{i\omega_{i_1}\Delta t}\right) + \left(1 + e^{-i\omega_{p_2}\Delta t}\right)^2 \left(1 + e^{i\omega_{i_2}\Delta t}\right)\right|^2$$
$$= \left[a_1 \cos\left((2\omega_{p_2} - 2\omega_{p_1})\Delta t\right) + a_2 \cos\left(\omega_{i_1}\Delta t\right) + a_3 \cos\left(\omega_{i_2}\Delta t\right) + \ldots\right], \quad (11)$$

where the a_n are coefficients and the indices 1 and 2 refer to the PFWM pathways cited above and illustrated in Figure 1. As we might expect, therefore, the variation of the signal wave intensity with Δt, I'_s, is modulated at several frequencies, one of which corresponds to the energy defect between the s and d states of the atom excited by two-photon absorption. This simplified treatment assumes that the nonlinear processes of Figure 1 are truly parametric in that no net population is transferred into any of the excited (intermediate) states. If, however, excited atoms in the 5d, 7s, or 6p states *appear from any external source*, the amplitudes of the Fourier domain representation of Equations (10) and (11) will be affected. Consequently, the temporal history of the wavepacket will be captured by recording $I'_s(\Delta t)$. The temporally-varying coefficients associated with each Fourier component (of which the wavepacket is comprised) can be regarded as reflecting the time (Δt) dependence of the population of either (or both) of the excited states responsible for that frequency. Expressed in different terms, the fundamental interaction in these experiments – that of a wavepacket with PFWM – results in the superposition of the temporal variation of the amplitude and phase for all frequency (Fourier) components of the wavepacket onto the PFWM signal wave intensity.

2.3 Experimental Arrangement and Data Acquisition

Two laser systems were employed in separate experiments to generate pump and probe wavelengths of ~620 nm and ~770 nm for Rb, and ~590 nm for Na excitation. Initial studies were conducted with a colliding pulse, mode-locked (CPM) oscillator/dye amplifier system (shown schematically in Figure 2) capable of generating 80–100 fs (sech^2)

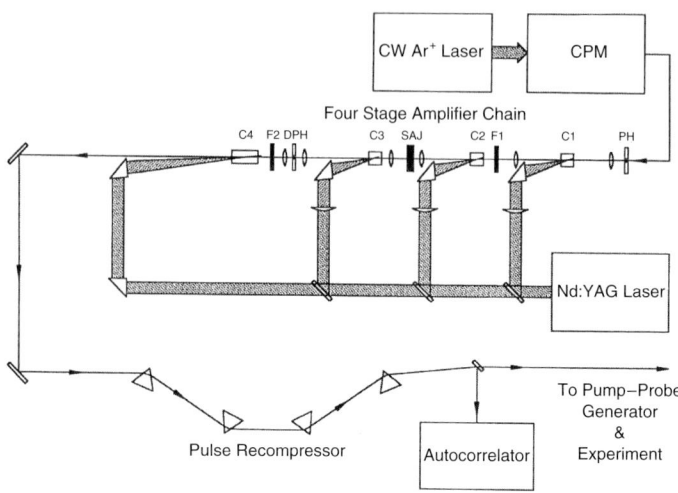

FIGURE 2 Diagram of the colliding pulse, mode-locked (CPM) oscillator/dye amplifier system for studies of 12s–11d and 11s–9d quantum beating in Rb. After pulse compression, the ~16 nm bandwidth of the dye amplifier output was ample to simultaneously photoexcite the two-photon 11s, 9d ←← 5s and 12s, 10d ←← 5s transitions. The acronyms SAJ, F, C, PH, and DPH denote (respectively) saturable absorber jet, filter, dye cell, pinhole, and diamond pinhole

pulses in the red ($\lambda_p \sim 620$ nm) with energies of 600 µJ. Pumped by 4.5 W (all lines) from an Ar ion laser, the CPM produced ~110 fs pulses (as measured by an autocorrelator) with a bandwidth of 8–10 nm, a peak wavelength of 620–625 nm, a pulse repetition frequency (PRF) of 90 MHz and single pulse energies of ~150 pJ, thus necessitating the addition of an amplifier. The four stage amplifier system [46,47] comprised quartz cells or custom machined stainless steel cells having a clear aperture of several cm and dye path lengths ranging from 2 cm to 7 cm (for the final amplifier). To maximize output beam quality, all of the sulforhodamine 640 amplifiers were pumped longitudinally by a pulsed (~8 ns), frequency-doubled Nd:YAG laser ($\lambda = 532$ nm), operating at 30 Hz and providing single pulse energies of typically 5 mJ for the first stage to 17 mJ for the saturated final stage. A malachite green absorber jet situated between the second and third amplifier stages served to suppress amplified spontaneous emission (ASE). Pulses emerging from the last amplifier were compressed from a temporal width of ~600 fs to sub-100 fs in a set of four flint glass (SF-10) prisms and the final pulse bandwidth of 16 nm (~12.5 THz) was more than ample to encompass all four of the 11s, 9d ←← 5s and 12s, 10d ←← 5s two-photon transitions of atomic Rb. Including reflection losses introduced by the prism compressor, the net overall gain of the amplifier chain was $\sim 4 \times 10^6$, yielding output

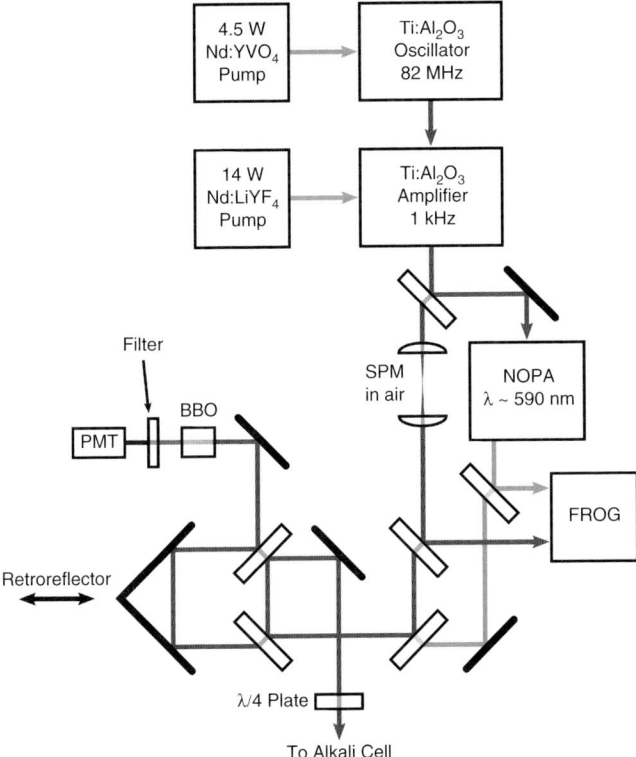

FIGURE 3 Schematic diagram of the experimental arrangement for Rb 7s–5d quantum beating experiments. The Ti:Al$_2$O$_3$ oscillator/amplifier system is operated at $\lambda \sim 770$ nm to drive the 7s, 5d ←← 5s transitions. A nonlinear optical parametric amplifier (NOPA) produces ~40 fs pulses centered at $\lambda \sim 590$ nm to generate Na 5s-4d quantum beating at ~40.4 THz (Section 2.5). pairs of identical pulses, separated in time by Δt, are produced by an interferometer and subsequently directed into a sapphire cell (not shown) containing Rb or Na vapor. The 1 m monochromator and the diagnostics for profiling the PFWM signal beam are also not illustrated

pulse energies of nominally 600 µJ, and the contribution of ASE to the overall pulse energy was maintained below 5%. Further details regarding the construction and performance of the CPM/dye amplifier system can be found in Refs. [46] and [47].

Pump–probe experiments at 770 nm, in which the 7s and 5d states of Rb were excited simultaneously by two-photon transitions, were conducted with the Ti:Al$_2$O$_3$ oscillator/amplifier system illustrated in Figure 3 (Refs. [44,48]). A Ti:sapphire oscillator, pumped continuously with 4.5 W from a frequency-doubled, diode-pumped Nd:YVO$_4$ laser, produces pulses having energies of ~10 nJ and a nominal duration of 100 fs at a PRF of 80 MHz, corresponding to an average output power of ~0.7 W.

Amplification of selected oscillator pulses is provided by a regenerative amplifier, which is driven by a diode-pumped Nd:LiYF$_4$ laser (14 W at 527 nm). The linearly-polarized output pulses from the amplifier, having a temporal width of ~120 fs and maximum energies above 1 mJ, are produced at a PRF of 1 kHz. In order to simultaneously excite the 7s and 5d states of Rb from ground (within the same pulse), it was necessary to: (1) tune the oscillator off the Ti:Al$_2$O$_3$ gain peak and slightly to the blue (to 770 nm), and (2) increase the bandwidth of the pulses exiting the amplifier. The latter was accomplished by self-phase modulation (SPM) in ambient air. After focusing the output of the amplifier with a lens having a 15 cm focal length, the beam was recollimated. With this simple arrangement, the breadth of the laser spectrum was adequate to access both the 5d$_{5/2}$ and 7s states by the 7s, 5d$_{5/2}$ ←← 5s transitions.

Following recollimation of the beam, pairs of pulses were produced by a beamsplitter and one pulse (probe) was delayed in time with respect to the other by a Michelson interferometer incorporating a retroreflector mounted onto a computer-controlled translation stage. The time delay Δt between the pump and probe pulses could be varied up to 300 ps with a resolution selectable from 2.7 to 10.7 fs, the choice depending upon the total scan length. Both pulses were focused by a lens having a focal length of 26 cm, and propagated collinearly through a 10 mm diameter cylindrical optical cell (not shown in Figure 3) constructed entirely of sapphire. Containing saturated, natural abundance Rb vapor at pressures between 2.5×10^{-3} and 20 Torr (corresponding to number densities, [Rb], between $\sim 6 \times 10^{13}$ cm^{-3} and 3×10^{17} cm^{-3}, respectively), the sapphire cell has a vapor column 1.7 cm in length. An uncertainty of $\lesssim \pm 1\%$ in the Rb cell temperature results in an estimated error in the Rb number densities of ~6%–8%. Throughout these experiments, the single pulse energy was typically maintained below 1 μJ, producing a peak intensity in the vapor cell estimated to be $\sim 3 \times 10^{10}$ W cm^{-2}, and the pump and probe pulse energies were kept at the same value. For most of the studies reported here, the pump and probe pulses were linearly polarized but the $\lambda/4$ plate and polarizer in the optical path (Figure 3) allow for the ellipticity (ε) of the pump and/or probe pulses to be varied continuously between $\varepsilon = 0$ (linear polarization) and $\varepsilon = 1$ (circular polarization). Autocorrelation measurements of the pump and probe pulse widths could be made with a $\beta - BaB_2O_4$ (BBO) crystal but the pulses were generally characterized with a frequency-resolved optical gating (FROG) [49,50] system that was available for these experiments. Measurements of the pulse chirp were made throughout each data set but small pulse-to-pulse variations in the pump and probe pulse intensities, in conjunction with the nonlinearity of SPM, complicated the control of the pulse chirp. Nevertheless, the duration and spectral width of the pump and probe pulses were maintained at nominally 120 fs and 20 nm, respectively. The relative intensity of the coherent signal wave, produced in the violet at $\lambda_s \sim 420$ nm, was detected by a 1 m spectrometer (having a reciprocal dispersion in first order of 8.2 Å/mm) and a photomultiplier.

Phase matching in PFWM is governed by the energy and momentum conservation conditions, [45] the latter of which is expressed as $2\mathbf{k}_p = \mathbf{k}_i + \mathbf{k}_s$, where $\mathbf{k}$ is the wavevector associated with a specific optical field and the subscripts p, i, and s again denote the pump, idler, and signal, respectively. In scalar form, the phase matching condition can be written as

$$n(\omega_s)\omega_s \cos\theta + \sqrt{n^2(\omega_i)\omega_i^2 - n^2(\omega_s)\omega_s^2 \sin^2\theta} = 2n(\omega_p)\omega_p \tag{12}$$

where θ is the angle between $\mathbf{k}_p$ and $\mathbf{k}_s$ and $n(\omega)$ is the index of refraction. If θ is small, (12) can be simplified to yield

$$\theta^2 = \frac{n^2(\omega_i)\omega_i^2 - [2n(\omega_p)\omega_p - n(\omega_s)\omega_s]^2}{n^2(\omega_s)\omega_s^2}. \tag{13}$$

Measurements of the phase matching angle (θ) for the violet signal beam in the 5s $\rightarrow\rightarrow$ 7s, 5d $\rightarrow$ 6p $\rightarrow$ 5s PFWM process were made downstream of the Rb vapor cell with a CCD camera. For [Rb] = 2.9×10^{16} cm^{-3}, θ is determined to be $0.9 \pm 0.1°$ which is consistent with Equation (13) and the measured peak signal wavelength of $\lambda_s = 419.8 \pm 0.1$ nm. It should be noted that the wavelength in air for the 6p$_{3/2}$ $\rightarrow$ 5s transition of Rb is 420.18 nm.

Similar comments could also be made for the Rb (12s–10d, 11s–9d) and Na (5s–4d) experiments. The corresponding PFWM processes by which quantum beating is generated and observed are also axially phase matched. Excitation of the Na 5s, 4d $\leftarrow\leftarrow$ 3s two-photon transitions was accomplished with the nonlinear optical parametric amplifier (NOPA) shown in Figure 3. Incorporating three BBO crystals, the NOPA produces $\sim$40 fs pulses with a bandwidth (30 nm) sufficiently large to accommodate the Na 5s–4d$_{5/2}$ energy defect of 1348 cm^{-1} (40.41 THz). For all of the Rb and Na experiments, data acquisition involved recording the relative intensity of the signal wave (311 nm, 316 nm, and $\sim$330 nm for the Rb (12s–10d), Rb (11s–9d), and Na (5s–4d) experiments, respectively) as Δt was scanned by the interferometer. It was also noted earlier that Zhu et al. [43] have observed 7s–5d quantum beating in Rb by monitoring the wavepacket with the PSWM process: $\omega_{s_2} = 2\omega_p - \omega_{s_1} - \omega_{s_3} - \omega_{i_{1,2}}$. The signal waves at 1.323 μm and 1.367 μm were upconverted into the visible ($\lambda \sim 494$ nm) with an angle-tuned LiIO$_3$ crystal. Quantum beating was evident in the Fourier domain representation of the dependence of the blue-green intensity on Δt.

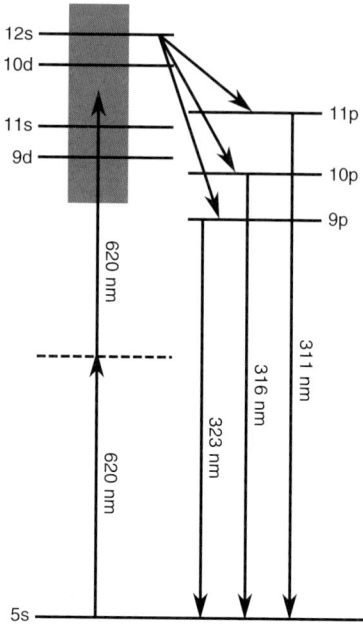

FIGURE 4 Diagram similar to Figure 1 but illustrating the 5s →→ 9d, 11s → 10p → 5s and 5s →→ 10d, 12s → 11p → 5s PFWM processes. Approximately twice the bandwidth of the ~110 fs, λ ~ 620 nm red pulses is indicated by the shaded box

2.4 Rb Wavepackets: Quantum Beating at 2–18.2 THz

2.4.1 Excitation in the Red (620 nm)

Data representative of the experiments in Rb are presented in this section. Let us first consider the simultaneous generation of quantum beating at several frequencies when Rb vapor is photoexcited in the red (λ ~ 620 nm) with the ≲100 fs pulses provided by the CPM oscillator/dye amplifier system of Figure 2. Figure 4 is a partial energy level diagram for Rb similar to that of Figure 1 but, in this instance, illustrating the PFWM processes culminating in signal waves generated in the vicinity of the $11p_{3/2}$ → 5s and $10p_{3/2}$ → 5s resonance transitions of the atom (~311 nm and ~316 nm, respectively). The shaded region represents the approximate two-photon bandwidth available from the laser system. An example of the data acquired by recording the relative intensity of the signal wave is given in Figure 5 for λ_s ~ 311 nm ($11p_{3/2}$ → 5s resonance transition) and [Rb] ≃ $5.7 \times 10^{16} cm^{-3}$. In the temporal domain (Figures 5(a) and (b)), this 200 ps scan in Δt exhibits the expected coherent transient at $\Delta t = 0$, and the expanded view of the ~10 ps scan segment in panel (b) of Figure 5 shows that the dominant periodicity is ~500 fs. The Fourier domain representation of the entire scan of Figure 5 (a), given

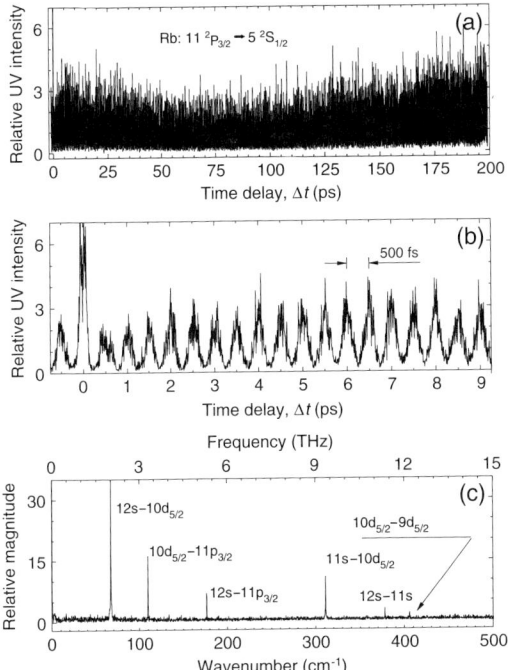

FIGURE 5 Representative data acquired by recording the variation of the Rb ($11p_{3/2} \rightarrow 5s$) signal intensity at 311.26 nm with the pump–probe pulse delay, Δt: (a) full 200 ps scan; (b) expanded view of a ~10 ps segment of the scan in (a), Showing the coherent transient at $\Delta t = 0$ and the primary periodicity of 500 fs; (c) Fourier domain representation of the full Δt-scan (part (a)), revealing several wavepacket frequency components in the ~67–406 cm^{-1} region

in part (c) of the figure, confirms the presence of the dominant feature at 66.88 cm^{-1} which is attributed to 12s–10d$_{5/2}$ quantum beating and corresponds to the ~500 fs period of Figure 5(b). Several other, higher frequency contributions to the wavepacket are also clearly observed at 109.6 cm^{-1}, 176.5 cm^{-1}, 310.9 cm^{-1}, 377.8 cm^{-1}, and 405.6 cm^{-1} which (as indicated in Figure 5(c)) correspond to the 10d$_{5/2}$–11p$_{3/2}$, 12s–11p$_{3/2}$, 11s–10d$_{5/2}$, 12s-11s, and 10d$_{5/2}$–9d$_{5/2}$ energy defects in Rb, respectively. Because of the length of the scans recorded in these experiments (100–300 ps), and the concomitant resolution of such scans in the Fourier domain (0.1–0.3 cm^{-1}), identifying unambiguously the fine structure level of the ^2D_J or ^2P_J states associated with a specific frequency component of a wavepacket is straightforward. For interpreting the dynamical data to be presented later, it is important to note here that the measured frequencies of the Fourier components of the wavepacket are within 0.08 cm^{-1} of their accepted values. With zero-padding of the temporal

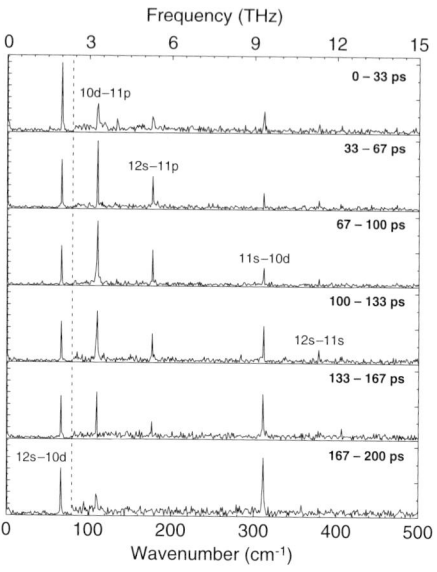

FIGURE 6 Fourier spectra for six Δt segments, each 33 ps in length, of the 200 ps experimental scan of Figure 5(a)

data (discussed later), the uncertainty in frequencies measured from the Fourier spectrum is decreased by an additional order of magnitude. As discussed in Ref. [51], this result suggests that the Stark effect has a negligible impact on the temporal history of the wavepacket [34]. On the basis of these experiments, it is evident that: **(1)** the Rb wavepackets generated with $\lambda_p \sim 620$ nm comprise several frequencies associated with quantum beating among states lying between $9d_{5/2}$ and 12s, and **(2)** the higher frequency components of these wavepackets are responsible for the structure observed in the temporal domain signal data of Figures 5(a) and (b). In addition to the reproducible Fourier components discussed above, other frequencies (such as the second and third harmonics of the strongest frequency components) were observed repeatedly but not in all spectra.

Dividing scans such as that of Figure 5(a) into 33 ps segments and calculating the fast Fourier transform (FFT) of each segment reveal, as illustrated in Figure 6, that the frequency composition of the wavepacket evolves in time. Specifically, the $12s-10d_{5/2}$ frequency component remains relatively strong throughout the scan but diminishes significantly after the first 33 ps segment. In contrast, the amplitude of the $10d_{5/2}-11p_{3/2}$ and $12s-11p_{3/2}$ Fourier terms of the wavepackets are initially quite weak but rise in intensity as Δt progresses, eventually peaking between $\Delta t = 33$ and 100 ps. A more detailed picture of the temporal dynamics of both the amplitude and phase of a specific wavepacket frequency component

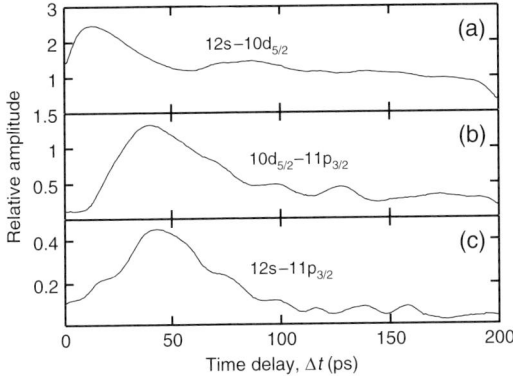

FIGURE 7 Temporal behavior of the amplitude of: (a) $12s-10d_{5/2}$; (b) $10d_{5/2}-11p_{3/2}$, and (c) $12s-11p_{3/2}$ quantum beating, calculated from the experimental data of Figure 5(a) with the STFT and a hanning window having a noise equivalent bandwidth of 2 cm^{-1}

can be extracted from pump–probe time delay scans, such as that of Figure 5(a), by means of the short time Fourier transform (STFT). An example of the results obtainable is provided by Figure 7 which shows the dependence on Δt of the relative Fourier amplitude for the three strongest frequency components of the Rb wavepacket ($12s-10d_{5/2}$, $10d_{5/2}-11p_{3/2}$, and $12s-11p_{3/2}$). These profiles were calculated from Δt data scans with the STFT and a Hanning ($\cos^2$) window having a noise equivalent bandwidth of 2 cm^{-1}. We have examined Hanning, Hamming, and Kaiser–Bessel windows and found each to be superior to the frequency-sampled ideal low pass filter. The Hanning and Hamming windows were utilized for all of the Fourier transients presented in this chapter because both are more readily generated than the Kaiser–Bessel window.

The profiles of Figure 7 confirm the qualitative behavior observed in Figure 6 – namely, that the $12s-10d_{5/2}$ quantum beating amplitude quickly (i.e., $\Delta t \lesssim 20$ ps) reaches its maximum value and its subsequent decay is mirrored by the rise in the amplitude of both $10d_{5/2}-11p_{3/2}$ and $12s-11p_{3/2}$ beating. In retrospect, this is not surprising since the $10p_{3/2}$ state is a participant in quantum beating at both frequencies. Numerous other Δt-scans recorded for Rb vapor pressures of 2–4 Torr show behavior virtually identical to that of Figure 7, with a slight dependence on the pump and probe pulse intensity. Throughout the entire $\Delta t \sim 200$ ps interval, the phase was observed to be essentially constant. These amplitude and phase data suggest that $10d_{5/2}-11p_{3/2}$ and $11s-11p_{3/2}$ quantum beating competes with $12s-10d_{5/2}$ oscillations, and they apparently exchange energy among them. Despite this interesting behavior, the number of quantum beating frequencies excited at this

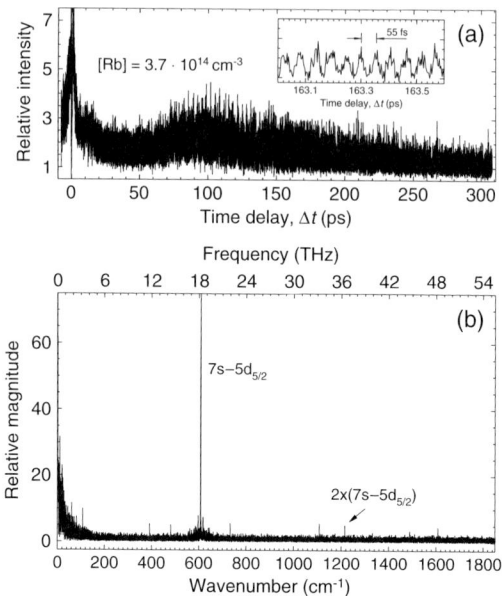

FIGURE 8 Dependence of the violet ($\lambda = 419.8$ nm) signal wave intensity on Δt for $\lambda_p \simeq 770$ nm and [Rb] $\simeq 3.7 \times 10^{14}$ cm^{-3}. The inset is an expanded view of a 0.6 ps segment of the full 300 ps scan, showing the $\sim$55 fs periodicity of the data which corresponds to the 7s–5d$_{5/2}$ energy defect of $\sim$607.9 cm^{-1}; (b) FFT of the scan of part (a). Quantum beating at 18.2 THz is dominant but weaker features, such as the second harmonic of the 7s–5d quantum beating frequency, are also noticeable

pump wavelength (620 nm), and the interactions between them, hamper the application of wavepackets comprising the Rb 12s, 11s, 10d$_{5/2}$, 9d$_{5/2}$ and 11p$_{3/2}$ states to the detection of atom–atom interactions. One further drawback of these experiments is the limited performance of the CPM/amplifier system with respect to PRF (30 Hz) and amplitude jitter. With these considerations in mind, all of the Rb experiments discussed in the remainder of this chapter involve quantum beating at a single frequency – 18.2 THz.

2.4.2 Rb (7s–5d) Quantum Beating

As discussed earlier in connection with Figure 1, two-photon excitation of Rb at $\sim$770 nm will yield quantum beating at 18.2 THz, corresponding to the Rb 7s–5d$_{5/2}$ energy defect of 607.94 cm^{-1}, if the bandwidth of the pump optical field is sufficient to encompass both the 7s $\leftarrow\leftarrow$ 5s and 5d$_{5/2}$ $\leftarrow\leftarrow$ 5s transitions. Data similar to those of Figure 5, for the $\lambda = 620$ nm photoexcitation experiments, are presented in Figure 8 for the 5s $\rightarrow\rightarrow$ 7s, 5d$_{5/2}$ $\rightarrow$ 6p$_{3/2}$ $\rightarrow$ 5s PFWM process. Part (a) of the figure shows the dependence of the violet signal wave intensity on Δt for a Rb number density of 3.7×10^{14} cm^{-3}. An expanded view of a $\sim$0.6 ps segment of the full scan is given by the inset, which illustrates that the observed periodicity (~55 fs) is consistent with the expected quantum beating frequency of 18.2 THz.

Panel (b) of Figure 8 is the FFT of the Δt-scan of Figure 8(a). Although quantum beating at 607.9 cm^{-1} dominates the Fourier spectrum, several weaker features are also apparent, including the second harmonic of the 7s–5d$_{3/2}$ beating frequency. Difference frequencies, such as that identified as (5d$_{5/2}$–5p$_{3/2}$) − (5p$_{3/2}$–5s) by Felinto et al., [52,53] have also been observed. A significant fraction of the Fourier spectrum also lies below 100 cm^{-1} and analysis confirms that the ultralow frequency ($<$10 cm^{-1}) and 608 ± 1 cm^{-1} portions of the spectrum together account for virtually all of the temporal domain profiles. For most of the experiments described in Sections 3–6, 7s–5d$_{5/2}$ quantum beating will serve as an *in situ* detector of atom–atom processes, responding to its environment through the dipole–dipole interaction.

2.5 Na 5s–4d$_{5/2}$ Quantum Beating at 1348 cm^{-1} (40.41 THz)

The highest quantum beating frequency observed to date in our laboratory is 40.41 THz (1348 cm^{-1}) which corresponds to the 5s–4d$_{5/2}$ energy defect in Na. In recent experiments, resonantly-enhanced two-photon excitation of the Na 5s and 4d$_{5/2}$ levels with $\lambda_p \sim$ 590 nm, spectrally-broad ($\Delta\lambda \simeq$ 30 nm) pulses produced by a Ti:Al$_2$O$_3$ oscillator/amplifier-driven NOPA system (described in Section 2.3) resulted in the generation of a coherent signal at $\lambda_s \sim$ 330 nm by the 3s $\rightarrow\rightarrow$ 5s, 4d$_{5/2}$ $\rightarrow$ 4p $\rightarrow$ 3s PFWM process. Data have been obtained over the past year by recording the relative 330 nm intensity for $0 \leq \Delta t \lesssim$ 100 ps. Figure 9 is the Fourier domain representation of one such scan. The presence of 5s–4d$_{5/2}$ quantum beating is clear and, again, the identification of the pertinent fine structure level of the 4d$_J$ state as $J = 5/2$ is unambiguous. As will be discussed in Section 3, experiments have been conducted in which many-body, dipole–dipole interactions have been measured in Na and the results will be compared with the more extensive body of data for Rb. Preliminary measurements of the nascent atomic fragment distributions resulting from Na$_2$ predissociation will also be mentioned in Section 4.

2.6 PFWM and PSWM in Other Atoms and Molecules

Before proceeding, a brief survey of the observations of parametric multi-wave processes in both atoms and molecules is in order. The literature on this subject is substantive, and the comments here are by no means intended to be exhaustive. PFWM in the alkali vapors was first reported by Lumpkin et al. [54,55] and was extensively characterized by Sorokin, Lankard, and Wynn [56,57] in the early 1970s. Since these early experiments, PFWM has been observed in a broad range of atoms and

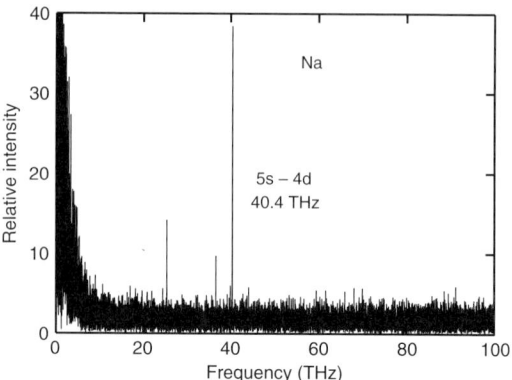

FIGURE 9 Data similar to those of Figure 8(b) but for quantum beating at 40.41 THz with the Na number density fixed at 10^{15} cm^{-3}. In this Fourier domain representation of a 100 ps scan obtained while monitoring the $\lambda_s \sim 330$ nm signal wave intensity, the strong feature at 1348 cm^{-1} (40.4 THz) is evident

molecules, among which are the alkalis, Xe, Kr, H_2 and NO (Refs. [58–65]). The $\chi^{(5)}$ counterpart of PFWM – parametric six wave mixing (PSWM) – was not reported until 1984 (Ref. [66]) but has since been observed in all of the alkali atoms except Cs [43,67–71]. The demonstrated ability to detect atomic wavepackets by both PFWM and PSWM (Ref. [43]) suggests that other coherent nonlinear optical processes, such as third harmonic generation or non-parametric FWM processes, [72,73] will also prove to be viable detection mechanisms.

Previous work with PFWM has generally been directed to its potential for efficiently generating coherent ultraviolet, vacuum ultraviolet or infrared radiation from a visible pump source [74]. Ishii et al. [65] comment on this trend when, in reporting PFWM in NO in 1996, they state: 'Although the PFWM process in the gas phase has attracted considerable attention as an effective frequency conversion technique, there have been few applications to the field of molecular spectroscopy.' This observation remains largely valid today but Sections 3 and 4 demonstrate that PFWM (and, possibly, other coherent nonlinear optical processes as well) is capable of being a powerful probe of chemical dynamics and fundamental interactions at the atomic level.

3. MANY-BODY, DIPOLE–DIPOLE INTERACTIONS AMONG EXCITED ALKALI ATOMS

3.1 Introductory Comments

Two seemingly disparate developments of the last two decades – preparing novel states of matter at low temperature, and fabricating nanostructured electronic and photonic devices – are driving a resurgence of interest in dipole–dipole interactions in the gas and solid phases. As examples of the latter, the resonant dipole–dipole interaction is known to be instrumental in the retention of coherence in an ensemble of quantum dots (with profound implications for quantum computing) [75,76] and experiments implicate dipole–dipole effects in the long-range quenching of excitons that has been observed in organic light-emitting diodes [77]. Such phenomena will undoubtedly become more prevalent as optoelectronic device fabrication ventures further into the nanometer region. Viewed more broadly, dispersive forces are (as noted by Amthor et al. [78]) of increased interest in a wide range of fields, from microelectromechanical systems (MEMS) to biology, [79] as interactions at the μm and sub-μm spatial scale are explored in greater detail with microfabrication-inspired technologies.

Emphasis is placed here on probing dipole–dipole interactions in the gas or vapor phase for several reasons, one of which is the large dynamic range in internuclear separation that is available with gases and plasmas. Another incentive is the optical-field–medium interaction lengths that are available, and the rapid advance, in recent years, of techniques for trapping and cooling atoms and molecules. Rydberg atoms and polar molecules are of particular interest, the former because of the quadratic variation of the atomic dipole moment with the principal quantum number n (Ref. [80]). With regard to polar molecules, it has been known for decades that long-range dipole–dipole interactions play a dominant role in molecule–molecule scattering, [81] and cooling polar molecules [82] offers the prospect for realizing quantum degenerate molecules and tuning the dipole–dipole interaction with external fields [83].

This section provides evidence for the detection of dipole–dipole interactions, and the measurement of absolute interaction energies, with wavepackets [84]. It is shown that the net interaction of a wavepacket with its atomic neighbors is evidenced by sidebands associated with the quantum beating frequency.

3.2 Estimates of Many-Body Interaction Energies

At long range (well beyond the LeRoy radius), [85] chemical effects can be neglected and the interaction energy between two atoms can be expressed in terms of the power series

$$V(R) = \pm \frac{C_3}{R^3} - \frac{C_6}{R^6} - \frac{C_8}{R^8} - \cdots \qquad (14)$$

in which $C_3, C_6, C_8, \ldots$ are known as London dispersion coefficients and R is the interatomic separation.[3] Determining C_3, the dipole–dipole potential constant, from diatomic molecular spectra associated with probing the region in R near the dissociation limit provides critical details concerning molecular structure but characterizing O_g^- dimer states also yields atomic alkali excited state lifetimes to within $\sim \pm\, 0.1\%$ [86]. Although it is accurate to state that photoassociative spectroscopy involving cold atoms has successfully isolated the dipole–dipole interaction between two atoms, considerably less is known of multiple atom interactions – dipole interactions within ensembles of excited atoms. Many-body Rydberg atom effects have been identified in cold atom experiments [15,16,78,80,87] but it is desirable to develop an experimental and theoretical capability to determine absolute values for the net dipole–dipole interaction of an atom with it neighbors, and to resolve the contribution of n-body interactions (where n is a given integer) to the whole.

Weiner et al. [1] have reviewed the expressions developed by King and van Vleck [88] for the dipole–dipole interaction between two identical atoms. Briefly, the interaction term $V^{(2)}$ for the Hamiltonian is written as

$$V^{(2)} = \frac{1}{4\pi\varepsilon_0 R^3}\left[\vec{\mu}_A \cdot \vec{\mu}_B - \frac{3(\vec{\mu}_A \cdot \vec{R})(\vec{\mu}_B \cdot \vec{R})}{R^2}\right] \tag{15}$$

$$= \frac{1}{4\pi\varepsilon_0 R^3}\left[\mu_{Ax}\mu_{Bx} + \mu_{Ay}\mu_{By} - 2\mu_{Az}\mu_{Bz}\right] \tag{16}$$

in which $\vec{\mu}_A$ and $\vec{\mu}_B$ are the dipole moments of atoms A and B. As discussed by Li et al., [80] atoms A and B have no permanent dipole moments if $R \to \infty$ and no external fields are present. If, however, A and B are in states $|a\rangle = |n_A, \ell_A, m_A\rangle$ and $|b\rangle = |n_B, \ell_B, m_B\rangle$, respectively, that are coupled by dipole-allowed transitions, then the dipole–dipole interaction energy between the atomic pair can be expressed by the relation:

$$\begin{aligned}E_{ab} &= \langle ba|\, H^{(2)}\,|ab\rangle \\ &= \frac{1}{4\pi\varepsilon_0 R^3}\langle ba|\, \mu_{Ax}\mu_{Bx} + \mu_{Ay}\mu_{By} - 2\mu_{Az}\mu_{Bz}\,|ab\rangle \\ &= \frac{1}{4\pi\varepsilon_0 R^3}\left[|\langle b|\,\mu_x\,|a\rangle|^2 + |\langle b|\,\mu_y\,|a\rangle|^2 - 2|\langle b|\,\mu_z\,|a\rangle|^2\right] \\ &\equiv \frac{C_3^{(ab)}}{R^3},\end{aligned} \tag{17}$$

where $H^{(2)}$ is the interaction Hamiltonian and $\langle ba|\,\mu_{Ax}\mu_{Bx}\,|ab\rangle = \langle b|\,\mu_{Ax}\,|a\rangle\langle a|\,\mu_{Bx}\,|b\rangle = |\langle b|\,\mu_x\,|a\rangle|^2$. Assuming A and B to be in a p and an s state, respectively, then Equation (17) yields

$$\left|C_3^{(sp)}(\Delta m = 0)\right| = 2\left|C_3^{(sp)}(\Delta m = \pm 1)\right| = d_{sp}^2, \quad (18)$$

where $\Delta m = m_B - m_A$, and d_{sp} is the electronic transition dipole moment. In calculating the atomic interaction potentials, C_3 for the Rb ($5d_{5/2}$–$6p_{3/2}$) interaction was assumed to be d_{5d-6p}^2 (Ref. [89]). Fine structure effects, estimated to represent a correction of <5%, were neglected and Hund's case (a) designations were adopted for the diatomic potentials. Although significant in Li$_2$ (Ref. [90]) and Na$_2$ (Ref. [4]), retardation effects introduce a correction [86] of <2% for Rb$_2$ in the present situation and have, therefore, been ignored in computing both the diatomic and multiple atom interactions. Taking the electric dipole matrix elements to be those reported by Safronova et al., [91] the (7s–6p$_{3/2}$)$_\beta$ and (5d–6p$_{3/2}$)$_\beta$ interaction potentials (where $\beta = \pm 1$ represents the symmetry of the molecular wave function) [89] were calculated and representative results are shown in Figure 10. The Σ^+ states are of particular interest because of their participation in the PFWM process discussed below. In the experiments discussed here, the primary diatomic dipole–dipole interactions are those between Rb (7s, 5d$_{5/2}$) and the 6p$_{3/2}$ states, and the unperturbed 7s–6p$_{3/2}$ and 5d$_{5/2}$–6p$_{3/2}$ molecular potentials are those of Figure 10. Accounting for secondary interactions such as 7s–6p$_{1/2}$ and 5d$_{3/2}$–6p$_{1/2,3/2}$ has the effect of perturbing the (7s, 5d$_{5/2}$)–6p$_{3/2}$ states and requires expansion of the Hamiltonian. However, these interactions are expected to play a minor role in interpreting the experimental data and will not receive further consideration here.

Equations (16)–(18) can be generalized to estimate the dipole–dipole interactions among three or more excited atoms. In the case of a three atom ensemble, for example, the interaction term for the Hamiltonian becomes

$$V^{(3)} = V_{AB} + V_{BC} + V_{CA} \quad (19)$$

and the overall dipole–dipole interaction energy is approximated by

$$E_{abc} = \frac{C_3^{(ab)}}{R_{AB}^3} + \frac{C_3^{(bc)}}{R_{BC}^3} + \frac{C_3^{(ca)}}{R_{CA}^3}. \quad (20)$$

If A, B, and C are assumed to be equidistant ($R_{AB} = R_{BC} = R_{CA} \equiv \langle R \rangle$), Equation (20) can be simplified for the states of interest in the present experiments (Rb 7s, 5d$_{5/2}$, 6p$_{3/2}$) by expressing individual terms in the

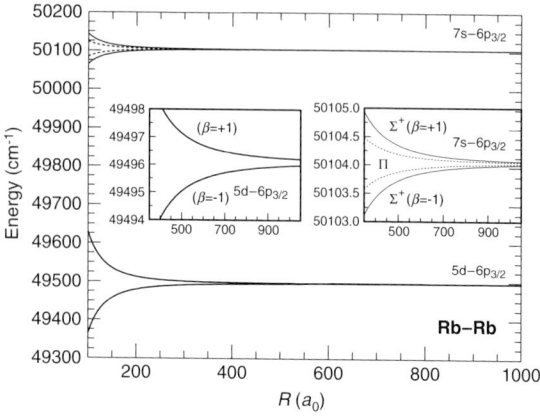

FIGURE 10 Calculated potentials for Rb $(7s-6p_{3/2})_\beta$ and $(5d-6p_{3/2})_\beta$ dipole–dipole interactions in the $100a_0 \leq R \leq 1000a_0$ region, where β denotes the symmetry of the molecular wavefunction (Ref. [89]). Expanded views of the $7s-6p_{3/2}$ and (selected) $5d-6p_{3/2}$ potentials in the $R = 350a_0-1050a_0$ interval are given by the insets

sum as

$$E_{7s6p6p} = \frac{2C_3^{(7s6p)}}{\langle R \rangle^3}, \quad E_{7s6p7s} = \frac{2C_3^{(7s6p)}}{\langle R \rangle^3},$$

$$E_{7s6p5d} = \frac{C_3^{(7s6p)} + C_3^{(5d6p)}}{\langle R \rangle^3}, \quad E_{5d6p5d} = \frac{2C_3^{(5d6p)}}{\langle R \rangle^3}. \tag{21}$$

Note that the interaction energy associated with three excited atoms in the $6p_{3/2}$, $6p_{3/2}$, and $7s$ states is twice that for the two atom $7s-6p_{3/2}$ interaction. Thus, the triatomic interaction effectively reduces to the diatomic case. Similar simplifications are valid for larger atomic ensembles. Furthermore, since the diatomic $7s-6p$ interaction gives rise to both Σ and Π states, then the triatomic $6p-6p-7s$ (or $5d$) interaction, for example, will normally include both the $6p-(6p-7s)_\Sigma$ and $6p-(6p-7s)_\Pi$ cases. For the sake of simplicity, calculations for ensemble sizes $n \geq 3$ considered only Σ states. The overall dipole–dipole interaction energies for ensembles as large as 13 excited atoms have been calculated as a function of $\langle R \rangle$ from expressions similar to Eqs. (20) and (21), and the results will be discussed in connection with Figure 13. Before concluding this section, it should be mentioned that the assumption made above of equidistance ($R_{AB} \equiv R_{BC} \equiv R_{CA}$) is an obvious oversimplification which will be discussed briefly in Section 3.4 with regard to the pair distribution function.

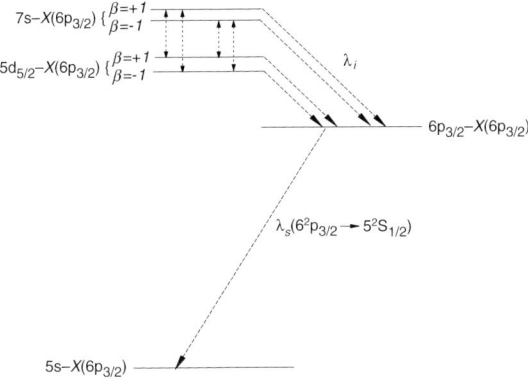

FIGURE 11 Qualitative diagram (not to scale) illustrating the idler and signal waves generated in the present experiments in which the dipole–dipole interaction between an Rb (7s or $5d_{5/2}$) atom and an ensemble of excited atoms [denoted X(6p) wherein at least one of the atoms is in the $6p_{3/2}$ state] is probed by PFWM. The diagram corresponds to a fixed value of interatomic spacing R and the vertical arrows represent quantum beating between the $[7s - X(6p)]_\beta$ and $[5d_{5/2} - X(6p)]_\beta$ interatomic potentials (reprinted by permission of the American Institute of Physics)

3.3 Experimental Results: Sideband Splittings Observed in the Fourier and Temporal Domains

As discussed previously (Section 2.4.2), $7s$–$5d_{5/2}$ quantum beating serves as an *in situ* detector of atom–atom interactions. For isolated atoms, quantum beating is generated at 607.94 cm^{-1} ($7s$–$5d_{5/2}$ energy defect) but, in the present experiments, each wavepacket is responsive to all of its excited neighbors. Thus, an Rb atom in the 7s or $5d_{5/2}$ states interacts simultaneously with ensembles of excited atoms, each ensemble of n atoms distributed over the surface of a sphere of radius $\langle R \rangle$. If X(6p) represents an ensemble of excited atoms in which at least one resides in the $6p_{3/2}$ state, then Figure 11 illustrates qualitatively the PFWM process for a fixed value of $\langle R \rangle$. Notice that quantum beating, induced by the absorption of two pump photons (not shown), between the 7s-X(6p) and 5d-X(6p) potentials gives rise to sidebands associated with the $7s$–$5d_{5/2}$ beating frequency.

Clear sidebands are, indeed, observed in the Fourier domain representation of Δt-scans such as those of Figure 8, and Figure 12 summarizes the results of 400 scans, each typically 100 ps in duration, in which the mean interatomic distance $\langle R \rangle \equiv \left(\frac{4\pi}{3} [\text{Rb}] \right)^{-\frac{1}{3}}$ has been varied from $\sim$300 a_0 to 2150 a_0. The minimum frequency splitting resolvable in the current experiments is $\sim$6 GHz ($\sim$0.2 cm^{-1}). Consequently, sidebands are not presently detectable for $\langle R \rangle \gtrsim 2000\ a_0$. Lengthening the scan period to several hun-

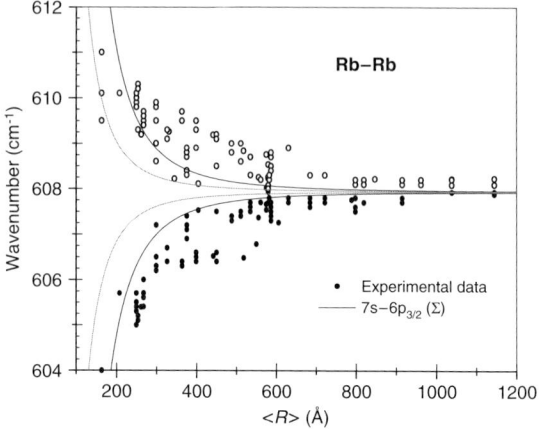

FIGURE 12 Comparison of the calculated diatomic $(7s-6p_{3/2})_\Sigma$ potentials of Figure 10 with the experimentally-observed dependence on $\langle R \rangle$ of the red and blue sideband peak positions. The minimum detectable offset from v_∞ (607.94 cm^{-1}) is approximately 0.2 cm^{-1} (6 GHz) with the present experiment (reprinted by permission of the American Institute of Physics)

dred ps and implementing heterodyning techniques in the experiment should permit the sensitivity to be improved.

Superimposed onto the data of Figure 12 are the Σ^+ potentials of Figure 10 calculated for the 7s–6p$_{3/2}$ interaction. It is clear that, over virtually all of the range of $\langle R \rangle$ examined, the diatomic dipole–dipole interaction potentials underestimate the measured sideband splittings by at least a factor of two. This discrepancy cannot be attributed to the spin–orbit effect or secondary interactions of the 7s, 5d$_{5/2}$ states with 6p$_{1/2}$, 5p, or other levels. As shown in Figure 13, however, the situation improves significantly when multiple atom dipole–dipole interactions are included in the calculations. Specifically, the three-body potentials [top, Figure 13] provide a much better match to the experimental data in the ~200–300 Å region. For the five atom ensemble predictions in the lower half of Figure 13, the calculations encompass ~90% of the measurements in the 250 Å $\lesssim \langle R \rangle \leq$ 900 Å interval and the potentials are most effective in describing the data in the $\langle R \rangle \gtrsim$ 550 Å region.

On the basis of Figure 13, therefore, one concludes that describing virtually all of the experimental data requires ensembles no larger than five atoms. In addition, the mean size of the ensemble ($\langle n \rangle$) that is the predominant contributor to the overall dipole–dipole interaction energy increases with $\langle R \rangle$. For $\langle R \rangle$ = 400 Å, for example, the measured sideband positions indicate that $\langle n \rangle \sim$ 3 atoms whereas, at 900 Å, $\langle n \rangle \sim$ 5. Any given excited atom participates simultaneously in ensembles of 2, 3, 4,

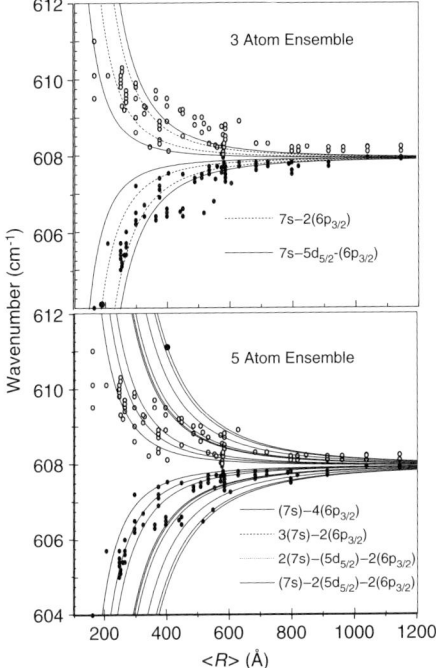

FIGURE 13 Comparison of the experimental data of Figure 12 with the estimated dipole–dipole interaction potentials for: (a) three atom and (b) five atom excited ensembles, each of which includes at least one atom in the $6p_{3/2}$ state. For clarity, only a few of the potentials associated with these ensembles are presented (reprinted by permission of the American Institute of Physics)

...atoms but the influence of the larger ensembles appears to be greatest at the higher $\langle R \rangle$ values.

Detecting atomic or molecular wavepackets by a coherent nonlinear optical process such as PFWM preserves both the phase and amplitude of the quantum beating process. As an example, Figure 14 presents representative temporal profiles for both the amplitude (dotted curve) and phase of $7s$–$5d_{5/2}$ quantum beating for a Rb vapor pressure of 1.2×10^{-2} Torr. The characteristics of amplitude transients will be discussed in detail in Section 5 but, for the purposes of the present discussion, we note that the phase φ remains essentially constant during much of the amplitude transient. An abrupt change in slope, however, is observed following a slight blueshift of the beat frequency. The dashed line in Figure 14 represents the least-squares fit to the declining portion of the phase trace, yielding a slope of $\dot{\varphi} = -970$ MHz $\simeq -0.03$ cm^{-1}. Such a redshift, observed here in the temporal domain, is precisely what one expects for a classical oscillator perturbed by a neighboring atom or molecule.

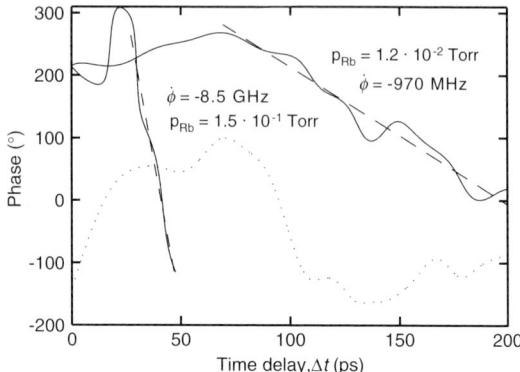

FIGURE 14 Temporal histories of the amplitude (dotted curve) and phase of the quantum beating at ~608 cm^{-1}, recorded for a Rb vapor pressure of 1.2×10^{-2} Torr. The dashed line, which represents the least-squares fit to the declining portion of the phase data, yields a redshift of the wavepacket oscillation frequency of ~970 MHz (~0.03 cm^{-1}). A phase trace for one other Rb vapor pressure is also shown, illustrating the increase in $\dot{\varphi} \equiv \Delta\omega$ with increasing [Rb]

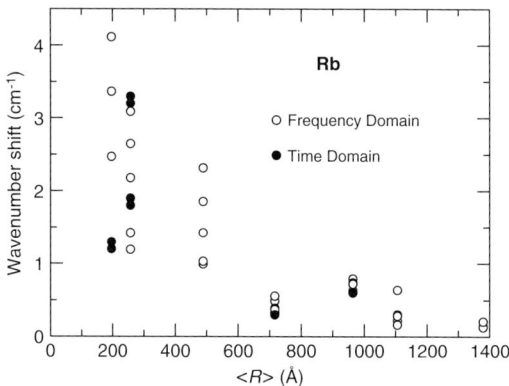

FIGURE 15 Data illustrating the magnitude of the redshift of the quantum beating frequency in Rb. Fourier (frequency) data, represented by open circles (○), are a portion of the red sideband results of Figure 12. The solid dots (●) are temporal domain data obtained from phase scans such as those of Figure 14

A phase transient for one other value of Rb number density (2.9×10^{15} cm^{-3}) is also illustrated in Figure 14 and it is clear that the downshift in the quantum beating frequency ($\Delta\omega = \dot{\varphi} = -8.5$ GHz) increases in magnitude with rising [Rb]. When these results (and similar measurements for other Rb number densities) are plotted as a function of p_{Rb} (Figure 15), one finds that the temporal data are consistent with those from the Fourier domain analysis. In Figure 15, the solid dots represent

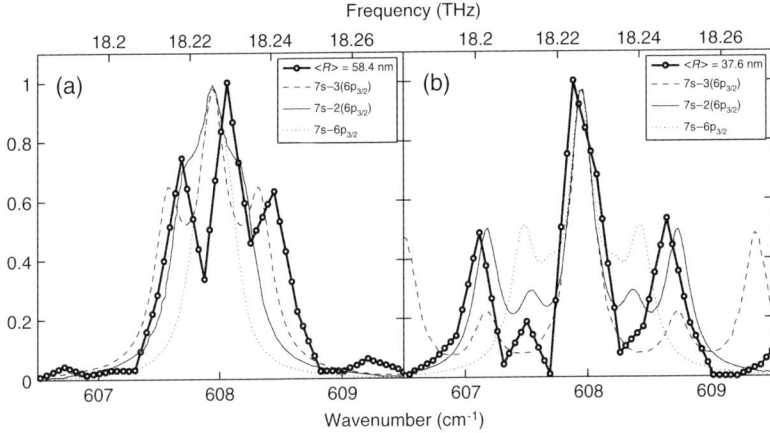

FIGURE 16 Representative experimental data (○), Showing the Fourier spectra associated with measurements of the PFWM signal intensity for $0 \leq \Delta t \leq 100$ ps. Results for: (a) [Rb] $\cong 1.9 \times 10^{15}$ cm^{-3} and (b) [Rb] $\cong 7.3 \times 10^{15}$ cm^{-3} are given which correspond to mean interatomic distances of $\langle R \rangle \cong 584$ Å and 376 Å, respectively. Predicted spectral profiles, calculated on the basis of two-, three-, and four-body $7s - n(6p_{3/2})$ dipole–dipole interaction potentials, are represented by the dotted, solid, and dashed curves, respectively (reprinted by permission of the American Institute of Physics)

the magnitude of the redshift of the Rb quantum beating frequency as determined from phase scans such as those of Figure 14. A portion of the red sideband data of Figure 12 (presented by the open circles) is also included. The corroboration of the Fourier domain redshifts by the temporal domain phase data provides confirmation of the reliability of the PFWM detection process, and demonstrates that the dipole–dipole interaction of a wavepacket with its neighbors is, indeed, being observed.

Further support for this interpretation of the data is provided by Figure 16 which presents representative Fourier spectra (FFTs) of measurements of the PFWM signal intensity, recorded for $0 \leq \Delta t \leq 100$ ps. The Rb number density was fixed at 1.9×10^{15} cm^{-3} and 7.3×10^{15} cm^{-3} for Figures 16(a) and (b), respectively, which correspond to values of $\langle R \rangle$ of 584 Å and 376 Å, respectively. Illustrated by the open circles (○), the experimental spectra display a single pair of sidebands at larger values of $\langle R \rangle$ [Figure 16(a)] but evidence for a second, weaker set emerges as the mean Rb–Rb distance is decreased [Figure 16(b)]. Simulations of the spectral profiles with the calculated $7s-n(6p_{3/2})$ potentials ($n = 1, 2, 3$) are also given by the dotted, solid, and dashed curves of Figure 16, respectively. Considering Figure 16(a), we note that two-body dipole–dipole interactions (dotted curve) do not match the data for any reasonable value of $\langle R \rangle$ and the same is true for the three-body ensemble

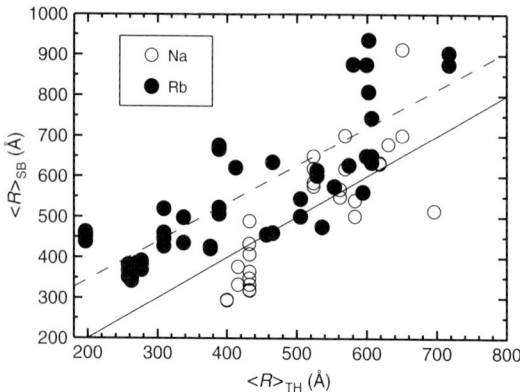

FIGURE 17 Variation of the experimental value of $\langle R \rangle$ ($\langle R \rangle_{\rm TH} \propto [{\rm Rb}]^{-\frac{1}{3}}$) with the value determined from Fourier sideband positions and calculated dipole–dipole interaction potentials ($\langle R \rangle_{\rm SB}$). Results for both Rb (•) and Na (o) are presented, and the solid and dashed lines are least-squares fits to the Na and Rb data, respectively, for $\langle R \rangle_{\rm TH} < 1000$ Å

(solid curve). However, the calculated four-body profile [dashed curve, Figure 16(a)] is essentially identical to the experimental spectrum when $\langle R \rangle = 616$ Å which represents a difference between the experimental and predicted $\langle R \rangle$ values of only ~5%. The two profiles differ primarily in a systematic redshift of the simulated spectrum by ~0.12 cm^{-1}, a value well within the experimental resolution of ~0.3 cm^{-1}. A similar result is obtained at other values of $\langle R \rangle$, as exemplified by the data of Figure 16(b) for $\langle R \rangle = 376$ Å. In this case, the calculated two-body interaction is again insufficient to account for the observed sideband offsets but now the predicted four-body interaction is too large. The three-body 7s–2(6p$_{3/2}$) simulation (at a best fit $\langle R \rangle$ of 402 Å) follows the experimental profile almost exactly except for a 0.08 cm^{-1} blueshift of the calculated spectrum with respect to experiment. This discrepancy is considerably smaller than the experimental resolution, and the difference between the measured and predicted (solid curve) $\langle R \rangle$ values is ~7%. Such errors lie within the uncertainty with which the Rb vapor pressure is presently controlled.

Experimental Fourier domain profiles similar to those of Figure 16 have been simulated over a range in Rb number density of more than 3 orders of magnitude. From experimental measurements of the sideband positions, simulations drawing upon the calculated n-body potentials (such as the $n = 2, 3$, and 5 curves of Figures 10 and 13) and the experimental resolution of ~0.3 cm^{-1} are able to predict the value of $\langle R \rangle$, denoted $\langle R \rangle_{\rm SB}$, at which the data were acquired. Comparisons of the calculated $\langle R \rangle_{\rm SB}$ with the experimental value ($\langle R \rangle_{\rm TH}$), determined from the measured

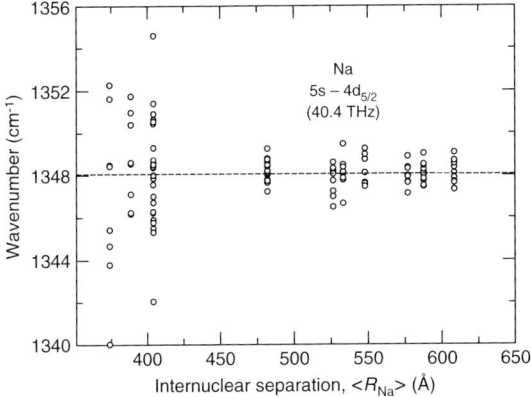

FIGURE 18 Measured energies of sidebands associated with Na quantum beating at ~ 1348 cm^{-1} (40.4 THz:5s–4d$_{5/2}$). The data, analogous to those for Rb in Figures 12, 13 and 16, were recorded for interatomic spacings in the range 375 Å $\lesssim \langle R \rangle \lesssim$ 610 Å

alkali cell temperature and the saturated vapor pressure relation, are summarized in Figure 17 for both Na and Rb. Linear least-squares fits to the Na and Rb data are represented by the solid and dashed lines, respectively. The Rb data acquired in the $\langle R \rangle_{TH} > 1000$ Å region were ignored for this calculation because the sideband splittings fall below the experimental resolution.

Over this extended range in alkali number density, the discrepancy between the predicted and measured values of $\langle R \rangle$ rises significantly above the 5–7% cited earlier in conjunction with Figure 16. For a few measurements, in particular, $\langle R \rangle_{SB}$ differs from R_{TH} by more than a factor of 1.5. Nevertheless, the least-squares fit for Rb, for example, deviates from ideal behavior ($\langle R \rangle_{TH} = \langle R \rangle_{SB}$ for all [Rb]) by <30% and <20% at $\langle R \rangle_{TH} = 400$ and 600 Å, respectively, a remarkable result considering the present uncertainty of ±30% in measurements of the alkali number density.

Preliminary data similar to that of Figures 12, 13 and 16 have been obtained for Na (5s-4d$_{5/2}$:40.41 THz) and the results for $\langle R \rangle \sim$ 375–610 Å are presented in Figure 18. For a given value of interatomic separation, the magnitude of the observed sideband splittings is considerably larger for Na than the corresponding values for Rb. A preliminary analysis of the data of Figure 18 indicates that ion-induced dipole interactions contribute significantly to the measured energy shifts.

3.4 Impact of the Pair Distribution Function

We would be remiss in not concluding this section with a few comments concerning the influence of the atom–atom pair distribution function on the interpretation of the data. The estimates of many-body, dipole–dipole interaction energies in Section 3.2 assumed that the excited atoms in an interacting ensemble are equally spaced. This simplification is tolerable for an initial analysis of the measurements of Figures 12 and 18, for example, but accounting for the spatial distribution of atoms in a gas or vapor is essential for interpreting the experimental data.

Pair distribution functions have been studied theoretically since the seminal work of Hertz [92] early in the last century. Torquato et al. [93] have proposed a formalism to calculate the nearest-neighbor probability density functions, and determined the density function for the specific case of impenetrable spheres. Recently, Marcassa et al. [94] have inferred the pair distribution function from cold (T $\sim$ 200 µK) Rydberg atom experiments in which the temporal dynamics of collisions between two Rb ($33p_{3/2}$) atoms were observed, but radiative lifetime effects complicate the analysis of the data from this experiment. A more rigorous treatment of the many-body, dipole–dipole interaction potentials discussed earlier will likely require convolving the nearest-neighbor distribution function with the calculated variation of the interaction energy with R. The clear involvement of n-atom ensembles in the present experiments will also mandate the determination, at a minimum, of the $n = 2\text{--}5$ spatial distribution functions. Finally, it should be noted that the reverse of the above statements may also be valid – namely, that simulations of the experimental data will enable at least the pair distribution functions to be extracted.

4. OBSERVATION OF MOLECULAR DISSOCIATION AND NASCENT PRODUCT STATE DISTRIBUTIONS BY THE DIPOLE–DIPOLE INTERACTION

4.1 Introduction

Predissociation occupies a prominent position in the study of molecular structure and dynamics [95]. In recent years, this subject has taken on renewed significance for several reasons, one of which is the centrality of predissociation in determining the lifetime of electronically-excited, ultracold alkali dimers produced in a magneto-optical trap [96]. Experimental tools introduced over the past several decades to probe direct dissociation or predissociation to ever greater levels of detail have generally focused on the asymptotic properties of the dissociation products, such as the angular and internal energy distributions. Laser photofragment spectroscopy, including absorption and Doppler spectroscopy, has frequently been combined with time-of-flight and coincidence imaging techniques to determine product state distributions and fragment kinetic energies in both diatomic [97–99] and polyatomic [100–102] systems.

A new dimension to examining intramolecular dynamics culminating in dissociation was provided by the sub-ps, pump–probe experiments of the late 1980s which opened a window onto the pathway from the initial molecular state(s) to the dissociation products. Zewail and co-workers [103–108] probed molecular wavepackets produced, with 150–200 fs pulses, [105] on dissociative or bound states of ICN (Ref. [103]), NaI (Refs. [104–106]), and I_2 (Ref. [107,108]). In 1988, Misewich et al. [109] at IBM reported the first of a series of experiments [110,111] in which the photodissociation at 248 nm of a metal-halide diatomic molecule (TlI) in the vapor phase was observed by monitoring optical absorption by the excited metal atom fragment. Subsequent experiments inspired by these pioneering studies have pursued the dynamics of rovibrational wavepackets in bound molecular electronic excited states, probed by wavepacket interferometry [112,113] or conventional pump–probe spectroscopy. Coupled with the detection of the molecular species or an atomic fragment by ion mass spectrometry, fluorescence, or photoelectron spectroscopy, both approaches have proven to be powerful tools for investigating the structure of electronic excited states in a broad range of diatomics, including the alkali dimers, [47,114–126] I_2, [107,108] and the van der Waals complex HgAr (Ref. [113]).

Five years ago, Senin et al. [51] reported the observation, with an atomic wavepacket, of the excited fragments produced by the dissociation of a diatomic molecule (Rb_2). Interaction of the wavepacket with an approaching excited atom occurs through the long-range dipole–dipole mechanism. Rubidium atoms generated by the dissociation of Rb_2 alter the spatially-averaged, excited state population distribution in the medium which, in turn, impacts the frequency composition of the wavepacket. Fourier analysis of the intensity of the coherent signal wave allows one to reconstruct the temporal history and product state distribution of Rb_2 dissociation. Clear molecular dissociation transients are observed and are readily identified with a specific dissociation channel.

This new approach to probing atomic and molecular dynamics is discussed in this section. Experiments are described in which interrogating atomic wavepackets by a coherent nonlinear optical process (PFWM) during and after the predissociation of electronically-excited Rb dimers provides for the determination of the nascent atomic product state distribution [48]. The product channels for excited atoms released from the molecule by predissociation are identified primarily by the transit time of the fragments, as determined by the observed interaction of the approaching atoms with wavepackets situated at $\langle R \rangle$. These results demonstrate that coupling of the atomic fragments of dissociation with an atomic wavepacket through the dipole–dipole interaction offers an alternative pathway to the

4.2 Electronic Structure and Predissociation of Rb_2: Generation of Excited Atomic Fragments

Demonstrating the detection of molecular dissociation fragments with an atomic wavepacket requires an auxiliary source of excited Rb atoms which is accomplished by having the pump pulse in ultrafast pump/probe experiments serve two purposes. Figure 19 is a partial energy level diagram of atomic Rb and its dimer counterpart Rb_2. The production of atomic wavepackets, comprising predominantly the 7s and $5d_{5/2}$ states of Rb, by two-photon excitation is portrayed on the right side of Figure 19. As discussed earlier in this chapter, the coherent superposition of Rb atomic states produced by the probe pulse interferes with that generated by the pump pulse and, therefore, this pump–probe sequence serves as a monitor of the amplitude and phase of the wavepacket at the moment the probe pulse arrives. Atomic wavepacket production will occur predominantly at an atom–atom separation of $\langle R \rangle$, the mean value.

However, as illustrated by the left-hand portion of Figure 19, the pump pulse will also excite colliding pairs of thermal ground state atoms in the Rb vapor. Photoassociation has assumed a pivotal role in the laser spectroscopy of alkali dimer excited states at long range and, in this regard, the photoexcitation of pairs of cold atoms moving along the dissociative $a\,^3\Sigma_u^+$ potential is well known [1]. In the present experiments, two-photon transitions of Rb–Rb colliding pairs from the vibrational and thermal continuum of the $a\,^3\Sigma_u^+$ ground state provide access at small R (<10 Å) to $^3\Lambda_u$ levels of Rb_2 in the ~26 000–28 000 cm^{-1} interval. Although two-photon-accessible ungerade states of Rb_2 in this energy range have not been characterized experimentally, calculations reported in Refs. [127] and [128] provide insight into the structure of the dimer for energies up to ~ 26 000 cm^{-1}. A few of the Rb dimer triplet states selected from Ref. [128] are presented in Figure 19. Following Refs. [127] and [128], the electronic states of Rb_2 are labeled by Hund's case (a) representations ($^{1,3}\Lambda_{u,g}$). As noted earlier, the $a\,^3\Sigma_u^+$ potential is essentially repulsive [despite a shallow (van der Waals) minimum at $R_e \simeq 6.1$ Å] [128] and the Franck–Condon region associated with the photoassociation process will be strongly dependent upon the precise contour (and R_e) of the $a\,^3\Sigma_u^+$ state in combination with the thermal energy of the colliding Rb–Rb pairs. Of primary importance, however, is the choice of 770 nm as the central wavelength for the ultrafast laser pulses which ensures that the molecular wavepacket established by the pump radiation extends at least partially into the vibrational continua of the $^3\Sigma_u^+$, $^3\Pi_u$, or $^3\Delta_u$ states correlated with Rb(5d) + Rb(5s) or Rb(7s) + Rb(5s) in the separated atom limit. This

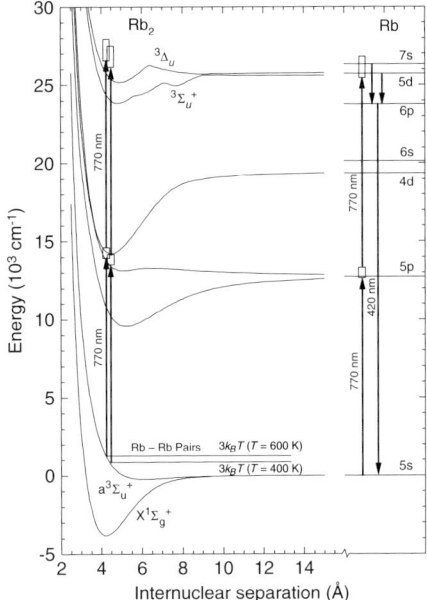

FIGURE 19 Selected electronic states of Rb and Rb_2 showing the simultaneous production of both molecular and atomic wavepackets in Rb vapor by near-infrared pump pulses ($\lambda_p \sim 770$ nm). Generation of molecular wavepackets in the 26 000–28 000 cm^{-1} region occurs by two-photon photoassociation of thermal Rb–Rb collision pairs moving along the $a\,^3\Sigma_u^+$ potential. The Rb_2 potentials, discussed in the text, are selected from Ref. [128] and the box at the terminus of each photon absorption represents the approximate laser bandwidth. The idlers and signal produced by the PFWM process are also indicated

situation, in combination with predissociation, is the means by which the requisite excited atomic Rb fragments are produced for the purpose of interacting with atomic wavepackets.

4.3 Molecular Dissociation Transients, Atomic Product State Distributions

From pump–probe time delay (Δt) scans such as that of Figure 8(a), the temporal dynamics of the amplitude and phase of a specific frequency (Fourier) component can be extracted by means of the short time Fourier transform (STFT) and a Hanning window having a noise equivalent bandwidth of 2 cm^{-1}. Representative examples of the temporal history of the 608 cm^{-1} relative amplitude are given in Figure 20 for four values of the Rb vapor pressure (p_{Rb}), spanning almost two orders of magnitude. For convenience, the amplitudes of all of the transients have been normalized.

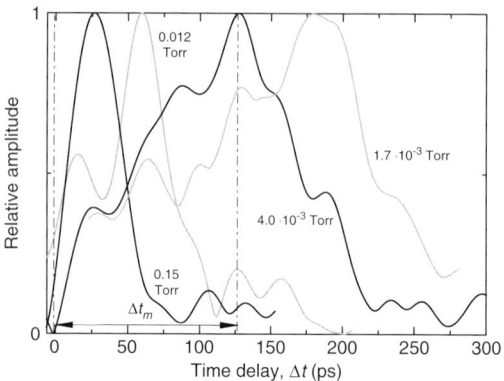

FIGURE 20 Temporal behavior of the 7s–5d$_{5/2}$ quantum beating amplitude for four values of Rb vapor pressure between 1.7×10^{-3} Torr and 0.15 Torr. For clarity, each of the profiles has been normalized and the value of Δt_m for the 4×10^{-3} Torr curve is indicated. To extract these transients, temporal (Δt) scans were analyzed with a hanning window having a noise equivalent bandwidth of 2 cm^{-1}

It is evident from Figure 20 that a transient in the quantum beating amplitude is observed for each value of Rb pressure. The maximum value is reached at a time (Δt_m) that becomes progressively smaller as the Rb number density is increased. Although the precise shape of the transient envelope varies slightly from run to run and with the bandwidth chosen for the STFT spectral window, the local maxima – observed in the lower pressure data to coincide in time with the absolute maxima in higher pressure transients – and the general contour are reproducible. We interpret the observed temporal histories of the amplitude and phase of the 7s–5d$_{5/2}$ quantum beating as reflecting the interaction of the atomic fragments of Rb$_2$ dissociation with nearby wavepackets. As demonstrated in Section 3, the perturbation of the amplitude and phase of the wavepackets by approaching Rb atoms is induced by the dipole–dipole interaction and readily detectable by PFWM. The observed behavior of the amplitude and phase transients of Figures 14 and 20, and literally hundreds more recorded over the past several years, are in agreement with a simple model of the interaction of molecular dissociation fragments with atomic wavepackets established by the pump pulse in nearby atoms. For example, the velocities of the excited atomic fragments can be estimated readily from the experimental data if it is assumed initially that each amplitude transient (such as those of Figures 8, 14 and 20) is dominated by a single exit channel. As discussed later, any given molecular dissociation transient is itself a sum of transients, each associated with a specific Rb excited state product. To a first approximation, however, the mean velocity v_f of an excited atomic fragment can be estimated from the relation:

$$v_f \approx \frac{\langle R \rangle - R_o - R_d}{2\Delta t_m} \quad (22)$$

which is clearly an approximation because it assumes the fragment velocity to be essentially constant, and that predissociation occurs on a time scale much shorter than the fragment transit time. In Equation (22), R_o is the value of internuclear separation at which atomic fragments are produced (~3–10 Å) from Rb_2, and R_d is the approximate interatomic spacing at which the wavepacket is dephased by the dipole–dipole interaction. From the data of Ref. [84], R_d is estimated to be ≈ 100 Å. Despite its simplicity, Equation (22) is remarkably accurate in identifying the primary dissociation channel associated with an experimental amplitude transient.

Expected values for the velocities of the atomic fragments can also be estimated by recalling that the average kinetic energy for a colliding pair of Rb ground state (5s) atoms is, in the center-of-mass frame, $E_{th} = 3k_B T$. Consequently, electronically-excited Rb dimers produced by photoassociation of Rb atomic pairs possess excess energy with respect to several separated atom limits of interest. Defining ΔE as the energy beyond that necessary to access a given exit (dissociation) channel, then

$$\Delta E \equiv 3k_B T + 2\hbar\omega_p - E^* \quad (23)$$

where the subscript p again denotes a pump photon, and E^* represents the energy of a given atomic Rb excited state. Assuming that all of ΔE appears as kinetic energy of the atomic fragments, then the expected average velocity of the dissociation products can be estimated from

$$v_c = \sqrt{\frac{2\Delta E}{\mu}} \quad (24)$$

where μ is the reduced mass of the dimer. For the Rb(7s)+Rb(5s) separated atom limit, for example, Equation (24) yields expected fragment velocities of ~5.5 Å/ps and 7.3 Å/ps for $T = 403$ K and 603 K, respectively. Table 1 summarizes the estimated atomic fragment velocities, calculated on the basis of Equation (24), for several Rb product channels (7s, $5d_{5/2}$, $6p_{3/2}$, ...) and two background gas temperatures.

Figure 22 presents histograms comparing the measured atomic fragment velocity distributions for two Rb cell temperatures (423 K and 510 K) with the calculated velocities associated with specific separated atom limits. The measured values of the velocity were determined from 30 amplitude transients similar to those of Figures 8, 14 and 20 and the

TABLE 1 Expected RMS velocities for Rb_2 dissociation fragments, estimated for several Rb product states and two ambient temperatures.

Product State	Energy $E_{Rb}(cm^{-1})$	v_f(Å/ps)-403 K	v_f(Å/ps)-603 K
7s	26 311.46	5.5	7.3
$5d_{5/2}$	25 703.52	8.0	9.4
$6p_{3/2}$	23 792.69	13.1	14.0
6s	20 133.60	19.4	20.0
$4d_{5/2}$	19 355.01	20.5	21.0
$5p_{3/2}$	12 816.56	28.0	28.4
5s	0	38.8	39.1

results are displayed in 2 Å/ps bins. Expected velocities for fragments exiting into a specific Rb excited state channel are indicated by the vertical arrows, and the notation 5p* denotes the Rb(5p) + Rb(5p) separated atom limit. The agreement between the calculated velocities (v_c, Equation (24)) and those determined from the maxima of the amplitude transients (v_f, Equation (22)) is remarkable. It is also evident that the low thermal energy and associated vapor pressure ($p_{Rb} \sim 4 \times 10^{-3}$ Torr) of Figure 22(a) favors the production of the highest energy Rb fragments that are accessible i.e., the 7s and 5d states. Few of the atomic fragments exit into the lower-lying asymptotes. Consider, however, the contrast with the $T = 510$ K ($p_{Rb} = 0.22$ Torr) data of panel (b) of Figure 22. The agreement between the calculated values of the Rb* velocities and distinct peaks in the histogram is again excellent but the excited state product distribution has shifted sharply to the lower-lying, separated atom limits (Rb 6s, 4d, and 5p). A more global view of this trend is provided by the three-dimensional histogram of Figure 21 that displays fragment velocity data recorded for four selected temperatures between 420 K and 515 K. These results are only a fraction of those obtained to date but demonstrate vividly the gradual shift of the atomic fragment distribution from the Rb(7s, 5d) + Rb(5s) limits to the higher kinetic energy products: 6s, 4d, 5p, and 5s. Over the course of a 100 K increase in the background vapor temperature (from 420 K to 515 K), the dominant exit channel clearly moves from 7s, 5d products to Rb(5p) + Rb(5s). The measurements of Figure 21 are distributed into 2 Å/ps and 6 K bins for the atomic velocities and vapor temperature, respectively. A composite histogram summarizing more than 100 transients recorded over the 410–515 K temperature range is shown in Figure 23. Incorporating an increased number of scans (relative to that of Figure 22) into the histogram has suppressed the background and accentuated the peaks associated with the 7s, 5d, 6s, 4d, and 5p product states. Displaying data acquired over such a wide temperature range, however, obscures the product distribution available at any specific temperature. Nevertheless, note that the production of 5p atoms via the 5p + 5s and 5p + 5p exit channels can be distinguished and the former

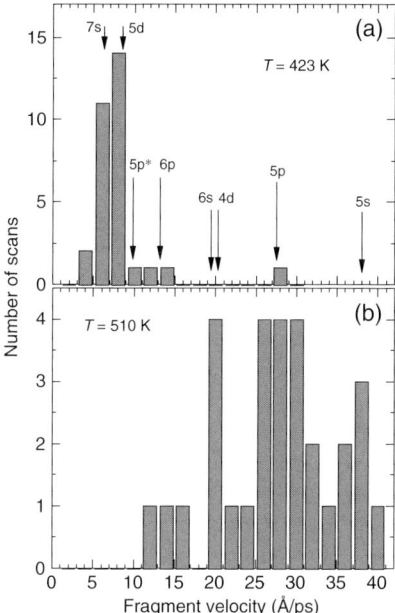

FIGURE 21 Three-dimensional histogram for rb atomic fragment velocity data obtained in four selected temperature ranges: 420 ± 3 K, 440 ± 5 K, 475 ± 3 K, and 515 ± 3 K. Data are sorted into 2 Å/ps and 6 K bins for the fragment velocity and temperature, respectively

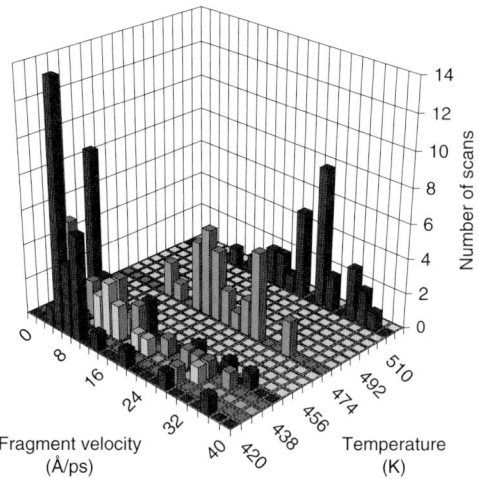

FIGURE 22 Histograms representing the atomic fragment velocity distribution measured for Rb cell temperatures of: (a) $T = 423$ K, and (b) $T = 510$ K. The velocity bin size is 2 Å/ps, and the expected values of the velocity for each atomic limit, calculated from Equation (24), are also indicated. The atomic asymptote denoted 5p* represents the Rb(5p) + Rb(5p) limit

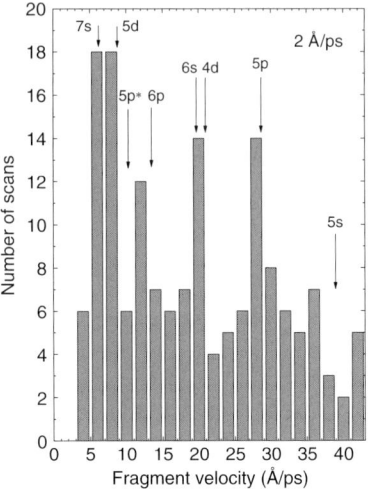

FIGURE 23 Summary of the Rb data obtained from more than 100 transients recorded over the 410–515 K temperature range. Velocities in this composite histogram are again distributed into 2 Å/ps Bins

is clearly dominant. Dissociation processes other than those described above may also be contributing to the velocity distribution of Figure 23. For example, dissociation of the electronically-excited molecule into the 5p exit channel may result either from the absorption of one pump photon by an Rb–Rb collision pair *or* two photons absorbed by Rb$_2$(X). The former will yield fragment velocities near the 5d value indicated in Figure 23 whereas the latter produces velocities between those for the 4d and 5p labels. Examination of the data suggests that the single-photon absorption process may be a significant contributor to the measured velocity distribution whereas Rb$_2$ absorption is not.

Preliminary data for Na similar to the Rb results of Figure 23 are shown in Figure 24. This composite histogram presents results acquired over the 590–650 K temperature range. The predicted (Equation (24)) velocity for the 5p fragment coincides with the measured value but the match is less satisfactory for the 3p* and 5s product channels.

4.4 Diffusion Description of Amplitude Transients

As noted earlier, the identification of each Fourier amplitude transient with a single dissociation channel is an obvious simplification. A more precise description of these scans in Δt must account for the contributions of multiple excited state fragments to the overall envelope of the data. To that end, an elementary model is introduced here in an effort to estimate the relative atomic excited state yields associated with each amplitude

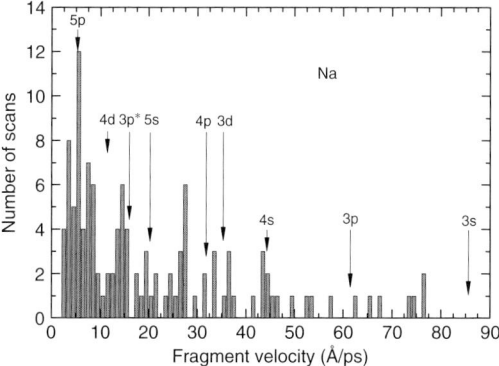

FIGURE 24 Data similar to those of Figure 23 but for Na. The velocities have been sorted into Bins 1 Å/ps in width, and 3p* denotes the Na(3p) + Na(3p) exit channel

transient. In the last section, a statistical approach was invoked to infer the excited state distribution from scores of Δt-scans.

The amplitude transients are interpreted as representing the distribution of transit times for atomic fragments moving from parent molecules toward nearby wavepackets. Although the average distance traveled by any given fragment to interact with the nearest wavepacket is given roughly by the mean atom–atom separation in the vapor, $\langle R \rangle$, the positions of the wavepackets are dispersed in a distribution about $\langle R \rangle$. This distribution is one of the factors determining the breadth of the transients in Figures 8, 14 and 20 (i.e., the spread in 'arrival times' of the fragments). An equivalent picture is that provided by a diffusion model in which the dissociation fragment spatial distribution broadens as the atoms propagate outward in R. From a quantum mechanical perspective, the partitioning of the molecular wavepacket into the various atomic exit channels is also expected to generate transients which, detected by atomic wavepackets, yield the quasi-Gaussian profiles portrayed in Figure 20, for example. We, therefore, express the experimentally-observed amplitude transient as a sum of Gaussians, each of which is associated with a specific Rb excited state. The contribution of the ith exit channel to the overall transient is given by

$$A_i(\Delta t) = C\delta t^{-1} \exp\left\{-\left(\frac{\Delta t - \Delta t_i}{\delta t}\right)^2\right\} \qquad (25)$$

where C is a constant, Δt_i is the time delay at which this transient reaches its maximum value and is given by $\Delta t_i \approx \langle R \rangle / v_c$, and $\delta t \propto \Delta t_i^{1/2}$, the half-width of the normalized Gaussian at the e^{-1} points, reflects the

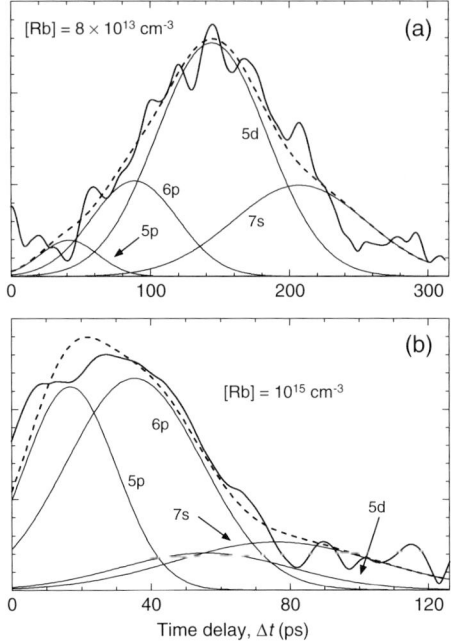

FIGURE 25 Representation of experimentally-observed Fourier amplitude transients as a sum of Gaussians, each representing a specific Rb exit channel (7s, 5d, 6p, 4d, or 5p): (a) [Rb] = 8×10^{13} cm^{-3}, and (b) [Rb] = 10^{15} cm^{-3}. The least-squares fit of the sum to the experimental profile is denoted by the dashed black curve

broadening of the transients observed in experimental scans. Combining the relations for Δt_i and δt indicates that $\delta t \propto [\text{Rb}]^{-1/6}$, a dependence that is confirmed by measurements of δt from the experimental transients.

Numerical simulations of the experimental Fourier amplitude scans entailed iterating the sum of the Gaussians associated with the Rb 7s, 5d, 6p, 4d, and 5p exit channels. Since the approximate position Δt_i of the peak in each Gaussian is specified by the calculated fragment velocities v_c (Equation (24)), the amplitude transient associated with each atomic excited state can be calculated. Assuming that the total response $A_T(\Delta t)$ is given by the RMS expression

$$A_T(\Delta t) = \sqrt{\sum_i A_i^2(\Delta t)}, \qquad (26)$$

then the contribution of each channel to the overall amplitude transient can be determined by an iterative least-squares fitting procedure in which the sum of Equation (26) is compared to the normalized

TABLE 2 Branching ratios for production of specific Rb fragments, estimated from amplitude transients. The uncertainty in each value given is ±10%.

[Rb], cm^{-3}	7s	5d	6p	5p
8×10^{13}	16	35	24	25
10^{15}	7	5	34	46

experimental profile. Representative results are given in Figure 25 in which experimental profiles recorded for [Rb] = $(8 \pm 2) \times 10^{13}$ cm^{-3} and $(1.0 \pm 0.2) \times 10^{15}$ cm^{-3} are decomposed into Gaussians associated with the 7s, 5d, 6p, and 5p exit channels. The solid black curves represent the measured transients, and the least-squares fit of the Gaussian sum to the experimental data is shown by the dashed curves. This process yields the branching ratio associated with atomic fragments exiting into a specific excited state channel, and Table 2 summarizes the results for the Rb number densities of Figure 25 ($(8 \pm 2) \times 10^{13}$ cm^{-3} and $(1.0 \pm 0.2) \times 10^{15}$ cm^{-3}). The analysis of transients from multiple Δt-scans determined the branching ratios for the 7s, 5d, 6p, and 5p states to be $(16 \pm 5)\%$, $(35 \pm 10)\%$, $(24 \pm 5)\%$, and $(25 \pm 5)\%$, respectively, for [Rb] = 8×10^{13} cm^{-3}. To within the estimated uncertainty of the analytical process, the 4d yield was found to be negligible which is in agreement with the nascent product distribution implied by the histogram of Figure 22(a). Also in accordance with Figure 22 is the observation that increasing the Rb number density (i.e., decreasing $\langle R \rangle$) skews the atomic excited state distribution toward the lower-lying levels. One concludes that resolving the contributions of individual exit channels to a single amplitude transient yields product distributions that are in qualitative agreement with the statistical picture provided by scores of 100–300 ps scans. These data suggest the feasibility of extracting nascent atomic product distributions directly in the temporal domain under collision-free conditions.

4.5 Dominant Rb$_2$ Predissociation Channels

The atomic fragment distributions presented in the preceding two sections exhibit a pronounced sensitivity to the vapor cell temperature. At the lowest temperatures studied, the 7s and 5d states are the favored products. With rising temperature, however, the excited state distribution moves steadily toward the lower-lying atomic asymptotes. This trend, which indicates that the dominant predissociation channel is a sensitive function of the internuclear separation at which photoassociation occurs, is in qualitative agreement with the calculated [127,128] electronic structure of Rb$_2$ in the 20 000–26 000 cm^{-1} region. A detailed discussion of the predissociative pathways that are most likely responsible for the production of 7s, 5d, 6p, and 5p fragments can be found in Ref. [48] and only the conclusions will be reviewed here.

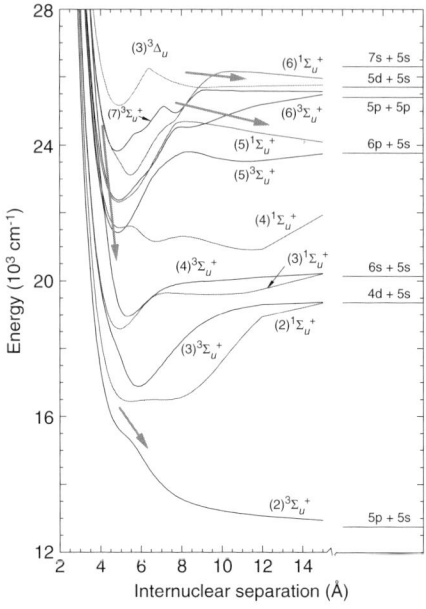

FIGURE 26 Summary of the channels expected to dominate the predissociation of Rb_2 wavepackets into specific atomic product states (molecular potentials adapted from Ref. [128])

Figure 26 summarizes the three primary predissociation channels by which the nascent atomic fragment distributions of Figures 21–23 are apparently produced. Assuming a major role in these processes are the $^{1,3}\Sigma_u^+$ states and each group of products is associated with predissociation of the molecule in a narrow interval in R. Spiegelmann et al. [127] and Park and co-workers [128] have identified the $(3)^3\Delta_u$ and $(7)^3\Sigma_u^+$ states of the dimer (illustrated in Figures 19 and 26) as correlated, in the separated atom limit, with Rb($5d_J$) + Rb(5s) and Rb(5p) + Rb(5p), respectively. Predissociation of the $(7)^3\Sigma_u^+$ state [128] by $(6)^1\Sigma_u^+$ in the 8–9 Å interval appears to be primarily responsible for Rb(7s) generation, for example. Similarly, a crossing between $(5)^1\Sigma_u^+$ and both $(7)^3\Sigma_u^+$ and $(6)^1\Sigma_u^+$ in the R ~7–8 Å region provides a convenient and presumably efficient mechanism for molecular fragmentation into the 6p asymptote, since the former is correlated with Rb(6p) + Rb(5s) in the separated atom limit. Furthermore, the inner branch of the $(7)^3\Sigma_u^+$ potential is the likely origin of 6p atom production at much smaller R (~3.5–4 Å) but coupling of the $(3)^3\Delta_u$ and $(7)^3\Sigma_u^+$ states to the 4d asymptote by predissociation appears to be weak. The presence of 5d atoms in the experimental product distributions is attributable simply to the direct dissociation of dimers generated in the vibrational continuum of the $(3)^3\Delta_u$ state which is derived from Rb(5d) + Rb(5s). These general considerations support the premise that the identity of the atomic fragment distribution produced by predissociation is determined by the initial structure of the molecular wavepacket and, therefore, the region of internuclear separation in which the wavepacket is born. Producing the wavepacket at larger R appears to increase the yield of those fragments (7s, 5d) with large internal energies. Conversely, wavepackets generated at smaller R, in the vicinity of the classical potential barrier, will favor the production of the 6p species. One concludes that the above scenario is in qualitative agreement with the experimental observation that reduction of the vapor temperature favors the generation of 7s atoms whereas heating the vapor provides access of Rb–Rb collision pairs to smaller R portions of the ground state potential. Photoassociation at short range establishes the molecular wavepacket in a region disposed to 6p fragment production. Indeed, experiments described in the next section, in which the chirp of the pump pulse is varied systematically, confirm that control over the Rb atom fragment distribution is accomplished by optically engineering the molecular wavepacket.

5. COHERENT CONTROL OF Rb_2 PREDISSOCIATION

Developing the ability to drive a chemical reaction to a predetermined outcome has long been a priority in chemistry. First proposed in the 1980s and subsequently realized through several basic approaches, coherent control seeks to achieve chemical selectivity optically by tailoring the temporal shape of ultrafast (sub-ps) laser excitation pulses, or by exploiting interferences between competing quantum pathways to discriminate against undesired products [129–131]. The literature of coherent control is extensive but several milestones should be mentioned, including quantum control of the temporal behavior of wavepackets, [112] molecular ionization yields, [132] and photodissociation branching ratios [133]. Perhaps the single advance having the most far-reaching impact has been the proposal [134] and implementation [135,136] of feedback-optimized quantum control which has been applied over the last twenty years to shaping quantum states,[137] enhancing the efficiency of high-order harmonic generation, [138] controlling energy flow within (or the formation of) biological molecules, [139,140] and directing the dissociation and reactivity of polyatomic molecules [141]. For further details of these and other seminal developments in the field, the reader is referred to the reviews of Refs. [142–145].

The last section demonstrated the ability to determine, with a wavepacket and PFWM, the nascent product state distribution resulting from the dissociation of a diatomic molecule. Experiments briefly described here exploit this capability to demonstrate the selectivity of quantum control in the predissociation of Rb$_2$. Specifically, the simultaneous detection of atomic fragments in several excited states allows for detailed observations of molecular fragmentation patterns under the influence of chirped pump pulses. Bardeen et al. [133] first employed chirped pulses in 1997 in the pursuit of quantum control of the photodissociation of NaI but the specificity at the experimental temperature of 1000 K was small. The experimental results and simulations presented here show that pump pulses with slightly negative or no chirp generate Rb(7s) fragments with near unit efficiency. Strongly positive chirp (1000 fs^2) produces primarily Rb(6p) fragments and negative chirp of the same magnitude exhibits no selectivity in fragment production. Quantum mechanical calculations of the structure of the molecular wavepacket established by the pump pulse are fully in agreement with the data.

These experiments were conducted with the PFWM scheme of Figure 1. Also, the experimental arrangement was that of Figure 3 with the exceptions that the NOPA was not required and a pair of SF-10 prisms was installed in the optical train for the purpose of varying the group velocity dispersion (GVD) of the pulses. The amplitude and phase of the pulses entering the Rb vapor cell were monitored by a FROG system and all experiments were conducted with the Rb vapor pressure fixed at 10^{-2} Torr, which corresponds to [Rb] = 2.3 × 10^{14} cm^{-3}. To minimize the influence of multiphoton ionization, the peak pump pulse intensity was maintained at or below 10^{11} W cm^{-2}.

Data illustrating the dependence of the measured Rb product state distribution, resulting from Rb$_2$ dissociation, on the pump pulse (linear) chirp are presented in Figure 27 in which the dashed horizontal lines denote the estimated velocity for specific atomic Rb fragments. Two sets of measurements were made for chirp values of ±1000 fs^2 and the laser pulse width set at 75 fs. A third group of experiments was conducted with a slight negative chirp of −400 fs^2. Removing all pulse chirp yields a transform-limited pulse 50 fs in duration which is insufficient to initiate Rb(7s–5d) quantum beating, presumably because the driving pulse is shorter than the beat frequency (∼ 608 cm^{-1}) period of ∼54.8 fs. For this reason, a data set was acquired for a pump pulse chirp and temporal width of −400 fs^2 and 55 fs, respectively. Under these latter experimental conditions, Rb(7s) atoms are produced from Rb$_2$ predissociation with near unit efficiency. However, similar experiments with larger chirp magnitudes (±1000 fs^2) yield dramatically different results. For positive chirp, the Rb product distribution is peaked at the velocity expected of Rb(6p) fragments but a tail extending to higher velocities is also observed. On the other hand, no selectivity with regard to the Rb$_2$ dissociation exit channel is observed in the −1000 fs^2 experiments as Rb(5d, 6p and 6s) atoms are detected with approximately equal probability.

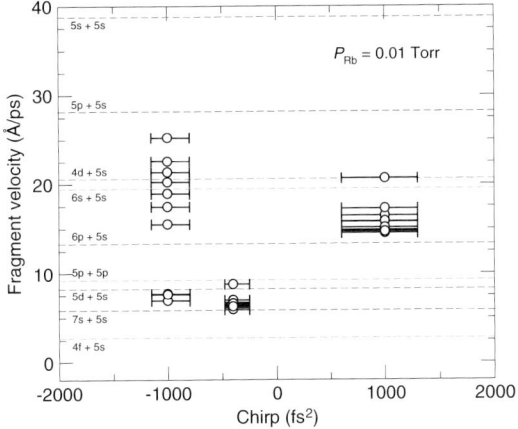

FIGURE 27 Dependence of measured Rb$_2$ fragmentation patterns (expressed as fragment velocities) on the magnitude of the pump/probe linear chirp. The dashed horizontal lines denote the expected velocities for specific atomic products

In an effort to simulate the results of Figure 27, the spatial characteristics of the molecular wavepacket produced on the terminal Rb$_2$ electronic state by the absorption of two photons were calculated from the expression:

$$|\Psi(R,t)\rangle \propto \sum_v \left| c_v \psi_v(R,t,v) e^{(-ik_v(R-R_e)-iG_v t/\hbar)} E(G_v) e^{-i\phi(G_v)} \right\rangle \quad (27)$$

where the c_v are coefficients associated with each vibrational state of quantum number v, R_e is the equilibrium internuclear separation for the electronic state in which the wavepacket is established, G_v is the vibrational term value, and $\exp\left\{-i\left[k_v(R-R_e)+\frac{G_v t}{\hbar}\right]\right\}$ is the temporal and spatial propagator for the wavepacket. A similar formalism has been described by several groups previously [47,146].

One significant unknown in the calculations introduced above is the identity and structural characteristics of the ungerade Rb$_2$ state in which the molecular wavepacket is formed. This electronic level is accessed from the dissociative $a\,^3\Sigma_u^+$ ground state by photoassociation, bound ← free absorption of two pump photons by ground state Rb atoms in the $a\,^3\Sigma_u^+$ vibrational continuum. The discussion of Rb$_2$ predissociation in Section 4.5 dealt exclusively with the $(3)^3\Delta_u$ and $(7)^3\Sigma_u^+$ states and, indeed, both appear to be immediate precursors to excited Rb fragments. However, the combined energies of two pump photons access an energy region lying above the Rb(7s, 5d) + Rb(5s) separated atom limits, a region in which experimental data are sparse. For the sake of convenience,

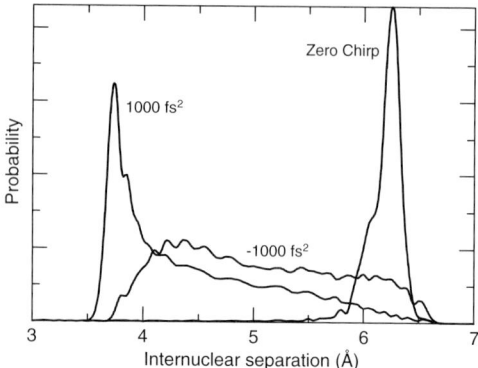

FIGURE 28 Spatial probability densities for molecular (Rb$_2$) wavepackets launched by two-photon absorption onto an Rb$_2$ $^3\Sigma_u^+$ state with pump pulses having a chirp of: (a) zero, (b) 1000 fs^2, and (c) −1000 fs^2

therefore, the wavepacket calculations assume that the terminal Rb$_2$ state is the $^3\Sigma_u^+$ level correlated with Rb(7p) + Rb(5s) in the separated atom limit, and its spectroscopic constants are the same as those calculated by Park et al. [128] for its counterpart derived from Rb(6p) + Rb(5s). Furthermore, chirped femtosecond laser pulses influence wavepacket generation and propagation through the amplitude and phase modulator term $E(G_v)e^{-i\phi(G_v)}$, where $\phi(G_v)$ is the spectral distributed phase. The simulations assumed $E(G_v)$ to be a Gaussian specified by the pump laser spectrum having a width and center wavelength of 20 nm and 769 nm, respectively. For these conditions, the wavepacket comprises the $v' \approx$ 26–49 levels of the $^3\Sigma_u^+$ state.

Figure 28 shows the predicted probability density, $\int |\Psi(R,t)|^2 dt$, integrated over the first 10 ps of the temporal history of the wavepacket. Results are given for three values of chirp: zero, 1000 fs, and −1000 fs^2. The localization of the wavepacket produced with zero chirp in the $R > 6$ Å region is consistent with the conclusion of Section 4.5 that the production of 7s and 5d atomic fragments by predissociation (or direct dissociation) of Rb$_2$ appears to occur in the 8–9 Å interval. Thus, the specificity exhibited by the chirp data of Figure 27 for Rb(7s) fragments is qualitatively in agreement with the calculated initial structure of the molecular wavepacket established by the pump on the terminal Rb$_2$ $^3\Sigma_u^+$ state. Similar comments could be made for the ±1000 fs^2 simulations of Figure 28. When the pulse chirp is +1000 fs^2, the wavepacket is launched onto the $^3\Sigma_u^+$ state at small R (<4 Å) which is precisely the region identified in Section 4.5 as most likely the origin of Rb(5p, 6p) fragments. In contrast, the wavepacket probability density for a pump pulse chirp of −1000 fs^2 is extended in R from ~4 to 6.5 Å, suggesting

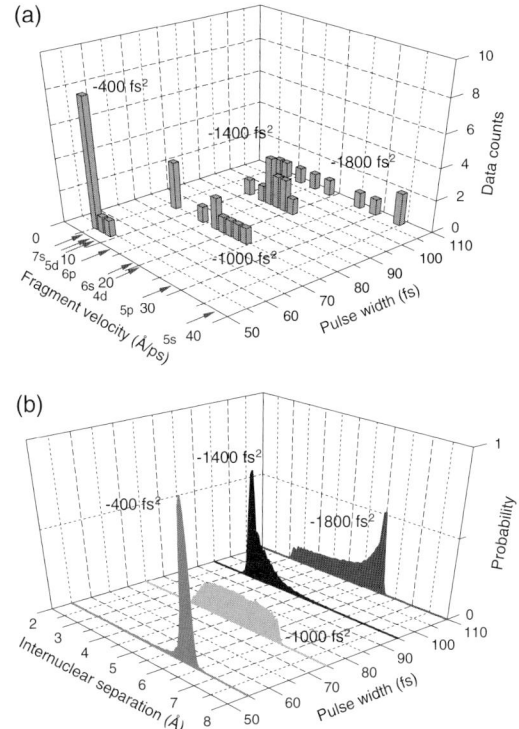

FIGURE 29 (a) Measured Rb fragment velocity distributions for four values of negative chirp (-400 fs^2, -1000 fs^2, -1400 fs^2, and -1800 fs^2); (b) calculated wavepacket probability densities, obtained by integrating $|\Psi(R,t)|^2$ over the $0 \leq t \leq 10$ ps interval

that a full spectrum of Rb fragments will be produced which is, indeed, the conclusion to be drawn from the experimental data of Figure 27.

The trends discussed above, although necessarily qualitative, are corroborated by further experimental data which are summarized in panel (a) of Figure 29. Measurements of Rb$_2$ dissociation fragment distributions (branching ratios) were extended to two additional values of negative chirp, -1400 fs^2 and -1800 fs^2. These larger chirp magnitudes yield results analogous to those discussed earlier – namely, the -1400 fs^2 chirp data shows fragment yields favoring Rb(6p) production. This is in agreement with Figure 29(b) which displays the calculated initial spatial probability density for the launched wavepacket. Once again integrated over 10 ps, the -1400 fs^2 simulation reaches its maximum value at R <3.5 Å which, as was the case for $+1000$ fs^2 chirp, favors the 6p exit channel. The -1800 fs^2 simulation is more broadly spread over R which is reflected well in the experimentally-determined fragment distribution.

On the basis of the measurements and simulations to date, it is reasonable to conclude that a strong correlation exists between the initial structure of the molecular wavepacket and the predissociation pathways that are favored. Localization of the wavepacket in the $R > 6$ Å region yields Rb(7s) production with near unit efficiency, whereas 6p atoms are favored when the wavepacket is born at small $R(\lesssim 4$ Å).

6. PHOTOASSOCIATION OF RARE-GAS–HALOGEN ATOMIC PAIRS AT AMBIENT TEMPERATURE

6.1 Introduction, Stationary Phase Approximation

Photoassociation has proven to be of inestimable value to cold atom and molecule physics by providing an avenue for probing spectroscopically the large R portions of interatomic potentials, as well as a means for cooling molecules. An example of an optically-assisted reaction in which a radiation field interacts with pairs of atoms to produce an electronically-excited molecule, photoassociation is expressed as

$$X + Y + n\hbar\omega \xrightarrow{\sigma_{PA}} XY(v', J'), \tag{28}$$

where n is an integer and, if $n = 1$, the reduced photoassociation cross section σ_{PA} is expressed in units of cm^5. It is known from both the semiclassical stationary phase approximation [147] and quantum calculations that process (28) principally occurs in the vicinity of the classical outer turning points for vibrational levels within the excited state potential. Because this bound $\leftarrow$ free process selects atomic pairs from the vibrational and thermal continua of the XY dissociative electronic ground state, photoassociation favors atomic pairs having the proper kinetic energies and interatomic separation R. For cold atoms, therefore, photoassociation occurs predominantly at large R, thereby offering access to the region of diatomic potentials near the separated atom limit. Information regarding other portions of the interatomic potentials must necessarily be gained from the photoassociation of atoms thermalized at temperatures well above those common to cold atom experiments.

Beyond its utility for exploring the physics of cold matter, photoassociation has several advantages over emission spectroscopy as an experimental technique for studying the chemical characteristics and optical properties of bound–free transitions in small molecules. The study of diffuse spectra and continua has a long and fruitful history but characterizing excited states by analyzing bound $\to$ free emission spectra alone presents a few limitations, among which is the difficulty in preparing well-defined vibrational state population distributions. Furthermore, since the difference potential is often not a single-valued

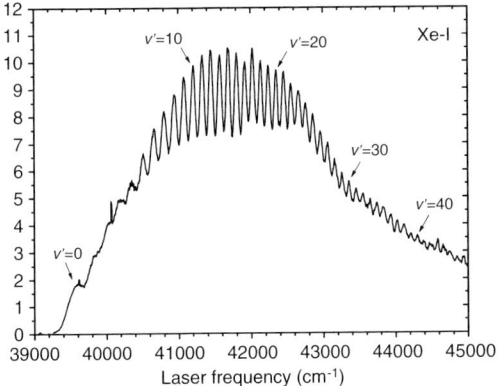

FIGURE 30 Excitation spectrum for the photoassociation of Xe–I atomic pairs thermalized at 300 K. Reported in Ref. [149], this spectrum is the composite of data acquired by monitoring both B → X ($\lambda \sim 253$ nm) and B → A ($\lambda \sim 319$ nm) emission. As indicated in the figure, each undulation is, to a first approximation, associated with bound ← free transitions of Xe–I pairs to a specific vibrational level (v') of the bound XeI ($B^2\Sigma_{1/2}^+$) excited state

function of R, little information is available regarding the dependence of the electronic transition moment μ on internuclear separation. Another drawback of interpreting bound → free spectra is that they generally display little structure other than weakly-modulated oscillations, thus complicating the determination of a unique set of molecular spectroscopic constants.

In the early 1990s, Schloss, Jones and Eden [148,149] reported the excitation spectra associated with the photoassociation of Kr–F and Xe–I atomic pairs thermalized at 300 K. Transitions of Kr($^1S_0$) – F(^2P) and Xe($^1S_0$) – I(^2P) pairs from their ground state thermal and vibrational continua to the lowest 27 and ≈40 vibrational levels of the $B^2\Sigma^+$ ion pair states of KrF and XeI, respectively, were shown to yield spectra that are characterized by deeply-modulated Franck–Condon structure. Figure 30 presents the experimentally-observed [149] excitation spectrum for the photoassociation of Xe–I atomic pairs thermalized at ∼300 K. Recorded in the ultraviolet over the 210–255 nm wavelength interval, this spectrum was acquired by first photolyzing I_2 (in mixtures of Xe and I_2 vapor) and photoassociating Xe–I pairs with a frequency-doubled dye laser while monitoring spontaneous emission from the XeI excimer at either $\lambda \sim$ 319 nm (B → A transition) or ∼253 nm (B → X transition). Because of scattered pump radiation, the spectrum of Figure 30 is a composite of data obtained through both XeI(B → X) and (B → A) excitation spectra. The experimental technique will not be discussed further but details can be found in Ref. [149].

The most prominent aspect of Figure 30 is the oscillatory Franck–Condon structure, superimposed onto an underlying continuum, which is considerably more pronounced than the weakly-modulated undulations generally displayed by the bound → free (emission) spectra of the rare-gas-halide molecules. The strongly-modulated spectra of Ref. [149] for Kr–F and Xe–I atomic pairs are reminiscent of the reflection (or Condon internal diffraction) spectra observed in a wide range of diatomics, including the Column IIB dimers Cd_2 and Zn_2 (Ref. [150]). However, reflection spectra result from bound → free emission from one or a limited number of v' states. Consequently, interference structure in emission spectra requires specificity in populating upper state vibrational levels. Several features of the spectrum of Figure 30 are, as discussed in Refs. [148] and [149], well described by the stationary phase approximation [147] which is an outgrowth of the conservation of nuclear position and momentum during an electronic transition. At the heart of this approximation is the premise that bound ⇌ free transitions occur predominantly in the vicinity of points of stationary phase – specific values of internuclear separation, denoted R^* – which satisfy the expression

$$E' - [V'(R^*) - V''(R^*)] = \varepsilon'' \qquad (29)$$

where ε'' is the energy of the ground state Xe–I atomic pair in the separated atom limit. The left-hand side of Equation (29) is the Mulliken difference potential where E' is the energy of a specific upper state [XeI(B)] vibrational level (v') and $V' - V''$ is the difference potential associated with the upper and lower electronic state interaction potentials, V' and V'' (respectively). If the ground state potential is sufficiently repulsive, the Xe–I pair distribution will be confined to large R (i.e., $> R_{eB}$), there will exist only one point of stationary phase for each v', and that point will lie close to the classical outer turning point R_2 for the upper state vibrational level in question.

Therefore, the excitation spectrum of Figure 30 represents the sum of single-photon transitions of collision pairs in the ground state's vibrational and thermal continua to individual vibrational levels of the XeI molecule's lowest excited state ($B^2\Sigma_{1/2}^+$). These transitions occur predominantly in the vicinity of the classical outer turning points and, if $V''(R)$ is sufficiently repulsive, the Franck–Condon (FC) region is confined to a narrow portion of the ground state potential. Since photoassociation occurs from a repulsive ground state in a 1–2 Å wide Franck–Condon region, there exists a roughly one-to-one mapping of internuclear separation onto excitation wavelength λ which results in a much more highly structured spectrum than is normally the case with bound–free spectra. The energy distribution of ground state collision pairs is thermalized and, therefore, is well characterized. Furthermore, the overall envelope for the spectrum is a direct reflection of the radial variation of the electronic transition moment $\mu_e(R)$, a parameter which is otherwise difficult to access experimentally. Since one is exciting atomic pairs from the vibrational continuum of the ground state, the envelope of the excitation spectrum is meaningful and can be interpreted directly in terms of $\mu_e(R)$.

Perhaps the most compelling feature of photoassociative excitation spectra, and the considerable structure they display, in particular, is the demanding test of theoretical interatomic potentials they provide. Toward that end, Jones et al. [149] simulated the experimental Kr–F and Xe–I photoassociative excitation spectra but adopted a perturbative description (discussed in the next section) of the $B^2\Sigma^+_{1/2}$ excited state potential. Spectroscopic constants derived by this approach differed from those obtained by the analysis of bound → free emission spectra [151]. The theme of this section is that the adoption of a more rigorous non-perturbative description of the XeI ($B^2\Sigma^+_{1/2}$) potential, as well as a segmented representation for the ground state, yields improved spectroscopic constants for this molecule.

6.2 Theoretical Considerations

6.2.1 Determination of Interatomic Potentials: Coupled VR Analysis

The central goal of molecular spectroscopy is the determination of interaction potentials. For diatomic molecules, the potential V(R) is, as indicated explicitly, dependent upon the internuclear separation R, and the Born–Oppenheimer approximation is generally invoked in an effort to separate the electronic and nuclear contributions to the wavefunction. The latter is responsible for describing the vibrational and rotational structure of the molecule and, in the center-of-mass frame, the time-independent Schrödinger's equation can be written

$$\left[-\frac{\hbar^2}{2\mu}\nabla_r^2 + V(R) \right] \Phi_{rv} = E_{rv}\Phi_{rv} \qquad (30)$$

in which Φ_{rv} is the rotational and vibrational portion of the wavefunction. Expressed in spherical coordinates, Equation (30) becomes

$$\left\{ -\frac{\hbar^2}{2\mu}\left[\frac{1}{R^2}\frac{\partial}{\partial R}\left(R^2 \frac{\partial}{\partial R} \right) + \hat{\mathbf{J}}^2 \right] + V(R) \right\} \Phi_{rv} = E_{rv}\Phi_{rv} \qquad (31)$$

where $\hbar^2 \hat{\mathbf{J}}^2/2\mu R^2$ is the rotational energy operator derived from the angular momentum operator $\hat{\mathbf{J}}$. Analytical eigensolutions for Equation

(31) are generally unavailable, and two paths are normally taken to pursue solutions to this relation. The approach most often adopted is a perturbative treatment of both vibration and rotation of the molecule, yielding the well-known series expansions:

$$G(v) = \omega_e \left(v + \frac{1}{2}\right) - \omega_e x_e \left(v + \frac{1}{2}\right)^2 + \omega_e y_e \left(v + \frac{1}{2}\right)^3 + \cdots \quad (32)$$

$$F(J) = B_e J(J+1) + D_e J^2 (J+1)^2 + \cdots, \quad (33)$$

where $G(v)$ and $F(J)$ are the spectroscopic term values for the vibrational and rotational states, respectively, of the molecule and the coefficients ω_e, $\omega_e x_e$, $\omega_e y_e$, ... and B_e, D_e, ... are the vibrational and rotational constants (respectively) for the electronic state in question.

The model represented by Equations (31)–(33) is strictly valid only for values of R near R_e, the equilibrium internuclear separation. A more rigorous description is required if the spectroscopic data are acquired over an extended range in R. In such a case, Φ_{rv} can be represented by the product $\Phi_{rv} = v(R) Y_{Jm}(\theta, \phi)/R$ and Schrödinger's equation in one spatial coordinate is expressed as

$$\left[-\frac{\hbar^2}{2\mu} \frac{d^2}{dR^2} + V(R) + \frac{J(J+1)\hbar^2}{2\mu R^2} \right] v(R)$$

$$= \left[-\frac{\hbar^2}{2\mu} \frac{d^2}{dR^2} + V^J(R) \right] v(R) = E_{v,J} v(R) \quad (34)$$

where $V^J(R) = V(R) + J(J+1)\hbar^2/2\mu R^2$ is known as the effective potential. Since the centrifugal term in Equation (34) is a function of both J and R, the energies $E_{v,J}$ are vibration–rotation(VR)-coupled values. Solutions to Equation (34) for a given value of the rotational quantum number J differ in several respects from those specified by the perturbative model. One such difference is that rotational constants are no longer relevant. Vibrational constants continue to be appropriate, however, and differ by less than a few per cent from those determined with the conventional perturbative model if $R \approx R_e$ and the electronic state being characterized is deeply bound. On the other hand, the discrepancies can be significant if the excited electronic state in question is shallow or the analyzed spectrum represents optical transitions over an extended range in R.

6.2.2 Electronic and Ground State Model Potentials

The functional forms adopted in these simulations for the excited ($B^2 \Sigma^+_{1/2}$) and ground ($X^2 \Sigma^+_{1/2}$) states of the XeI molecule are, with the exception of certain features of the ground state, drawn largely from prior experimental and theoretical work. As described in Refs. [149–151] and [152], for example, the $B^2\Sigma^+$ ion pair state of this diatomic excimer was assumed to have the form of a truncated Rittner potential:

$$V_B(R) = T_i + be^{-\beta R} - \frac{C_1}{R} - \frac{C_3}{R^3} - \frac{C_4}{R^4} - \frac{C_6}{R^6}, \tag{35}$$

where T_i is the $Xe^+(^2P_{3/2}) + I^-$ limit at 73 161 cm^{-1} and C_1, C_3, C_4, and C_6 are the constants associated with the Coulombic, ion-quadrupole, ion-induced dipole, and induced dipole-induced dipole interactions, respectively. Since expressions to calculate the C_n coefficients are provided by Refs. [153] and [154], only b and β in Equation (35) are free parameters.

Modeling the ground state of the XeI molecule is more complex owing to the existence of a van der Waals minimum in addition to the dissociative region at small R. Considerable computational testing has shown that no existing expression in the literature provides a fully satisfactory treatment of the $X^2\Sigma^+_{1/2}$ potential over the entire R range of interest. Accordingly, a three-segment potential describing separately the inner wall, the van der Waals region, and the long-range portion of the ground state has been developed. The inner wall and van der Waals regions are represented by a conventional Morse potential, whereas the asymptotic (longer-range) portion is described by the dispersion terms $-(C_6/R^6 + C_8/R^8)$ in which the dispersion coefficients C_6 and C_8 are given by Refs. [155–157]. Transitioning from one potential segment to the next is facilitated by Fermi-like 'switching' [156] (or damping) [157] functions, expressed as

$$f(R) = \frac{1}{\left(1 + e^{a(R-R_s)}\right)^b} \quad (b > 0), \tag{36}$$

where a and b are constants and R_s is the value of interatomic separation at which two potential segments are joined.

6.2.3 Photoassociation Spectrum Calculations

Given the model potentials proposed in the last section, calculating the photoassociation spectrum follows the procedure discussed in detail in Ref. [149]. Briefly, the reduced photoassociation cross section $\kappa_{PA}(\nu)$ can be expressed as

$$\kappa_{PA}(\nu) = \frac{8\pi^3\nu}{3c}e^2 a_0^2 \frac{1}{g_e Q_t} S(\nu) n_{Xe} n_I, \tag{37}$$

where n_{Xe} and n_I are the Xe and I atomic number densities, respectively, Q_t is the translational partition function, and g_e is the electronic degeneracy. The transition profile $S(\nu)$ can be written as

$$S(\nu) = \sum_{v',J}(2J+1)e^{-\frac{\varepsilon_J''}{kT}}\left|\langle v',J|\mu_e(R)|\varepsilon_J''\rangle\right|^2 \qquad (38)$$

$$= \sum_{v',J} P_{v',J}\left(\varepsilon_J''\right), \qquad (39)$$

where $P_{v',J}\left(\varepsilon_J''\right)$, known as the partial wave, represents the contribution to $S(\nu)$ from all bound $\leftarrow$ free transitions terminating at a specific rovibrational level (v', J'). It should be emphasized that Equations (38) and (39) account only for $\Delta J = 0$ transitions and the discussion to follow assumes that $J' = J'' \equiv J$. The term $\left|\langle v', J|\mu_e(R)|\varepsilon_{J'}''\rangle\right|^2$ in Equation (38) is the matrix element and $\mu_e(R)$ is the electronic transition moment. If μ_e is a slowly-varying function of the interatomic separation, then

$$\left|\langle v',J|\mu_e(R)|\varepsilon_J''\rangle\right|^2 \cong \mu_e^2\left|\langle v',J|\varepsilon_J''\rangle\right|^2. \qquad (40)$$

Once the partial waves are calculated, the lineshape function is determined by summing the $P_{v',J}$ over the appropriate vibrational–rotational states. All calculations also require that the rovibrational energies $E_{v',J}$ and the atomic collision pair energies ε_J'' are related to the laser (excitation) frequency by the relation

$$h\nu = E_{v',J} - \varepsilon_J''. \qquad (41)$$

Examples of partial waves and Franck–Condon factors ($\left|\langle v', J | \varepsilon_J''\rangle\right|^2$) for bound $\leftarrow$ free transitions in which $v' = 5$ and $J = 120$ or $J = 0$ are illustrated in panels (a) and (b) of Figure 31, respectively. For the simulations reported here, wavefunctions were calculated in R increments of 5×10^{-3} Å and the interval in ε_J'' was chosen to be 4 cm^{-1}. Also, the computational ranges for J and v' were increased, relative to the values of Ref. [149], to maxima of $J_{MAX} = 320$ and $v' = 60$. This necessarily also raises the span in R examined from 3.0–4.5 Å (Ref. [149]) to 2.7–4.7 Å.

6.3 Simulations and Derived Xenon Monoiodide Spectroscopic Constants

The simulation strategy centered on matching as precisely as possible the calculated photoassociation spectrum with specific features of the experimental profile. Systematic alterations in the spectroscopic parameters for the B and X state potentials were correlated with

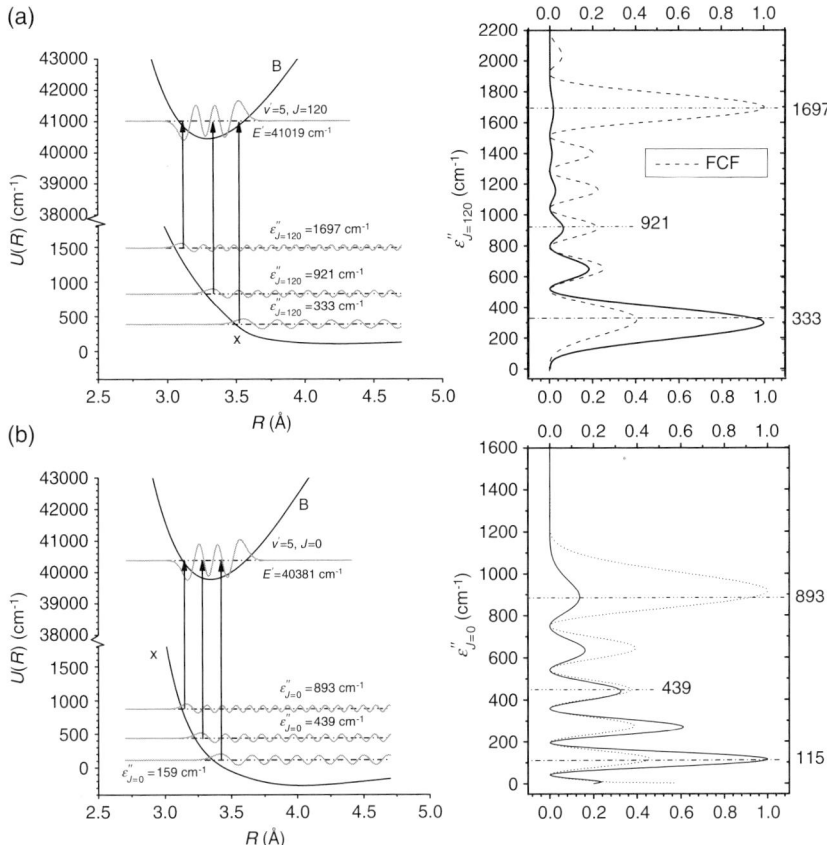

FIGURE 31 Examples of partial waves (solid curves) and franck–condon factors (dotted curves) calculated for Xe–I atomic pairs and $v' = 5$: (a) $J = 120$, and (b) $J = 0$. Notice the vertical displacement of both the ground and excited state potentials in (a)

observable characteristics in the experimental spectrum, including the modulation depth of each undulation, the peak positions and spectral widths of the oscillations, and the contour of the red edge of the spectrum. As an example, the modulation depth associated with the highest frequency peaks (i.e., $v' > 25$) and spectral positions of the longer-wavelength undulations ($v' < 10$) were found to be quite sensitive to the value of $T'_e \equiv T_e(B^2\Sigma^+_{1/2})$. Choosing a value too large rapidly shifts the most strongly-modulated peaks to larger v'. In a similar manner, all of the primary spectroscopic parameters of the B and X states were observed to have distinct effects in the calculated spectra. Narrowing the allowable range in each spectroscopic constant was accomplished by a series of simulations in which an approximate value for T'_e was first

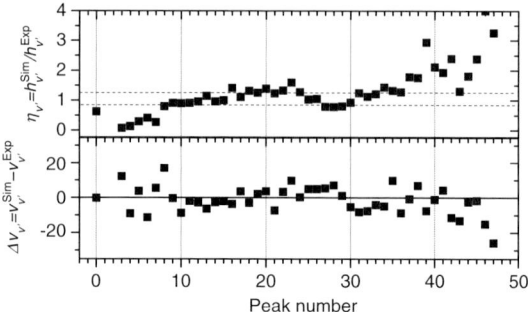

FIGURE 32 Comparison between the simulated and experimental values of the relative peak heights (top) and peak spectral positions (bottom) for the 47 observable oscillations in the Xe–I photoassociation spectrum

established. Subsequently, the value for ω'_e was identified by ensuring that the $v' = 0\text{–}9$ peaks of the simulations are spectrally aligned with experiment. At the same time, the ground state potential and its slope at R'_e ($V_X(R'_e)$ and $V'_X(R'_e)$, respectively) were adjusted such that the heights of the $v' = 0\text{–}9$ undulations have the correct relative values. After a final adjustment in ω'_e so as to align the $v' \sim 8\text{–}20$ peaks with experiment, D''_e and R''_e were determined by matching the $v' \gtrsim 20$ peak positions with their measured values. The overall quality of the fit of each simulation to the experimental spectrum was quantified by defining two parameters, $\eta_{v'}$ and $\Delta v_{v'}$, written as

$$\eta_{v'} = h_{v'}^{\text{sim}}/h_{v'}^{\text{exp}}, \qquad (42)$$

$$\text{and} \quad \Delta v_{v'} = v_{v'}^{\text{sim}} - v_{v'}^{\text{exp}}, \qquad (43)$$

where the terms *sim* and *exp* denote simulated and experimental values, respectively, $h_{v'}$ is the relative height (intensity), with respect to the underlying continuum, of the peak associated with vibrational level v', and $\Delta v_{v'}$ is the difference in frequency between the calculated and experimental values of the v' peak position.

Figure 32 presents the values of $\eta_{v'}$ and $\Delta v_{v'}$ for the simulation to date having the lowest RMS values of both parameters. It is evident that the peak positions and modulation levels for the oscillations associated with $v' \sim 8\text{–}33$ match the experiment quite well but the fit is less satisfactory beyond $v' > 35$. Perhaps this is to be expected because of the increasing spectral congestion in the $\tilde{v} > 43\,000$ cm^{-1} region. Also, the spectral positions of most of the $v' = 0\text{–}8$ peaks are in agreement with observed values but the peak heights are not accurately reproduced between $v' \sim 4$ and 8.

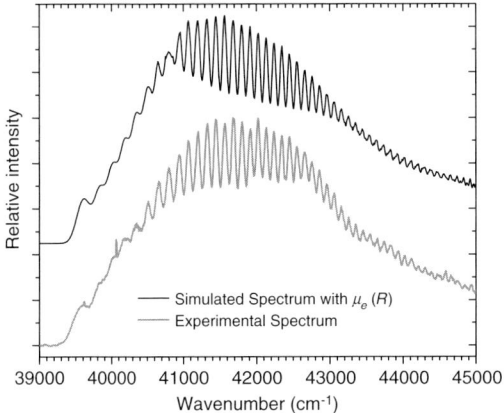

FIGURE 33 Comparison of the experimental Xe–I photoassociation spectrum with the best fit simulation to date

TABLE 3 Spectroscopic parameters for the $B^2\Sigma^+_{1/2}$ and $X^2\Sigma^+_{1/2}$ states of XeI derived from photoassociation spectra.

State	Parameter	
$B^2\Sigma^+_{1/2}$	T'_e (cm^{-1})	40 080 ($\pm$70)
	D'_e (cm^{-1})	33 081 ($\pm$70)
	R'_e (Å)	3.284 ($\pm$0.01)
	ω'_e (cm^{-1})	105.2 ($\pm$1.0)
	$\omega'_e x'_e$ (cm^{-1})	0.20
$X^2\Sigma^+_{1/2}$	D''_e (cm^{-1})	150.5 ($\pm$30)
	R''_e (Å)	4.04 ($\pm$.1)

Figure 33 provides a comparison of the experimental photoassociation spectrum with the best fit simulation. For clarity, the two profiles are offset vertically and the simulated spectrum includes the variation of μ_e with R that was ascertained from the calculations by a process similar to that described in Ref. [149]. The spectroscopic constants for the simulated spectrum of Figure 33 are presented in Table 3. Although further details regarding the computational procedure, as well as a comprehensive overview of the XeI spectroscopic constants in the literature, will be reported elsewhere, [F. Shen and J.G. Eden (unpublished)] several comments should be made here. The first is that the value of ω'_e determined in these simulations (105.2 $\pm$ 1.0 cm^{-1}) lies within the stated uncertainty of the values suggested in Refs. [149] (103 $\pm$4 cm^{-1}) and [151] (110.54 $\pm$ 10 cm^{-1}). Secondly, the most obvious impact on the XeI(B) spectroscopic constants of a more rigorous treatment of the ground state potential and the wavefunction calculations is to lower T'_e from the $40\,540 \pm 25$ cm^{-1} value of Jones et al. [149] to $40\,080$ cm^{-1} (Table 3). The latter is within 300 cm^{-1} (<0.8%) of the constant proposed by Radzykewycz and Tellinghuisen [151]. The overriding point to be made, however, is that, despite being recorded at ambient temperature, photoassociation spectra such as those of Figure 33 and Refs. [149] and [158] offer a rich array of structural features that provide a formidable challenge to theoretical potentials. Consequently, it is our conviction that such bound ← free excitation spectra constitute a significant new tool with which to elucidate the electronic and vibrational structure of diatomic and triatomic molecules having shallow or dissociative ground states.

7. APPLICATION OF PHOTOASSOCIATION TO HIGH INTENSITY DISCHARGE LIGHTING

High intensity discharge lighting, including metal-halide, sodium, and mercury lamps, continue to be among the most efficient lighting sources available and currently provide several GW of visible radiation worldwide on a continuous basis [159]. Introduced in the 1960s (Ref. [160]), metal-halide lamps are of particular economic value at present owing to luminous efficacy values above 85 lumens/W and a favorable emission spectrum as exemplified by color rendering indices (CRI) as large as 95 (Ref. [161]).

In the context of this chapter, and photoassociation, in particular, interest in metal-halide lamps stems from the chemical composition of the vapor within an operating lamp and the interaction of that vapor with the intense, broadband optical field produced by the thermal arc. It is well known from both theory and experiment that the most prevalent species in metal-halide lamps are atomic mercury and iodine, the latter of which is provided by the thermal dissociation of metal-iodide precursors introduced to the lamp. Thermochemical calculations demonstrate that Hg and I number densities in a lamp typically exceed 10^{19} cm^{-3} and 10^{18} cm^{-3}, respectively, which naturally raises the question of the potential influence of photoassociation on the optical power (and its spectral distribution) emitted by the lamp. Specifically, we might expect the spectrum produced by the arc on the axis of the lamp to be modified by photoassociative absorption losses as the optical radiation propagates outward toward the lamp envelope. Although models of HID lamp properties have been available for more than three decades, [162,163] this process has (to our knowledge) not previously been considered. To that end, Hg–I photoassociation spectra were calculated at several discrete temperatures between 1000 K and 6000 K from the *ab initio* configuration interaction potentials for the X, A, and B states of HgI reported by Salter et al. [164]. Figure 34 presents the results of the simulations for photon

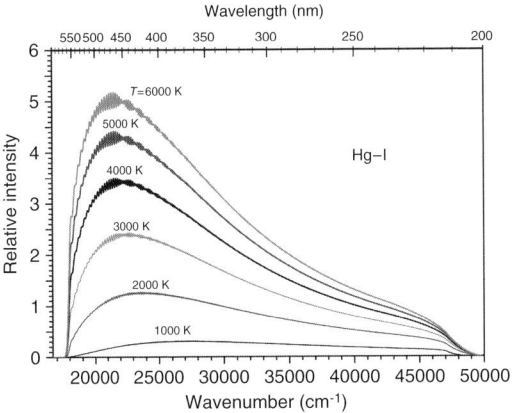

FIGURE 34 Spectrum for the photoassociation of Hg–I atomic pairs thermalized at temperatures between 1000 K and 6000 K

wavelengths in the ∼200–570 nm region ($\tilde{\nu} \sim 17\,500 - 50\,000$ cm^{-1}). Maximum absorption shifts monotonically to the red with increasing gas temperature and, at $T = 5000$ K, a broad peak is observed at ∼470 nm. As is the case for the Xe–I and Kr–F photoassociation spectra, Franck–Condon structure is evident in the vicinity of maximum absorption but the degree of modulation is low. The calculated profiles of Figure 34 clearly extend well into the visible – for a gas temperature of 6000 K, 50% of peak absorption on the long-wavelength side of the spectrum occurs at $\lambda \sim 560$ nm. Hg–I pair absorption in the visible is dominated by $A \leftarrow X$ transitions. At room temperature, visible absorption owing to Hg–I pairs is negligible owing to the steeply repulsive character of the HgI(A) ground state but lamp operating temperatures beyond 1000 K raise dramatically the impact of this process.

In combination with the spectra of Figure 34, the broadband radiation field, large Hg and I number densities, and temperatures of several thousand K existing in metal-halide lamps suggest that photoassociation is a potentially significant loss mechanism. Estimates based on Figure 34, the values of [Hg] and [I] cited earlier and calculated radial temperature profiles indicate that photoassociation is capable of significantly altering the blue-green portion of the spectrum of the lamp, and absorbing at least a few per cent of the power emitted by the arc. Of course, the calculations described above do not account for the photoassociation of Hg–Hg, Hg–Hg–Hg, and I–I atomic pairs or ensembles (forming electronically excited Hg_2, Hg_3, and I_2, respectively) which may also impact negatively the performance of the lamp.

8. SUMMARY AND CONCLUSIONS

Experiments have been described in which dipole–dipole interactions between excited Rb atoms have been detected at interatomic distances >1100 Å (>$2000a_0$) with wavepackets and PFWM. Sidebands associated with the 7s–$5d_{5/2}$ beating frequency are observed, enabling the measurement of the overall interaction energy of an excited atom with its electronically excited neighbors. Estimates of the many-body, dipole–dipole interaction energies for ensembles comprising as many as 13 excited Rb atoms indicate that groups of five or fewer atoms account for all of the data. For $\langle R \rangle \sim 1000$ Å, the primary interaction appears to be that involving 5–6 excited atoms whereas at 600 Å ensembles of four atoms are dominant. At $\langle R \rangle \sim 400$ Å, $\langle n \rangle \sim 3$ atoms. Similar (preliminary) results have been obtained for Na. The data and calculations demonstrate that fundamental aspects of interactions among excited atoms at internuclear distances of <200 Å to well beyond 1000 Å can be studied in gases or vapors at room temperature and beyond with a new form of laser spectroscopy based upon the coupling of a classical THz oscillator (wavepacket) with a coherent nonlinear optical process. The ability to determine absolute dipole–dipole interaction energies is a result of freezing the atom–atom interaction with $\sim$120 fs laser pulses, restricting the measurement interval to 100–200 ps, and monitoring offsets (sideband positions) relative to the standard provided by the Rb(7s–$5d_{5/2}$) or Na(5s–$4d_{5/2}$) quantum beating frequencies. Simulations of the Fourier domain (spectral) profiles predict the experimental Rb number density to within 30% and confirm that dipole–dipole interactions dominate the overall measured interaction energy. Extending this approach to probing atom–atom interactions at longer range appears to be feasible by adopting optical heterodyning techniques to detect sideband splittings in the sub-GHz domain. The pursuit of similar phenomena in small molecules having a permanent dipole moment, such as NO, is an attractive direction for future work. This ultrafast technique provides access to the investigation of a range of chemical dynamic and collisional processes, and is of particular value for those atomic and molecular species that have not thus far been confined successfully in a trap.

An immediate application of the ability to detect optically the dipole–dipole interaction is the observation of molecular dissociation. Indeed, it has been demonstrated that predissociation of a diatomic molecule (Rb_2) and the excited state distribution of the nascent atomic fragments can be detected and characterized by wavepacket interferometry in tandem with a coherent nonlinear optical process (PFWM). Femtosecond pump–probe experiments were described in which the predissociation of electronically-excited Rb_2 states in the $\sim$24 000–28 000 cm^{-1} interval, and the distribution of nascent atomic fragments into Rb excited states (7s, 5d, 6s, 4d, and 5p) spanning an energy range >1.25 eV, have been observed in Rb vapor with atomic number densities of $\sim 6 \times 10^{13} - 3 \times 10^{17}$ cm^{-3}. Quantum beating at 18.2 THz (corresponding to the Rb 7s–5d ($J = 5/2$) energy defect of $\sim$608 cm^{-1}) is superimposed onto the axially-phase-matched PFWM signal wave generated at $\lambda_S \sim 420$ nm (Rb $6^2P_J \rightarrow 5^2S_{1/2}$ transitions) and recovered by Fourier analysis of the signal wave intensity as the pump–probe time delay (Δt) is scanned. Stated succinctly, the molecular dissociation transient is detected by the dipole–dipole interaction and PFWM produces a macroscopic response in the form of a coherent, modulated beam of UV or visible radiation.

The dominant exit channels for Rb$_2$ predissociation are found to be sensitive to the interval of internuclear separation R in which the molecular wavepacket, produced by the pump pulse through two-photon association of Rb–Rb collision pairs, is localized. Generating Rb$_2$ wavepackets on $^3\Lambda_u$ ($\Lambda = \Sigma, \Pi$) potential surfaces at large R ($\sim$7–9 Å) favors the Rb 7s and 5d dissociation channels. In contrast, producing dimer wavepackets localized in the R$\sim$3–4 Å region suppresses Rb (7s, 5d) generation and favors the production of Rb atomic fragments (Rb (6p), in particular) with less internal energy but maximum velocities of $\sim$25–35 Å/ps. The sensitivity of this technique to individual dissociation channels and yet its multichannel detection capability are attributable to the sensitivity of PFWM to the distribution of population among the excited states involved in the four wave mixing process. It must also be emphasized that radiative and collisional effects are negligible owing to the $\sim$100 ps time scale of the experiments. It appears that this spectroscopic tool can readily be extended to studying the dissociation of polyatomic molecules in which diatomic (or larger) molecular fragments are produced.

Coherent control of Rb$_2$ predissociation has also been demonstrated with chirped pump pulses. For zero chirp, Rb(7s) atoms are observed to be produced with near unit efficiency because of the localization of the initial (launched) wavepacket beyond R$\sim$6 Å. Similarly, the Rb(6p) exit channel is predominant when the Rb$_2$ wavepacket is established by the pump at smaller values of internuclear separation.

Photoassociation of thermalized pairs of rare gas and halogen atoms has been observed by laser spectroscopy. Excitation spectra of the photoassociation of Xe–I atomic pairs in the ultraviolet, for example, display strongly-modulated Franck–Condon structure that imposes a stringent test on theoretical interatomic potentials. Analysis of such spectra by coupled vibration–rotation (VR) calculations yields spectroscopic constants for the $B^2\Sigma^+$ ion pair state of the molecule that are more precise than those extracted from the gently undulating, bound $\rightarrow$ free emission spectra of the rare-gas-halide excimers. Furthermore, bound ← free transitions originating from the ground state of the diatomic complex are of applied interest in those situations, such as that characteristic of the operating environment of high pressure metal-halide arc lamps, in which atoms at high number densities are immersed in an intense, broadband radiation field.

ACKNOWLEDGMENTS

The authors are indebted to numerous former colleagues in the Laboratory for Optical Physics and Engineering at the University of Illinois for their contributions to this research. Among them are J. Schloss, R. Jones, H. Tran, J. Gao, G. Rodriguez, A. Oldenburg, Z. Lu, and C. Zhu. Also, the support of this work by the U.S. Air Force Office of Scientific Research (H. R. Schlossberg) is gratefully acknowledged.

REFERENCES

[1] J. Weiner, V.S. Bagnato, S. Zilio and P.S. Julienne (1999), *Rev. Mod. Phys.*, **71**, 1.
[2] K.M. Jones, E. Tiesinga, P.D. Lett and P.A. Julienne (2006), *Rev. Mod. Phys.*, **78**, 483.
[3] J.O. Hirschfelder, C.F. Curtiss and R.B. Bird (1954), *Molecular Theory of Gases and Liquids*, John Wiley and Sons, New York.
[4] K.M. Jones, P.S. Julienne, P.D. Lett, W.D. Phillips, E. Tiesinga and C.J. Williams (1996), *Europhys. Lett.*, **35**, 85.
[5] See, for example K. Enomoto, M. Kitagawa, K. Kasa, S. Tojo and Y. Takahashi (2007), *Phys. Rev. Lett.*, **98**, 203201.
[6] D. Tong, S.M. Farooqi, J. Stanojevic, S. Krishnan, Y.P. Zhang, R. Côté, E.E. Eyler and P.L. Gould (2004), *Phys. Rev. Lett.*, **93**, 063001.
[7] K. Singer, M. Reetz-Lamour, T. Amthor, L.G. Marcassa and M. Weidemüller (2004), *Phys. Rev. Lett.*, **93**, 163001.
[8] T.C. Killian, S. Kulin, S.D. Bergeson, L.A. Orozco, C. Orzel and S.L. Rolston (1999), *Phys. Rev. Lett.*, **83**, 4776.
[9] M.P. Robinson, B. Laburthe Tolra, M.W. Noel, T.F. Gallagher and P. Pillet (2000), *Phys. Rev. Lett.*, **85**, 4466.
[10] M. Mayle, B. Hezel, I. Lesanovsky and P. Schmelcher (2007), *Phys. Rev. Lett.*, **99**, 113004.
[11] C.H. Greene, A.S. Dickinson and H.R. Sadeghpour (2000), *Phys. Rev. Lett.*, **85**, 2458.
[12] C.H. Greene, E.L. Hamilton, H. Crowell, C. Vadla and K. Niemax (2006), *Phys. Rev. Lett.*, **97**, 233002.
[13] C. Boisseau, I. Simbotin and R. Côté (2002), *Phys. Rev. Lett.*, **88**, 133004.
[14] J.-H. Choi, B. Knuffman, T. Cubel Liebisch, A. Reinhard and G. Raithel (2007), Cold Rydberg Atoms. *Adv. Atomic, Molecular, and Opt. Phys.*, **54**, 131–202.
[15] W.R. Anderson, J.R. Veale and T.F. Gallagher (1998), *Phys. Rev. Lett.*, **80**, 249.
[16] I. Mourachko, D. Comparat, F. de Tomasi, A. Fioretti, P. Nosbaum, V.M. Akulin and P. Pillet (1998), *Phys. Rev. Lett.*, **80**, 253.
[17] E.B. Alexandrov (1964), *Opt. Spectrosc.*, **17**, 957.
[18] J.N. Dodd, R.D. Kaul and D.M Warrington (1964), *Proc. Phys. Soc. (London)*, **84**, 176.
[19] S. Haroche, J.A. Paisner and A.L. Schawlow (1973), *Phys. Rev. Lett.*, **30**, 948.
[20] H.J. Andrä (1974), *Phys. Scripta*, **9**, 257.
[21] M.M. Salour (1976), *Opt. Comm.*, **18**, 377.
[22] R. Beach and S.R. Hartmann (1984), *Phys. Rev. Lett.*, **53**, 663.
[23] N. Morita and T. Yajima (1984), *Phys. Rev. A*, **30**, 2525.
[24] J.E. Golub and T.W. Mossberg (1986), *J. Opt. Soc. Am. B*, **3**, 554.

[25] H. Bitto and J.R. Huber (1990), *Opt. Comm.*, **80**, 184.
[26] W. Demtröder (2003), *Laser Spectroscopy*, 3rd edn., Springer, Berlin, 692–699.
[27] G. Alber, H. Ritsch and P. Zöller (1986), *Phys. Rev. A*, **34**, 1058.
[28] J. Parker and C.R. Stroud Jr. (1986), *Phys. Rev. Lett.*, **56**, 716.
[29] A. ten Wolde, L.D. Noordam, A. Lagendijk and H.B. van Linden van den Heuvell (1988), *Phys. Rev. Lett.*, **61**, 2099.
[30] A. ten Wolde, L.D. Noordam, A. Lagendijk and H.B. van Linden van den Heuvell (1989), *Phys. Rev. A*, **40**, 485.
[31] J.A. Yeazell, M. Mallalieu, J. Parker and C.R. Stroud Jr. (1989), *Phys. Rev. A*, **40**, 5040.
[32] J.A. Yeazell, M. Mallalieu and C.R. Stroud Jr. (1990), *Phys. Rev. Lett.*, **64**, 2007.
[33] G.M. Lankhuijzen and L.D. Noordam (1995), *Phys. Rev. A*, **52**, 2016.
[34] M. Strehle, U. Weichmann and G. Gerber (1998), *Phys. Rev. A*, **58**, 450.
[35] C.C. Hayden and D.W. Chandler (1995), *J. Chem. Phys.*, **103**, 10465.
[36] M. Motzkus, S. Pedersen and A.H. Zewail (1996), *J. Phys. Chem.*, **100**, 5620.
[37] M. Schmitt, G. Knopp, A. Materny and W. Kiefer (1997), *Chem. Phys. Lett.*, **270**, 9.
[38] M. Schmitt, G. Knopp, A. Materny and W. Kiefer (1997), *Chem. Phys. Lett.*, **280**, 339.
[39] M. Schmitt, G. Knopp, A. Materny and W. Kiefer (1998), *J. Phys. Chem. A*, **102**, 4059.
[40] T. Siebert, M. Schmitt, A. Vierheilig, G. Flachenecker, V. Engel, A. Materny and W. Kiefer (2000), *J. Raman Spectrosc.*, **31**, 25.
[41] S. Meyer and V. Engel (2000), *J. Raman Spectrosc.*, **31**, 33.
[42] H.C. Tran, P.C. John, J. Gao and J.G. Eden (1998), *Opt. Lett.*, **23**, 70.
[43] C.J. Zhu, Y. Xiao, A.A. Senin, J. Gao, J.G. Eden, T.S. Varzhapetyan and D.H. Sarkisyan (2007), *Phys. Rev. A*, **75**, 053405.
[44] Z.H. Lu, C.J. Zhu, A.A. Senin, J.R. Allen, J. Gao and J.G. Eden (2004), *IEEE J. Select. Topics Quantum Electron.*, **10**, 159.
[45] J.F. Reintjes (1984), *Nonlinear Optical Parametric Processes in Liquids and Gases*, Academic, New York.
[46] R.L. Fork, C.V. Shank and R.T. Yen (1982), *Appl. Phys. Lett.*, **41**, 223.
[47] G. Rodriguez, P.C. John and J.G. Eden (1995), *J. Chem. Phys.*, **103**, 10473; A.M. Farkas and J.G. Eden (1993), *IEEE J. Quantum Electron.*, **29**, 2923.
[48] Y. Xiao, A.A. Senin, B.J. Ricconi, R. Kogler, C.J. Zhu and J.G. Eden (2008), *J. Phys. B: At. Mol. Opt. Phys.*, **5** (in press).
[49] D.J. Kane and R. Trebino (1993), *J. Quantum Electron.*, **29**, 571.
[50] D.J. Kane and R. Trebino (1993), *Opt. Lett.*, **18**, 823.
[51] A.A. Senin, H.C. Tran, J. Gao, Z.H. Lu, C.J. Zhu, A.L. Oldenburg, J.R. Allen and J.G. Eden (2003), *Chem. Phys. Lett.*, **381**, 53.
[52] D. Felinto, L.H. Acioli and S.S. Vianna (2000), *Opt. Lett.*, **25**, 917.
[53] D. Felinto, C.A.C. Bosco, L.H. Acioli and S.S. Vianna (2001), *Phys. Rev. A*, **64**, 063413.
[54] O.J. Lumpkin, P.P. Sorokin and J.R. Lankard (1967), *Bull. Am. Phys. Soc.*, **12**, 1054.
[55] O.J. Lumpkin Jr. (1968), *IEEE J. Quantum Electron.*, **QE-4**, 226.
[56] P.P. Sorokin and J.R. Lankard (1973), *IEEE J. Quantum Electron.*, **QE-9**, 227.
[57] P.P. Sorokin, J.J. Wynne and J.R. Lankard (1973), *Appl. Phys. Lett.*, **22**, 342.
[58] W. Hartig (1978), *Appl. Phys.*, **15**, 427.
[59] F.S. Tomkins and R. Mahon (1981), *Opt. Lett.*, **6**, 179.
[60] J. Bokor, R.R. Freeman, R.L. Panock and J.C. White (1981), *Opt. Lett.*, **6**, 182.
[61] R. Hilbig and R. Wallenstein (1983), *IEEE J. Quantum Electron.*, **QE-19**, 194; **QE-19**, 1759.
[62] M. Alden, U. Westblom and J.E.M. Goldsmith (1989), *Opt. Lett.*, **14**, 305.
[63] R.K. Wunderlich, W.R. Garrett, R.C. Hart, M.A. Moore and M.G. Payne (1990), *Phys. Rev. A*, **41**, 6345.
[64] U. Czarnetzki and H.F. Döbele (1991), *Phys. Rev. A*, **44**, 7530.
[65] J. Ishii, Y. Ogi, Y. Tanaka and K. Tsukiyama (1996), *Opt. Comm.*, **132**, 316.
[66] P.-L. Zhang, Y.-C. Wang and A.L. Schawlow (1984), *J. Opt. Soc. Am. B*, **1**, 9.
[67] Z.G. Wang, H. Schmidt and B. Wellegehausen (1987), *Appl. Phys. B: Photophys. Laser Chem.*, **44**, 41.

[68] B.K. Clark, M. Masters and J. Huennekens (1988), *Appl. Phys. B: Photophys. Laser Chem.*, **47**, 159.
[69] M.A. Moore, W.R. Garrett and M.G. Payne (1989), *Phys. Rev. A*, **39**, 3692.
[70] Z.J. Jabbour, M.S. Malcuit and J. Huennekens (1991), *Appl. Phys. B: Photophys. Laser Chem.*, **52**, 281.
[71] F.Z. Chen, X.F. Han and C.Y.R. Wu (1993), *Appl. Phys. B: Photophys. Laser Chem.*, **56**, 113.
[72] C.H. Skinner and H.P. Palenius (1976), *Opt. Comm.*, **18**, 335.
[73] R.R. Freeman, J. Bokor and W.E. Cooke (1982), *Phys. Rev. A*, **26**, 3029.
[74] See, for example C.H. Muller III, D.D. Lowenthal and M.A. DeFaccio (1988), *Opt. Lett.*, **13**, 651.
[75] H. Matsueda (1999), *Superlatt. Microstruct.*, **26**, 211;
H. Matsueda (2002), *Superlatt. Microstruct.*, **31**, 87.
[76] H. Matsueda, K. Leosson, Z. Xu, J.M. Hvam, Y. Ducommun, A. Hartmann and E. Kapon (2004), *IEEE Trans. Nanotechnol.*, **3**, 318.
[77] Y. Divayana and X.W. Sun (2007), *Phys. Rev. Lett.*, **99**, 143003.
[78] T. Amthor, M. Reetz-Lamour, S. Westermann, J. Denskat and M. Weidemüller (2007), *Phys. Rev. Lett.*, **98**, 023004.
[79] K. Autumn, Y.A. Liang, S.T. Hsieh, W. Zesch, W.P. Chan, T.W. Kenny, R. Fearing and R.J. Full (2000), *Nature (London)*, **405**, 681.
[80] W. Li, P.J. Tanner and T.F. Gallagher (2005), *Phys. Rev. Lett.*, **94**, 173001.
[81] R.B. Bernstein (1982), *Chemical Dynamics via Molecular Beam and Laser Techniques*, Clarendon Press, Oxford, p. 78.
[82] J.M. Sage, S. Sainis, T. Bergeman and D. DeMille (2005), *Phys. Rev. Lett.*, **94**, 203001.
[83] H.P. Büchler, A. Michelli and P. Zöller (2007), *Nat. Phys.*, **3**, 726.
[84] F. Shen, J. Gao, A.A. Senin, C.J. Zhu, J.R. Allen, Z.H. Lu, Y. Xiao and J.G. Eden (2007), *Phys. Rev. Lett.*, **99**, 143201.
[85] R.J. LeRoy (1973), in: R.F. Barrow, D.A. Long and D.J. Millen (Eds.), *Molecular Spectroscopy*, Vol. I, Chemical Society, London.
[86] H. Wang, J. Li, X.T. Wang, C.J. Williams, P.L. Gould and W.C. Stwalley (1997), *Phys. Rev. A*, **55**, R1569.
[87] C. Ates, T. Pohl, T. Pattard and J.M. Rost (2007), *Phys. Rev. Lett.*, **98**, 023002.
[88] G.W. King and J.H. van Vleck (1939), *Phys. Rev.*, **55**, 1165.
[89] M. Marinescu and A. Dalgarno (1995), *Phys. Rev. A*, **52**, 311.
[90] W.I. McAlexander, E.R.I. Abraham and R.G. Hulet (1996), *Phys. Rev. A*, **54**, R5.
[91] M.S. Safronova, C.J. Williams and C.W. Clark (2004), *Phys. Rev. A*, **69**, 022509.
[92] P. Hertz (1909), *Math. Ann.*, **67**, 387.
[93] S. Torquato, B. Lu and J. Rubenstein (1990), *Phys. Rev. A*, **41**, 2059.
[94] L.G. Marcassa, A.L. de Oliveira, M. Weidemüller and V.S. Bagnato (2005), *Phys. Rev. A*, **71**, 054701.
[95] H. Katô and M. Baba (1995), *Chem. Rev.*, **95**, 2311.
[96] V. Kokoouline, O. Dulieu, R. Kosloff and F. Masnou-Seeuws (2000), *Phys. Rev. A*, **62**, 032716.
[97] H. Lefebvre-Brion and R.W. Field (2004), *The Spectra and Dynamics of Diatomic Molecules*, Elsevier, pp. 493-549.
[98] K.H. Meiwes and F. Engelke (1982), *Chem. Phys. Lett.*, **85**, 409.
[99] M. Carré, M. Druetta, M.L. Gaillard, H.H. Bukow, M. Horani, A.L. Roche and M. Velghe (1980), *Mol. Phys.*, **40**, 1453.
[100] See, for example Z. Xu, B. Koplitz and C. Wittig (1987), *J. Chem. Phys.*, **87**, 1062.
[101] M.D. Person, K.Q. Lao, B.J. Eckholm and L.J. Butler (1989), *J. Chem. Phys.*, **91**, 812.
[102] I. Ben-Itzhak, P. Wang, J. Xia, A.M. Sayler, M.A. Smith, J.W. Maseberg, K.D. Carnes and B.D. Esry (2005), *Nucl. Instrum. Methods Phys. Res. B*, **233**, 56.
[103] M. Dantus, M.J. Rosker and A.H. Zewail (1987), *J. Chem. Phys.*, **87**, 2395.
[104] T.S. Rose, M.J. Rosker and A.H. Zewail (1988), *J. Chem. Phys.*, **88**, 6672.
[105] T.S. Rose, M.J. Rosker and A.H. Zewail (1989), *J. Chem. Phys.*, **91**, 7415.

[106] P. Cong, G. Roberts, J.L. Herek, A. Mohktari and A.H. Zewail (1996), *J. Phys. Chem.*, **100**, 7832.
[107] R.M. Bowman, M. Dantus and A.H. Zewail (1989), *Chem. Phys. Lett.*, **161**, 297.
[108] M. Gruebele, G. Roberts, M. Dantus, R.M. Bowman and A.H. Zewail (1990), *Chem. Phys. Lett.*, **166**, 459.
[109] J.A. Misewich, J.H. Glownia, J.E. Rothenberg and P.P. Sorokin (1988), *Chem. Phys. Lett.*, **150**, 374.
[110] J.H. Glownia, J.A. Misewich and P.P. Sorokin (1990), *J. Chem. Phys.*, **92**, 3335.
[111] R.E. Walkup, J.A. Misewich, J.H. Glownia and P.P. Sorokin (1990), *Phys. Rev. Lett.*, **65**, 2366.
[112] N.F. Scherer, R.J. Carlson, A. Matro, M. Du, A.J. Ruggiero, V. Romero-Rochin, J.A. Cina, G.R. Fleming and S.A. Rice (1991), *J. Chem. Phys.*, **95**, 1487;
N.F. Scherer, A. Matro, L.D. Ziegler, M. Du, R.J. Carlson, J.A. Cina and G.R. Fleming (1992), *J. Chem. Phys.*, **96**, 4180.
[113] K. Ohmori, Y. Sato, E.E. Nikitin and S.A. Rice (2003), *Phys. Rev. Lett.*, **91**, 243003.
[114] T. Baumert, B. Bühler, R. Thalweiser and G. Gerber (1990), *Phys. Rev. Lett.*, **64**, 733.
[115] T. Baumert, B. Bühler, M. Grosser, R. Thalweiser, V. Weiss, E. Wiedenmann and G. Gerber (1991), *J. Phys. Chem.*, **95**, 8103.
[116] T. Baumert, V. Engel, C. Röttgermann, W.T. Strunz and G. Gerber (1992), *Chem. Phys. Lett.*, **191**, 639.
[117] G. Rodriguez and J.G. Eden (1993), *Chem. Phys. Lett.*, **205**, 371.
[118] A.L. Oldenburg, P.C. John and J.G. Eden (2000), *J. Chem. Phys.*, **113**, 11009.
[119] V. Blanchet, M.A. Bouchène, O. Cabrol and B. Girard (1995), *Chem. Phys. Lett.*, **233**, 491.
[120] V. Blanchet, M.A. Bouchène and B. Girard (1998), *J. Chem. Phys.*, **108**, 4862.
[121] R. de Vivie-Riedle, B. Reischl, S. Rutz and E. Schreiber (1995), *J. Phys. Chem.*, **99**, 16829.
[122] S. Rutz, R. de Vivie-Riedle and E. Schreiber (1996), *Phys. Rev. A*, **54**, 306.
[123] J.M. Papanikolas, R.M. Williams, P.D. Kleiber, J.L. Hart, C. Brink, S.D. Price and S.R. Leone (1995), *J. Chem. Phys.*, **103**, 7269.
[124] R.M. Williams, J.M. Papanikolas, J. Rathje and S.R. Leone (1997), *J. Chem. Phys.*, **106**, 8310.
[125] B. Zhang, L.-E. Berg and T. Hansson (2000), *Chem. Phys. Lett.*, **325**, 577.
[126] N. Gador, B. Zhang, H.O. Karlsson and T. Hansson (2004), *Phys. Rev. A*, **70**, 033418.
[127] F. Spiegelmann, D. Pavolini and J.-P. Daudey (1989), *J. Phys. B*, **22**, 2465.
[128] S.J. Park, S.W. Suh, Y.S. Lee and G.-H. Jeung (2001), *J. Mol. Spectrosc.*, **207**, 129.
[129] P. Brumer and M. Shapiro (1986), *Chem. Phys. Lett.*, **126**, 541.
[130] D.J. Tannor, R. Kosloff and S.A. Rice (1986), *J. Chem. Phys.*, **85**, 5805.
[131] M. Shapiro, J.W. Hepburn and P. Brumer (1988), *Chem. Phys. Lett.*, **149**, 451.
[132] A. Assion, T. Baumert, J. Helbing, V. Seyfried and G. Gerber (1996), *Chem. Phys. Lett.*, **259**, 488.
[133] C.J. Bardeen, J. Che, K.R. Wilson, V.V. Yakovlev, P. Cong, B. Kohler, J.L. Krause and M. Messina (1997), *J. Phys. Chem. A*, **101**, 3815.
[134] R.S. Judson and H. Rabitz (1992), *Phys. Rev. Lett.*, **68**, 1500.
[135] C.J. Bardeen, V.V. Yakovlev, K.R. Wilson, S.D. Carpenter, P.M. Weber and W.S. Warren (1997), *Chem. Phys. Lett.*, **280**, 151.
[136] A. Assion, T. Baumert, M. Bergt, T. Brixner, B. Kiefer, V. Seyfried, M. Strehle and G. Gerber (1998), *Science*, **282**, 919.
[137] T.C. Weinacht, J. Ahn and P.H. Bucksbaum (1999), *Nature*, **397**, 233.
[138] R. Bartels, S. Backus, E. Zeek, L. Misoguti, G. Vdovin, I.P. Christov, M.M. Murnane and H.C. Kapteyn (2000), *Nature*, **406**, 164.
[139] J.L. Herek, W. Wohlleben, R.J. Cogdell, D. Zeidler and M. Motzkus (2002), *Nature*, **417**, 533.
[140] V.I. Prokhorenko, A.M. Nagy, S.A. Waschuk, L.S. Brown, R.R. Birge and R.J.D. Miller (2006), *Science*, **313**, 1257.
[141] R.J. Levis, G.M. Menkir and H. Rabitz (2001), *Science*, **292**, 709.
[142] R.J. Gordon and S.A. Rice (1997), *Ann. Rev. Phys. Chem.*, **48**, 595.

[143] H. Rabitz, R. de Vivie-Riedle, M. Motzkus and K. Kompa (2000), *Science*, **288**, 824.
[144] S.A. Rice and M. Zhao (2000), *Optical Control of Molecular Dynamics*, John Wiley and Sons, New York.
[145] M. Dantus (2001), *Ann. Rev. Phys. Chem.*, **52**, 639.
[146] V. Engel and H. Metiu (1989), *J. Chem. Phys.*, **91**, 1596; **90**, 6116.
[147] J. Tellinghuisen (1985), in: K.P. Lawley (Ed.), *Photodissociation and Photoionization*, In: Advances in Chemical Physics, Vol. LX, John Wiley and Sons, New York, pp. 299–369.
[148] J.H. Schloss, R.B. Jones and J.G. Eden (1992), *Chem. Phys. Lett.*, **191**, 195.
[149] R.B. Jones, J.H. Schloss and J.G. Eden (1993), *J. Chem. Phys.*, **98**, 4317.
[150] See, for example G. Rodriguez and J.G. Eden (1991), *J. Chem. Phys.*, **95**, 5539. and refs. cited therein.
[151] D.T. Radzykewycz and J. Tellinghuisen (1996), *J. Chem. Phys.*, **105**, 1330.
[152] J. Tellinghuisen, A.K. Hays, J.M. Hoffman and G.C. Tisone (1976), *J. Chem. Phys.*, **65**, 4473.
[153] A. Sur, A.K. Hui and J. Tellinghuisen (1979), *J. Mol. Spectrosc.*, **74**, 465.
[154] J. Tellinghuisen (1993), *Can. J. Chem.*, **71**, 1645.
[155] K. Tamagake, D.W. Setser and J.H. Kolts (1981), *J. Chem. Phys.*, **74**, 4286.
[156] P. Casavecchia, G. He, R.K. Sparks and Y.T. Lee (1982), *J. Chem. Phys.*, **77**, 1878.
[157] P. Huxley, D.B. Knowles, J.N. Murrell and J.D. Watts (1984), *J. Chem. Soc. Faraday Trans. 2*, **80**, 1349.
[158] V.S. Zuev, J.G. Eden and H.C. Tran (2001), *Opt. Spectrosc.*, **90**, 516.
[159] J.F. Waymouth (1991), *IEEE Trans. Plasma Sci.*, **19**, 1003.
[160] See, for example G.H. Reiling (1964), *J. Opt. Soc. Am.*, **54**, 532.
[161] *High Intensity Discharge Lighting Technology*, U.S. Department of Energy (2006). (www.eere.energy.gov/buildings/tech/lighting/).
[162] R.O. Shaffner (1971), *Proc. IEEE*, **59**, 622.
[163] R.J. Zollweg, J.J. Lowke and R.W. Liebermann (1975), *J. Appl. Phys.*, **46**, 3828.
[164] C. Salter, P.C. Tellinghuisen, J.G. Ashmore and J. Tellinghuisen (1986), *J. Mol. Spectrosc.*, **120**, 334.

CHAPTER 3

Bose–Einstein Condensates in Disordered Potentials

Leonardo Fallani, Chiara Fort and Massimo Inguscio

LENS European Laboratory for Nonlinear Spectroscopy, Italy
Dipartimento di Fisica, Università di Firenze, Via Nello Carrara 1, 50019 Sesto Fiorentino (FI), Italy

Contents
1. Introduction 120
2. How to Produce a Disordered Potential 121
   2.1 Speckle Patterns 122
   2.2 Multi-chromatic Lattices 124
   2.3 Other Methods 126
3. Weakly-interacting Regime 127
   3.1 A Bose–Einstein Condensate in a Disordered Potential 130
   3.2 The Quest for Anderson Localization 133
   3.3 Further Directions 141
   3.4 Observing Anderson Localization 144
4. Strongly-interacting Regime 145
   4.1 The Quest for Bose Glass 148
   4.2 Experiments with Atomic Mixtures 152
5. Conclusions 154
Acknowledgments 154
References 155

Abstract The interplay between disorder and interactions is a *leitmotiv* of condensed-matter physics, since it constitutes the driving mechanism of the metal–insulator transition. Bose–Einstein condensates in optical potentials are proving to be powerful tools to quantum simulate disordered systems. We will review the main experimental and theoretical results achieved in the past few years in this rapidly developing field.

Advances in Atomic, Molecular, and Optical Physics, Volume 56 © 2008 Elsevier Inc.
ISSN 1049-250X, DOI 10.1016/S1049-250X(08)00012-8 All rights reserved.

1. INTRODUCTION

In nature many processes occur in an ordered way. Indeed, ordered configurations are often the ones minimizing the total energy of the system. A prominent example of this tendency towards order is given by the growth of a crystal, where the atoms arrange themselves in a spatially periodic configuration building up an ordered lattice. The physics of transport of electrons in a metal heavily relies on the periodicity of this lattice. However, when crystalline solids are studied on a sufficiently small length scale, one realizes that impurities and defects are always present, which may affect in a substantial way the transport of the electrons. Disorder is indeed an intrinsic property of all the real systems. In the last 50 years the effects of disorder on transport phenomena have been extensively studied in the context of both statistical and condensed-matter physics.

Despite the very general interest in understanding the physics of disorder in condensed-matter systems, still many questions remain open and unsolved, even from the theoretical point of view. As a matter of fact, the theoretical description of periodic systems, as perfect crystals, is much easier than the one for disordered system as disordered lattices or glasses. The problems that arise are related to the fact that the effects of disorder cannot be theoretically treated in a perturbative way: even a small amount of disorder can produce dramatic changes in the physical properties of the system under investigation. In 1958, P.W. Anderson published a seminal paper (Anderson, 1958) in which he showed under which conditions non-interacting electrons in a disordered metal can either move through the system, or be localized. It was soon realized that Anderson localization is a much more general phenomenon holding for the propagation of generic classical waves in disordered media. Localization is a coherent effect that arises from multiple scattering of a wave from randomly-distributed impurities and from the resulting destructive interference in the direction of propagation.

Also interactions are well known to induce localization effects, as pointed out by N.F. Mott who was able to explain the anomalous insulator behavior of some materials when electron–electron interactions were included in the band theory. In 1977 P.W. Anderson and N.F. Mott were awarded with the Nobel Prize in Physics for their fundamental theoretical investigations of the electronic structure of magnetic and disordered systems (Anderson, 1978; Mott, 1978). Following these pioneering works, a strong theoretical effort has been devoted in the last decades to investigate the combined role of disorder and interactions in the superfluid–insulator transition observed in many condensed-matter systems, such as ^{4}He adsorbed on porous media (Crowell et al., 1995), thin superconducting films (Goldman and Marković, 1998), arrays of

Josephson junctions (van der Zant et al., 1992) and high-temperature superconductors (Jiang et al., 1994; Budhani et al., 1994).

Ultracold atoms in optical lattices (Morsch and Oberthaler, 2006) represent an extremely powerful tool for engineering simple quantum systems with a broad tunability of the parameters, thus serving as 'quantum simulators' (Feynman, 1982) to reproduce the physics of different systems. The striking advantage offered by such atomic systems resides in the unprecedented possibility to work with perfectly isolated samples at quasi-zero temperature and to have experimental control on most of the Hamiltonian parameters, e.g. the lattice depth or the strength of the atom–atom interactions, that can be precisely tuned even in real time. One spectacular demonstration of this opportunity has been given by the observation of the superfluid (SF) to Mott insulator (MI) transition in a three-dimensional optical lattice (Greiner et al., 2002), which pioneered the investigation of strongly quantum correlated regimes with ultracold atoms (Bloch et al., 2008).

A natural extension of these experiments is the realization of disordered systems using ultracold atoms in optical potentials. In this paper we will review the recent progresses in this field, that was experimentally initiated in 2004 with the first investigation of atomic Bose–Einstein condensates in disordered potentials. Different possibilities can be followed to produce disordered ultracold atomic systems. Disordered or quasi-disordered potentials can be created optically by using speckle patterns (Lye et al., 2005) or multi-chromatic incommensurate optical lattices (Fallani et al., 2007). These methods allow the production of disordered potentials in which both the spectral properties and the amount of disorder are known with very good accuracy and can be easily controlled. In addition to the optical way, disordered systems could also be created by using atomic mixtures (Gavish and Castin, 2005) or inhomogeneous magnetic fields (Gimperlein et al., 2005; Courteille et al., 2006). We will review these different possibilities together with the illustration of the diverse interaction regimes that can be investigated. The first experimental results obtained with ultracold bosons in disordered potentials will be presented, discussing the state-of-the-art of this newborn field and the perspectives for future breakthroughs.

2. HOW TO PRODUCE A DISORDERED POTENTIAL

In this section we will present different experimental approaches to the production of disordered potentials for neutral atoms. We will mostly focus on two methods allowing the production of complex optical potentials: speckle patterns and multi-chromatic lattices.

2.1 Speckle Patterns

The first realization of disordered potentials for cold atoms has been obtained with speckle patterns (Boiron et al., 1999). Speckles are produced whenever light is reflected by a rough surface or transmitted by a diffusive medium (Goodman, 2006). We will mostly consider the case of transmission, sketched in Figure 1(a), and we will refer to the scattering device as a *diffusive plate*. Such a device can be modeled as being made up of many randomly-distributed impurities by which the illuminating light is scattered. Since the scattering of laser light is mainly a coherent process, the partial waves emerging from the scattering interfere and produce a complex distribution of light, called a *speckle pattern*, an example of which is shown in Figure 1(b). This disordered distribution of light can be imaged onto the atoms, producing a disordered potential $V(\mathbf{r})$ proportional to the local laser intensity $I(\mathbf{r})$.

In general, if the wavelength of the light is detuned far from the atomic resonance, no absorption is involved and the resulting mechanical effect can be described by a potential energy of the form

$$V(\mathbf{r}) = \frac{3\pi c^2}{2\omega_0^3} \left(\frac{\Gamma}{\Delta}\right) I(\mathbf{r}), \tag{1}$$

where c is the speed of light, ω_0 is the frequency of the atomic resonance, Γ its radiative linewidth, $\Delta = \omega - \omega_0$ the detuning, and $I(\mathbf{r})$ the intensity distribution. This potential is often called *dipole potential* (Grimm et al., 2000). It is worth noting that the sign of this potential depends on Δ, which is the only quantity which can take either positive or negative values. In particular, when $\Delta < 0$ (*red detuning*) $V(\mathbf{r})$ is negative, hence maxima of light intensity correspond to potential minima: atoms will move towards higher-intensity regions. Instead, when $\Delta > 0$ (*blue detuning*) $V(\mathbf{r})$ is positive, hence maxima of light intensity correspond to potential maxima: atoms will move towards lower-intensity regions.

Speckle patterns represent a valuable way to produce a disordered potential in a controlled way. The possibility to accurately measure the statistical and correlation properties of the disordered potential comes from the fact that the intensity of the speckle pattern can be directly recorded by a CCD camera (typically the same one used to image the BEC atoms). In Figure 2(a) we show the cross section of a typical speckle pattern used at LENS for the first investigation of disordered Bose–Einstein condensates (Lye et al., 2005). Among the different quantities characterizing the properties of the speckle field, one can define an *average speckle height* V_S. Different definitions are used in literature, however one of the most used in the context of BEC experiments corresponds to taking twice the standard deviation of the speckle potential $V(x)$ (supposed one-dimensional)

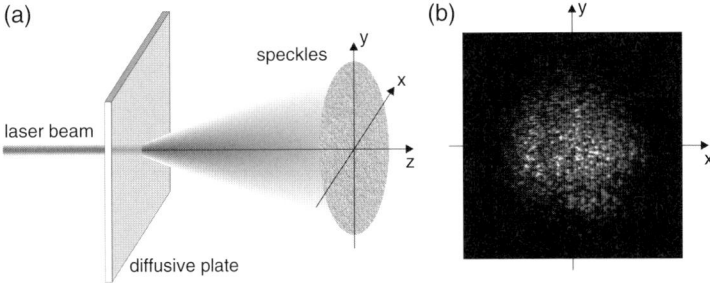

FIGURE 1 Production of speckle patterns. (a) A laser beam is shone through a diffusive plate and the resulting speckle pattern is then imaged onto the BEC. (b) Intensity distribution of a typical speckle pattern recorded with a CCD camera

around its mean value $\overline{V}$ (Lye et al., 2005):

$$V_S = 2 \left[\frac{1}{L} \int_{-L/2}^{L/2} \left(V(x) - \overline{V} \right)^2 dx \right]^{1/2}. \qquad (2)$$

An even more important quantity, as we shall see in the following, is the *autocorrelation length* σ, giving information on the speckle grain size (Goodman, 2006). This quantity is defined as the rms width of the autocorrelation integral $G(d)$ of the speckle potential

$$G(d) = \int_{-L/2}^{L/2} V(x)V(x+d) dx \approx e^{-\frac{d^2}{2\sigma^2}}, \qquad (3)$$

an example of which is shown in Figure 2(b). The autocorrelation length σ depends on the wavelength of the light, on the nature of the diffusive medium producing the speckle pattern and, most importantly, on the optical resolution of the lens system used to image the speckle pattern onto the atoms. As a matter of fact, the typically determined autocorrelation length is set by the diffraction limit spot size of the imaging system. A detailed description of the speckle potential is given in Clément et al. (2006), where the statistical properties of the speckle field are discussed both from a theoretical point of view and with the introduction of experimental methods which allow their precise determination.

The random potential produced by a speckle pattern is static. This means that the atoms experience just one realization of disorder, which can be reproduced in the same way from one experiment to another. However, shifting the position of the diffuser leads to a different realization of the speckle pattern that preserves the same spectral and statistical proper-

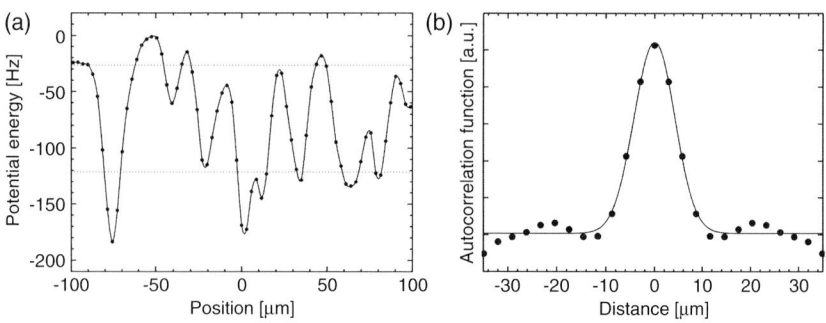

FIGURE 2 Production of speckle patterns. (a) Cross section of a typical speckle potential (the energy difference between the horizontal lines is the average speckle height V_S). (b) Autocorrelation integral of the same potential

ties. Thus, averages on multiple realizations of disorder can be achieved in a simple way.

Before concluding this section, we note that speckles are intrinsically two-dimensional in the plane perpendicular to the propagation axis. Actually, a speckle pattern also varies along the direction of propagation of the light. However, the typical correlation size along this direction is much larger. Nonetheless, speckle potentials with different dimensionality can be produced: one-dimensional speckles can be produced by using cylindrical lenses stretching the speckle pattern along one direction, while three-dimensional speckles could be obtained by adding speckle patterns coming from different directions.

2.2 Multi-chromatic Lattices

As we have seen in the previous section, speckle patterns are a powerful and easy-to-implement method to produce random potentials. We have pointed out that a crucial parameter of such potentials is the autocorrelation length σ, which gives an estimate of the minimum length scale below which the potential loses its random nature and becomes correlated. Typically, this length is connected with the diffraction limit dimension at which optical speckles are imaged onto the atomic sample. For this reason, the random potential produced by speckles is often too coarse grained (with σ of the order of several microns), unless one builds a dedicated setup to overcome the usual optical access restrictions. Recently, progresses in the realization of speckle potentials with autocorrelation length below 1 μm have been achieved by the experimental groups of A. Aspect (Clément et al., 2006) and B. DeMarco (White et al., 2008).

On the other side, having in mind many years of exciting physics with cold atoms in optical lattices, we know that optical standing waves can be easily created providing spatial periodicities that can also be smaller

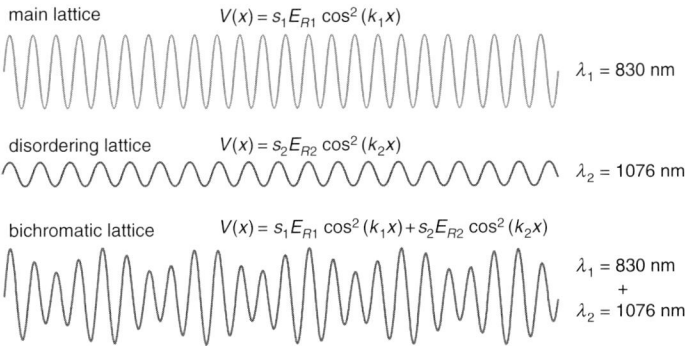

FIGURE 3 A bichromatic optical lattice. The discrete translational invariance of the main lattice is perturbed by the addition of a secondary lattice with incommensurate wavelength

than half a micron (roughly speaking, one order of magnitude smaller than the autocorrelation length of the speckle pattern shown in Figure 2). This suggests the idea that, by combining several optical standing waves with different incommensurate spacings, it is possible to produce complex potentials with very small 'grain size'.

The simplest example is given by a bichromatic lattice resulting from the addition of two lattices with incommensurate wavelengths. Bose–Einstein condensates in potentials of this kind have been first investigated in experiments at LENS (Fallani et al., 2007), where a main lattice with wavelength λ_1 was perturbed by a weaker secondary lattice with wavelength λ_2, as sketched in Figure 3. The resulting potential can be written in the form

$$V(x) = s_1 E_{R1} \cos^2(k_1 x) + s_2 E_{R2} \cos^2(k_2 x) \quad (4)$$

where $k_1 = 2\pi/\lambda_1$ and $k_2 = 2\pi/\lambda_2$ are the lattice wavenumbers and s_1 and s_2 are adimensional numbers indicating the heights of the two lattices in units of the recoil energies $E_{R1} = h^2/(2m\lambda_1^2)$ and $E_{R2} = h^2/(2m\lambda_2^2)$, respectively. In the limit $s_2 \ll s_1$ the height of the optical barriers is roughly constant across the whole lattice and it is possible to define a tunneling rate J which only depends on the main lattice height s_1. In this limit, the effect of the secondary lattice reduces to an inhomogeneous and non-periodic shift of the potential energy at the bottom of the lattice wells (see Sections 3.2.2 and 4).

As we will further discuss in the following, bichromatic incommensurate lattices are not truly disordered potentials. They differ from both purely random potentials and speckle potentials, which exhibit different statistical and correlation properties. Strictly speaking, they are *quasi-*

periodic potentials, since their spectrum is made up of a set of discrete frequencies. However, because of the lack of any translational invariance, they can be used to investigate the physics of finite-sized disordered systems and study the emergence of quantum localization effects, as we shall see in Section 3.2.2.

We note that, since one always deals with finite-sized atomic samples, the notion of incommensurability (i.e. the wavelength ratio being an irrational number) is a rather sophistic concept, and should be substituted with a more practical definition. From an experimental point of view, since the lattice wavelengths are known with finite precision, the measurement of the ratio λ_2/λ_1 always gives a rational number. From a theoretical point of view, it is important to consider that the finite size of the systems under investigation releases the constraints on the incommensurability: even a periodic potential (resulting from a commensurate ratio) does not show any periodicity if the system size is smaller than the period. The bichromatic lattice is thus *effectively* incommensurate provided that the ratio between the wavelengths is far from a ratio between *simple* integer numbers. More precisely, a bichromatic lattice can be considered incommensurate whenever the resulting periodicity (if any) is larger than the system size.

2.3 Other Methods

In solids disorder is often caused by the presence of impurities, i.e. atoms of a different kind that randomly occupy the sites of the crystalline lattice where atoms of different species were expected. This kind of disorder can be simulated also in cold gases by using a mixture of two different atomic species, as proposed in Gavish and Castin (2005). The atoms of one of the two species are trapped in the sites of a deep optical lattice. If the filling factor (i.e. the average number of atoms per site) is less than unity, only some of the sites will be occupied by one atom and the other ones will be empty, as schematically shown in Figure 4. The atoms of the other species, that could be weakly affected by the presence of the lattice, feel the collisional interaction with the randomly-distributed atoms of the first species. This kind of disorder is spectrally different from both the speckle and bichromatic potentials, since it is a binary kind of disorder (yes/no) on top of a periodic backbone. In Gavish and Castin (2005) the authors have theoretically investigated the possibility to study one-dimensional Anderson localization of matter waves with this system and this work has then been extended to 3D in Massignan and Castin (2006).

The possibility of using mixtures of atomic species to study disordered systems had been previously discussed in Vignolo et al. (2003) for the case of bosonic atoms flowing in an optical lattice partially filled by fermionic impurities. A similar approach was also proposed in Paredes et al. (2005) as a method to quantum-simulate disordered strongly interacting systems. The main ideas of this work have been developed for hard-core

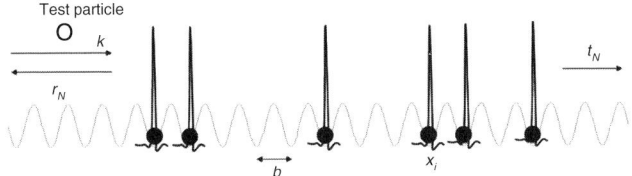

FIGURE 4 Disordered potential produced by the interaction of the atoms with a different species. An incoming test particle (white) experiences the collisional potential produced by randomly-distributed (black) atoms trapped in an optical lattice. Reprinted figure with permission from Gavish and Castin (2005) © 2005, of the American Physical Society

bosons in Horstmann et al. (2007) and in Roscilde and Cirac (2007) for the study of metastable glassy states of the mixture. The schemes proposed in all these theoretical works have not been experimentally realized yet, although first experiments with binary ultracold mixtures have been performed (see Section 4.2).

Another way to introduce disorder in the system has been proposed in Gimperlein et al. (2005) by using inhomogeneous magnetic fields, e.g. by exploiting the magnetic field fluctuations in the proximity of a microtrap caused by imperfections in the chip fabrication (Wang et al., 2004). If the bias magnetic field is kept close to a Feshbach resonance (Inouye et al., 1998), small field fluctuations on top of it produce spatial fluctuations in the scattering length characterizing the interactions between the atoms. Therefore, this technique allows introducing disorder on the atom–atom interaction strength, rather than on the external potential. In Gimperlein et al. (2005) the phase diagram of interacting bosons in the presence of such disorder has been derived, evidencing novel features with respect to the phase diagram with disorder in the external potential (that will be presented in Section 4).

3. WEAKLY-INTERACTING REGIME

One of the most fascinating phenomena characterizing the transport of waves in random systems is *Anderson localization*. This effect takes its name after the seminal work of P.W. Anderson in 1958, who identified the fundamental role of disorder in the metal–insulator transition observed in solid-state systems (Anderson, 1958). Anderson first formulated his localization theory for a simple model of particles hopping on a lattice with random on-site energies, arguing that above a critical disorder amplitude the quantum states had to change from extended to spatially localized. This intuition, together with the mathematical tools developed to describe the localization transition, led to the award of the Nobel Prize in Physics in 1977 (Anderson, 1978).

In the following decades, however, it was realized that Anderson localization is a much more general phenomenon, holding for propagation of generic linear waves in disordered media. Indeed, it has been observed for sound waves and light waves (Wiersma et al., 1997; Schwartz et al., 2007), whereas a direct observation for matter waves has not yet been possible.

In the language of wave propagation, Anderson localization arises because of interference effects in the scattering of a wave by disordered defects. When studying wave propagation in disordered systems, different localization regimes can be identified. A precursor effect of Anderson localization is *weak localization*, which arises from interference effects in multiple scattering events: an example of weak localization is given by *coherent backscattering*, i.e. the enhanced probability of backdiffusion for light incident on a disordered sample, owing to the interference between the forward and backward scattering paths (Wiersma et al., 1995). In the strong scattering limit $kl \simeq 1$ (with k wave number and l mean free path between scattering events), these interferences can add up to completely halt the waves inside the random medium, resulting in *strong localization*, or Anderson localization.

Anderson-localized states are characterized by the typical exponential decay of their tails in the space distribution. In the case of light, this means that the intensity is an exponentially decreasing function of the distance traveled in the disordered medium. For quantum-mechanical wavefunctions, this means that a localized state $\Psi(x)$ can be written as

$$\Psi(x) \sim \exp\left(-\frac{x}{\zeta}\right), \qquad (5)$$

where ζ is the *localization length*. Generally speaking, the stronger the disorder the smaller the localization length.

In the physics of Anderson localization an important role is played by the dimensionality of the system. After the first Anderson conjecture, scaling arguments have been proposed which predict different scenarios with changing dimensionality d of the system (Abrahams et al., 1979). For $d < 2$ all the states are localized. For $d > 2$ a localization transition exists, with a mobility edge separating extended states for weak disorder from localized states above a critical value. The two-dimensional case is marginal, since the states are localized for any amount of disorder as in the one-dimensional case, but the localization length at weak disorder can be exponentially large.

A Bose–Einstein condensate is characterized by long-range coherence and can be described with a classical order parameter which corresponds to the wavefunction of the Bose-condensed atoms. In the non-interacting case all the atoms are described by the same single-particle wavefunction which obeys the Schrödinger equation

$$i\hbar \frac{d\Psi}{dt} = -\frac{\hbar^2}{2m}\nabla^2 \Psi + V(\mathbf{r})\Psi. \quad (6)$$

In the presence of disorder, this wavefunction can be Anderson localized. Bose–Einstein condensates represent an appealing system where it is possible to directly study the effect of localization. By using the techniques described in the previous section, one is able to create disordered potentials in an extremely controlled way, knowing precisely the kind and amount of disorder. Furthermore, the wavefunction (more precisely, its squared modulus) can be directly observed by imaging the condensate with a CCD camera. As a result, the typical exponential tails of Anderson-localized states could be observed (if the imaging resolution and sensitivity are good enough), allowing the detection of localization.

An extremely interesting, and still open, problem regards the effect of interactions on localization. Originally, Anderson formulated his theory for non-interacting quantum particles. If one considers real interacting particles, however, the scenario could be significantly different. In the case of electrons, repulsive long-range interactions are present due to the Coulomb electric force. In the case of ultracold neutral atoms long-range dipolar interactions can be nearly always neglected and the dominant interaction mechanism is represented by s-wave elastic collisions. These short-range interactions can be either attractive or repulsive, although only repulsive interactions allow for the existence of stable BECs with arbitrarily large number of atoms (Dalfovo et al., 1999). When the BEC density is sufficiently small, as happens in many experimental situations, the effect of the collisional forces can be described within a mean-field approach by adding a nonlinear term in Equation (6), which becomes the well-known Gross–Pitaevskii equation (GPE) (Dalfovo et al., 1999):

$$i\hbar \frac{d\Psi}{dt} = -\frac{\hbar^2}{2m}\nabla^2 \Psi + V(\mathbf{r})\Psi + g|\Psi|^2 \Psi. \quad (7)$$

The interaction strength, described by $g = 4\pi \hbar^2 a/m$, is parametrized as a function of one single scalar parameter a, which takes the name of *scattering length*. A similar kind of cubic nonlinearity is present also in the Maxwell equations describing the propagation of light in a nonlinear optical medium where the index of refraction depends on the light intensity (Kerr effect). This term is responsible for many effects of nonlinear dynamics, such as solitonic propagation (Burger et al., 1999), four-wave mixing (Deng et al., 1999) and instabilities (Wu and Niu, 2001; Fallani et al., 2004). The presence of interactions can

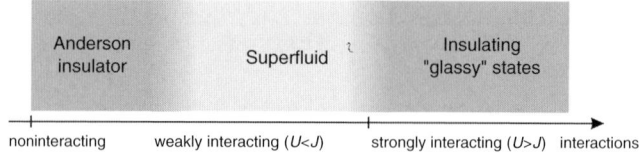

FIGURE 5 Pictorial representation of the different interaction regimes for a bosonic gas in the presence of disorder

heavily affect the physics of localization, which is intrinsically a single-particle effect, holding for linear waves. From a naive point of view, negative nonlinearities ($a < 0$, arising from attractive interactions between particles, or self-focusing behavior of the wave) could play a role in favor of localization. On the contrary, positive nonlinearities ($a > 0$, induced by repulsive interactions, or self-defocusing behavior) are expected to disfavor localization, making the problem much more interesting to study, both theoretically and experimentally.

The interplay between disorder and interactions in the physics of localization has been the object of a very intense theoretical investigation. It was soon realized that repulsive interactions can compete with disorder and eventually destroy the localization. In strongly-interacting systems, however, different regimes can be achieved and new quantum phases can be reached in which interactions and disorder co-operate in localizing the system in glassy states (see Section 4 for a discussion of these phases). An oversimplified picture of the different regimes for an interacting bosonic gas in a disordered potential is sketched in Figure 5. Real phase diagrams, of course, are much more complicated than this pictorial representation, the details depending e.g. on the kind of disorder and on the dimensionality of the system. Actually, there are still many open questions to be solved, which the experimental study of interacting Bose–Einstein condensates in disordered potentials could address. As a matter of fact, these systems offer the advantage of a broad tunability of the Hamiltonian parameters, including the kind and amount of the disorder (as seen in the previous section) and the interaction strength between the atoms (e.g. by using Feshbach resonances (Inouye et al., 1998)).

3.1 A Bose–Einstein Condensate in a Disordered Potential

3.1.1 Static Properties

A natural starting point to gather information on the behavior of the BEC in the disordered potential is the shape of the atomic density distribution after release from the confining potential. This time-of-flight detection technique has been used since the first experimental realization of BEC (Anderson et al., 1995) as a precious tool to study its ground state

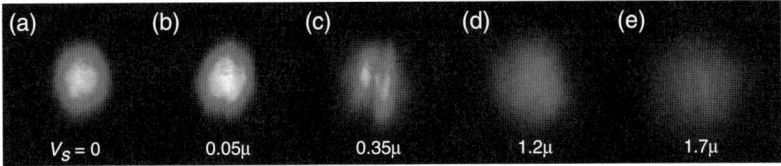

FIGURE 6 Ground state of a BEC in a harmonic + disordered potential. Absorption images of the atomic density distribution following a time-of-flight after the release from the confining potential. The numbers in the bottom indicate the average speckle height V_S in units of the BEC chemical potential $\mu \simeq 1$ kHz in the harmonic trap. The data are adapted from Lye et al. (2005)

properties. We start considering the case of disordered potentials created with optical speckles, first investigated with ^{87}Rb in Lye et al. (2005).

Basically, one can observe three different regimes, depending on the ratio between the speckle height V_S and the BEC chemical potential μ. For very small optical potentials $V_S \lesssim 0.1\mu$ one does not observe any significant deviation from the ordinary Thomas–Fermi shape of the BEC expanding from the harmonic trap (Figure 6(a),(b)). For higher speckle heights $0.1\mu \lesssim V_S \lesssim \mu$ one observes that the density distribution is strongly modified by the appearance of complex structures in the form of elongated stripes (Figure 6(c)). Finally, further increasing the speckle height to $V_S \gtrsim \mu$, the expanded density profile ceases to be characterized by stripes and one can detect only a broad unstructured Gaussian distribution (Figure 6(d),(e)).

The appearance of density modulations in the regime of weak disorder has recently attracted large interest. The structures observed after expansion could arise either from real in-trap density modulations or from phase fluctuations converted into density modulations after time-of-flight. More recent experimental works (Chen et al., 2008; Clément et al., 2008) systematically investigated such structures, measuring the fringe visibility as a function of the speckle height and evidencing that the fringe pattern is stable for the same realization of disorder. This observation, together with the comparison with GPE simulations (Clément et al., 2008), suggests that for the actual experimental parameters the most plausible scenario is the one in which small in-trap density fluctuations are amplified during the time-of-flight. The problem is studied in detail in Clément et al. (2008) with a thorough analysis of the mechanisms involved during the BEC expansion.

Further increasing the intensity of the disordered potential to $V_S \gtrsim \mu$ the condensate is split up into many condensates localized in the randomly-spaced minima of the speckle potential. The absence of interference structures in the observed density distribution is due to the fact that the spacing between different condensates is not uniform

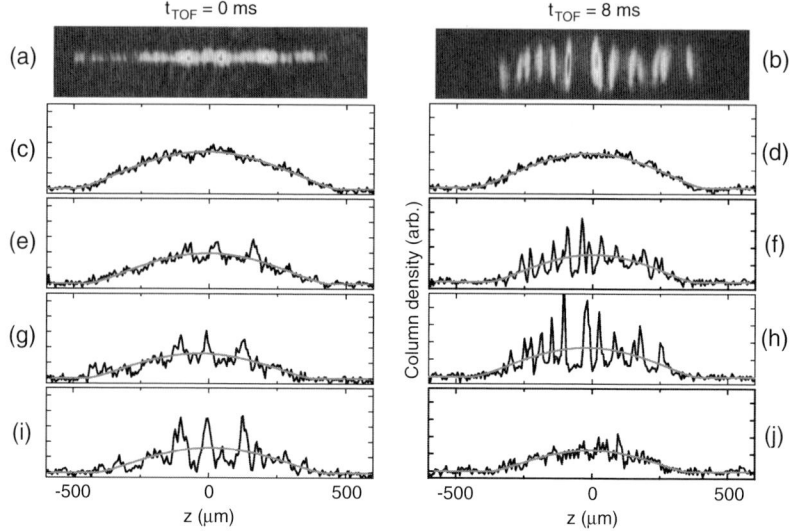

FIGURE 7 Ground state of a BEC in a harmonic + disordered potential. The left panel shows the *in-situ* density distribution, while the right panel refers to the density distribution observed after time-of-flight. The false-color images refer to a speckle potential height $V_S \approx 0.5\,\mu$, while the density plots refer to $V_S \approx 0, 0.3\,\mu, 0.5\,\mu, 1.0\,\mu$, increasing from top to bottom. Reprinted figure with permission from Chen et al. (2008)

and gives rise to an interference pattern that, averaged over the optical resolution of the system, is almost flat. This *fragmentation* scenario has been confirmed in Chen et al. (2008) by using direct *in-situ* imaging of the trapped ^{7}Li atoms (see Figure 7).

3.1.2 Collective Excitations

Low-energy collective excitations of the BEC have been studied in the presence of a weak speckle disorder producing a corrugation of the harmonic trap potential. Since the first production of BECs, frequency measurements of collective modes have provided precious information for the identification of superfluidity and, more in general, for the characterization of quantum fluids. In Lye et al. (2005) the frequency and the damping rate of the dipole and quadrupole modes of a BEC in a speckle potential have been measured as a function of the strength of disorder. With the term of 'dipole mode' one usually refers to rigid center-of-mass oscillations of the BEC in the parabolic trap, while the 'quadrupole mode' for a highly anisotropic trap (or 'axial breathing mode') indicates a shape oscillation in which the center-of-mass does not move, but the size of the BEC along its long axis is periodically changing in time (Dalfovo et al., 1999).

In Lye et al. (2005) a damping of both modes has been observed and measured as a function of the strength of the speckle potential.

Damping of the dipole mode has been recently observed also in Chen et al. (2008), in which the study of the different dynamic regimes has been accompanied by *in-situ* detection of the atomic density distribution. Indeed, Chen et al. (2008) demonstrated that the halting of the center-of-mass motion for strong disorder has to be connected with the creation of a fragmented BEC. In this regime different condensates are trapped in the different speckle potential wells and no global phase coherence is present, due to the extremely long tunneling times between different fragments.

Lye et al. (2005) measured also a frequency shift of the quadrupole mode. The frequency of the quadrupole mode is particularly important since its value depends not only on the strength of the parabolic potential but also on the Bose–Einstein equation of state. Hence it depends on the nature of the system, whether it is superfluid or not. The frequency changes measured in Lye et al. (2005), however, just reflected the change in the effective potential curvature induced by the corrugation produced by the disordered potential. The problem has been addressed theoretically in Modugno (2006) by numerical solution of the Gross–Pitaevskii equation combined with a sum-rules approach, confirming the frequency shift as an effect due to the change in the effective trap frequency.

The effect of the speckle potential produced in these experiments is mostly classical, and it does not really produce a change in the nature of the quantum fluid. We shall discuss more about this point in the next section, evidencing how the correlation length of the potential plays a crucial role in the observation of truly disordered-induced localization effects. In particular, concerning the measurements of collective excitations, the presence of disorder with short correlation length can modify the superfluid equation of state leading to non-trivial frequency shifts, as recently studied in Falco et al. (2007).

Very recently, the Rice group has studied dipole oscillations of solitons in the presence of optical speckles, observing that solitons are robust against damping of the oscillation amplitude but that the number of atoms comprising the soliton rapidly diminishes (Hulet, R. 2008. Private Communication).

3.2 The Quest for Anderson Localization

In this section we will review the advances of the experiments aiming to observe Anderson-like localization for Bose–Einstein condensates propagating in disordered optical potentials.

3.2.1 Localization in a Speckle Potential

Out of the condensed-matter systems for which it has been originally proposed, Anderson localization has been widely searched, and eventually demonstrated, in classical wave propagation experiments (Wiersma et al., 1997; Schwartz et al., 2007). In these kinds of experiments, an electromagnetic wave undergoes multiple scattering from the randomly-distributed scatterers of the disordered medium. Strong (Anderson) localization sets in when the multiple scattered waves interfere destructively in the propagation direction and localized states become populated. According to the Ioffe–Regel criterion (Ioffe and Regel, 1960) this happens when the mean free path of the wave becomes as small as its wavelength.

Experiments performed in 2005 at LENS (Florence) and by the group of A. Aspect at Institut d'Optique (Orsay) aimed to realize such scattering configuration with Bose–Einstein condensates propagating in disordered optical potentials produced with speckle patterns (Clément et al., 2005; Fort et al., 2005). The idea behind these two works was quite similar: an initially trapped Bose–Einstein condensate of ^{87}Rb was left free to expand in a one-dimensional disordered waveguide. In Clément et al. (2005) this waveguide was produced by a highly elongated magnetic trap, while in Fort et al. (2005) by a single beam optical trap. The propagation of the Bose-condensed matter wave in the waveguide was studied as a function of the height of the disordered speckle potential.

In Figure 8, taken from Fort et al. (2005), the density distribution of the condensate, imaged *in-situ* after a fixed expansion time in the optical waveguide, is shown for different speckle potential heights (ranging from $V_S = 0$ to $V_S = 0.7\mu$, with μ the BEC chemical potential) together with the picture of the actual speckle field used. Without speckles the condensate freely expands, while in the presence of the speckles both the expansion and the center-of-mass motion (induced by a small acceleration along the waveguide) start to be suppressed for $V_S \gtrsim 0.3\mu$. A closer look shows that actually two different components can be distinguished: while a low-density cloud expands without stopping, a few localized density peaks become observable when increasing the speckle height. In Figure 9, taken from Clément et al. (2005), the rms size of the BEC expanding in a disordered magnetic waveguide is plotted as a function of time for different heights of the disordered potential: one can clearly see the transition from a diffusive regime in the absence of disorder to a 'localization' regime when disorder is present.

Further investigations have demonstrated that this suppressed expansion is not Anderson localization, but a classical localization that can be explained with simple energetic arguments. The expanding condensate is not a monochromatic flux of atoms all moving with the same velocity: since the momentum distribution of the sample has a finite width (mainly caused by the atom–atom repulsive interactions which initially drive the expansion), a low velocity component of the cloud is always present and get trapped in the speckles since it has not sufficient energy to escape the

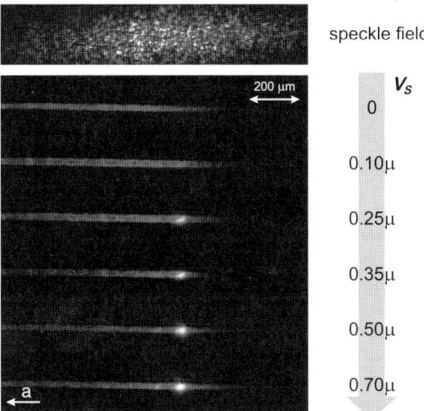

FIGURE 8 Expansion of a BEC in a disordered optical guide. (Top) Intensity profile of the speckle field used in the experiment. (Bottom) Density profiles of the condensate after expansion in the disordered optical guide for different speckle heights V_S, here expressed in units of the BEC chemical potential $\mu = 2.5$ kHz in the initial trap. The data are adapted from Fort et al. (2005)

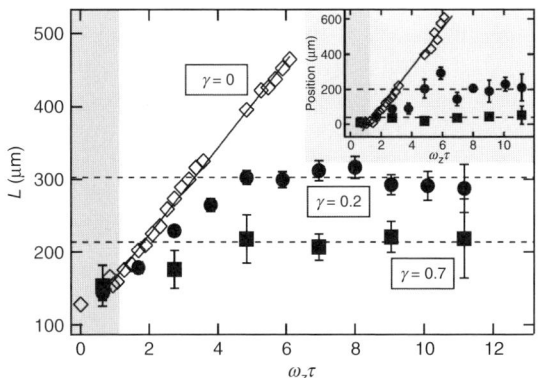

FIGURE 9 Expansion of a BEC in a disordered magnetic waveguide. Time evolution of the axial BEC rms size L and center-of-mass position (inset) for various amplitudes γ of the random speckle potential (in units of the BEC chemical potential). Reprinted figure with permission from Clément et al. (2005) © 2005, of the American Physical Society

deepest potential wells (in the case of red-detuned speckles, as in Fort et al. (2005)) or to tunnel through the highest potential barriers (in the case of blue-detuned speckles, as in Clément et al. (2005)).

Several theoretical works have studied the expansion of a Bose–Einstein condensate in a disordered potential (Clément et al., 2006; Modugno, 2006; Shapiro, 2007; Sanchez-Palencia et al., 2007; Akkermans et al., 2008;

Sanchez-Palencia et al., 2008; Skipetrov et al., 2008). In particular, in Clément et al. (2005, 2006) and Modugno (2006) it has been shown that the expanded BEC density profile is actually made up of two spatially separated parts. In the center of the cloud interaction energy is dominating over kinetic energy and the BEC density profile exactly follows the shape of the potential, as expected from the Thomas–Fermi approximation for an interacting Bose gas: no Anderson localization is expected to appear in this region. In the wings of the cloud the density is much smaller and kinetic energy is dominating over interaction energy: here the BEC almost behaves as a non-interacting gas and the density profile has deviations from the Thomas–Fermi approximation. However, numerical studies based on the Gross–Pitaevskii equation evidenced that no Anderson localization is present even in this region.

There are two possible physical reasons impeding the observation of Anderson localization. The first is indeed the presence of interactions: from an intuitive point of view, repulsive interactions between the atoms force them to spread more in space, contrasting localization. The second reason is the finite correlation length of the disorder: even in the absence of disorder, Anderson localization could not be observable because the disordered potential is not 'good' enough to produce the scattering strength which is necessary to have a localization length smaller than the system size. It is well known that, in the pure random case, any infinitesimal amount of disorder leads to localization in 1D (Abrahams et al., 1979), but in finite-sized systems (like a trapped Bose–Einstein condensate) the localization length plays an important role.

The further experiments reported by Fort et al. (2005) have evidenced that, apart from the problem of interactions, the typical speckle potentials employed so far in the experiments were not 'fine grained' enough to produce quantum reflection/transmission, which is at the basis of one-dimensional Anderson localization. This has been observed by studying the collision of a BEC with a potential defect created with a tightly focused laser beam, that mimics the effects of one single speckle grain. The absence of quantum reflection from the potential well created with this optical (red-detuned) defect indicated that, in order to have quantum scattering, one should use much steeper potentials, i.e. speckle potentials in which the autocorrelation length σ is smaller (the typical correlation length of the speckles used in Fort et al. (2005); Clément et al. (2005) was around 5 μm).

This problem has been theoretically addressed in Modugno (2006), where the coefficients of quantum reflection by a potential well and of quantum transmission from a potential barrier have been calculated as a function of the potential steepness and of the velocity of the incident matter waves. Of course, quantum reflection/transmission is fundamental in the case of one-dimensional localization, in which scattering just happens along a line: hence, for multiple scattering to appear, only a fraction of the incident wave has to be reflected/transmitted. In higher dimensions interference due to multiple scattering could happen also for classical reflection from potential hills, for which the requirements are less stringent. However, in higher dimensions localization itself is more difficult to achieve, owing to the larger localization lengths (if localization is present).

Despite the obstacles discussed above, Anderson localization can be observed in one-dimensional diffusion experiments similar to the ones reported in Clément et al. (2005); Fort et al. (2005) provided that the speckle autocorrelation length is made small enough. In Sanchez-Palencia et al. (2007) the density profile of the BEC expanding in a weak speckle potential has been analytically worked out, evidencing exponential localization in the dilute tails of the wavefunction, where density is very low and interactions can be neglected. In the case of interacting ^{87}Rb BECs this could happen for very small disorder correlation length, one order of magnitude less than the ones achieved in Clément et al. (2005) and Fort et al. (2005). The crucial parameter, as we shall see in the following section, is the ratio between the disorder correlation length and the *healing length*, which is the length scale associated to the effect of interactions. In Sanchez-Palencia et al. (2007) a mobility edge was also found as a maximum wavenumber of the expanding BEC above which localization cannot be observed, which is a peculiar characteristic of speckle potentials with finite correlation length.

3.2.2 Localization in a Bichromatic Lattice

Quasi-periodic lattices, introduced in Section 2.2, are a particular class of potentials which exhibit properties common to both periodic and disordered systems (Diener et al., 2001). As in the case of periodic lattices, their spectra show reminiscence of energy bands. On the other hand, owing to the lack of any translational invariance, they support the existence of localized states, which behave very similarly to the ones supported by truly disordered systems (Grempel et al., 1982). Therefore they can be used as a tool to study quantum localization, as a valid alternative to speckle patterns, with the experimental advantage of an effortless production of short length-scale potential fluctuations.

Localization in incommensurate bichromatic potentials is a well-known topic. This problem has been studied in detail in the framework of the Harper model (Harper, 1955) and of the one-dimensional tight-binding Aubry–André model (Aubry and André, 1980), which is described by the Hamiltonian

$$\hat{H} = -J \sum_n (|n\rangle\langle n+1| + |n+1\rangle\langle n|) + \Delta \sum_n \cos(2\pi\beta n)|n\rangle\langle n|, \quad (8)$$

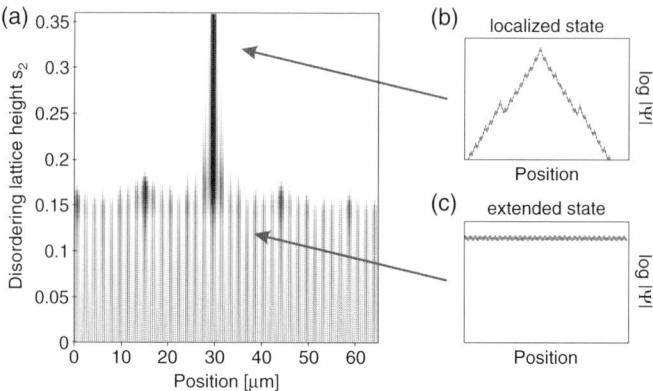

FIGURE 10 Transition from extended to localized states in a bichromatic incommensurate potential. (a) The square modulus of the ground state wavefunction is plotted in grayscale as a function of position and disordering lattice strength. (b), (c) Logarithmic plots of the ground state wavefunction below and above the localization transition

where J is the tunneling rate between next-neighboring sites and Δ is the amplitude of the quasi-periodic modulation of the potential energy, β being an irrational number. The Aubry–André model can be experimentally realized when the primary lattice height s_1 is much larger than the secondary lattice height s_2 (see Section 2.2). The primary lattice discretizes the system and produces a renormalization of the effective mass $m^* = mE_{R1}/J\pi^2$, while the height of the secondary lattice $s_2 = \Delta/E_{R2}$ is the control parameter which drives the localization transition. As a matter of fact, differently from what happens with a pure disordered potential, in the quasi-periodic case a localization transition exists even in 1D, with a critical value $\Delta \approx 2J$ of the quasi-disorder amplitude for producing a localized ground state.

This behavior is illustrated in Figure 10(a), in which we plot the lowest-energy single-particle eigenstate in the incommensurate lattice, as obtained by numerical integration of the one-dimensional Schrödinger equation (6), which holds in the continuum. The figure shows, in grayscale, the squared modulus of the ground state wavefunction as a function of position (horizontal axis) and disordering lattice strength (vertical axis). One clearly sees that for low values of disorder the ground state is an extended state (see Figure 10(c)), i.e. the wavefunction extends across the entire lattice (in the limit $\Delta = 0$ one recovers the extended Bloch state describing the ground state in a periodic lattice). Increasing disorder above a threshold value the wavefunction suddenly localizes around few lattice sites, with exponentially decreasing tails, in the same way as in

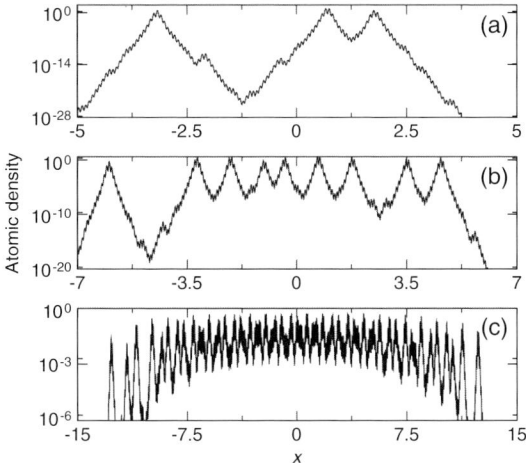

FIGURE 11 Ground state of the Gross–Pitaevskii equation for a Bose–Einstein condensate in a three-color optical lattice for different effective interaction strengths $\tilde{g} = 0.2$ (a), $\tilde{g} = 8$ (b), $\tilde{g} = 256$ (c). The latter value corresponds to the interaction strength for a Bose-condensed sample of $N = 10^4$ ^{87}Rb atoms in an elongated harmonic trap with frequencies $\omega_\perp = 2\pi \times 40$ Hz and $\omega_\parallel = 2\pi \times 4$ Hz. Reprinted figure with permission from Schulte et al. (2005) © 2005, of the American Physical Society

the case of Anderson-localized states in δ-correlated disordered potentials (Figure 10(b)).

So far, we have shown that the incommensurate lattice supports the existence of localized states for non-interacting particles. What happens when one introduces interactions? This problem has been discussed in Schulte et al. (2005, 2006), where the ground state of the system has been calculated for a three-color optical lattice. Introducing repulsive interactions between the atoms, the numerical integration of the one-dimensional Gross–Pitaevskii equation shows that the ground state wavefunction becomes a superposition of many single-particle localized states, which add up to form an overall extended state, as shown in Figure 11 for different interaction strengths. Similar results for a bichromatic lattice have been presented in Lye et al. (2007).

This behavior can be interpreted in terms of a *screening effect* induced by interactions (Sanchez-Palencia, 2006; Schulte et al., 2006). The nonlinear term in the Gross–Pitaevskii equation (7) can be treated as an effective potential canceling the spectral components of the original potential varying on length scales larger than the *healing length* $\xi = 1/\sqrt{8\pi a n}$, where a is the scattering length and n the atom density. The healing length is the typical length scale that is associated to the variation of the BEC wavefunction around sharp potential jumps (Dalfovo et al., 1999). More generally, it is the typical length scale below which the condensate

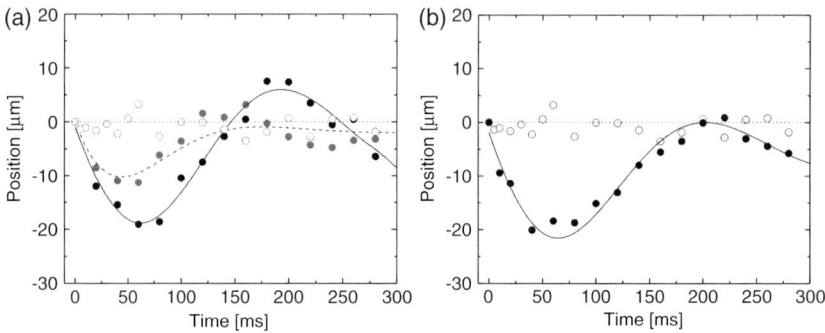

FIGURE 12 Oscillations of a BEC in a parabolic + incommensurate two-color lattice. (a) Center-of-mass position for $N = 1.5 \times 10^4$ atoms, $s_1 = 10$ and different disordering lattice heights $s_2 = 0.1$ (filled black circles), $s_2 = 0.17$ (filled gray circles), $s_2 = 0.25$ (empty circles). (b) Center-of-mass position for $s_1 = 10$, $s_2 = 0.25$ and different atom numbers $N = 1.5 \times 10^4$ (empty circles) and $N = 2 \times 10^5$ (filled circles). The data are adapted from Lye et al. (2007)

wavefunction is able to behave quantum-mechanically. As a consequence, in order to observe localization, the interference effects producing localization should take place on a distance smaller than the healing length, otherwise the BEC wavefunction would behave classically over longer distances. The healing length can be made larger by reducing the amount of interactions in the system, that could be achieved either by reducing the scattering length a or by reducing the atom density n. If the healing length is smaller than the disorder localization length ($\xi < \zeta$, see Equation (5)) no localized states can be observed. If the healing length is larger than the localization length, but smaller than the system size ($\zeta < \xi < L$), one could observe a superposition of many localized states. Finally, if the healing length is the largest length scale in the system ($\xi > L > \zeta$) the BEC wavefunction collapses in a single localized state. This crossover is illustrated in Figure 11, where the ground state in the three-color lattice is plotted for the same lattice heights but different interaction strength (hence same ζ but different ξ): while in panel (a) $\xi > \zeta$ and a few localized states can be easily detected with clearly exponentially decreasing tails, in panel (c) $\xi < \zeta$ and an overall extended state forms.

The existence of localized states can be probed with transport experiments, similar to those presented in the previous section. In Lye et al. (2007) the transport of an [87]Rb BEC has been studied in the presence of a bichromatic incommensurate potential. Localization of the center-of-mass motion has been observed, the stronger the smaller is the strength of interactions (tuned by changing the atomic density), as shown in Figure 12. This density-dependent behavior, with interactions pushing to delocalize the system, is reminiscent of Anderson-like localization.

However, in the regime of parameters studied in this work, no simple Anderson-like localization has to be expected, the eigenstates of the system being similar to the state dominated by interactions shown in Figure 11c. The suppression of the center-of-mass motion shown in Figure 12 was mainly caused by the strong modulation of the BEC wavefunction on the length scale of the beating between the two lattice periods, which resulted in very low tunneling times across the lattice and, consequently, in an extremely slow dynamics.

Effects of nonlinear dynamics have also been considered as possible mechanisms to damp the motion. As a matter of fact, suppression of transport is expected to appear for mechanisms alike the interaction-induced dynamical instability observed in Cataliotti et al. (2003), Fallani et al. (2004) and Cristiani et al. (2004) for monochromatic optical lattices. In the system studied in these works the interplay between repulsive nonlinearities and band structure resulted in fast-growing excitations dephasing the system and halting the motion. The same effect can also be observed in bichromatic optical lattices, in which a band structure can still be identified (Diener et al., 2001), with a multitude of energy gaps opening in the spectrum and getting denser and denser with increasing height of the secondary incommensurate lattice.

From what was discussed above we can draw a preliminary conclusion. Bichromatic potentials do allow solving the problem of the correlation length of the speckles produced in Lye et al. (2005), Clément et al. (2005), Fort et al. (2005) and Schulte et al. (2005), which causes the localization length to be too large to be observable. In bichromatic potentials the localization length can easily be smaller, however the presence of too strong interactions still remains and makes it impossible to observe a clear localization of the wavefunction in a few localized states. In order to achieve clear signatures of Anderson localization one needs to work with extremely weakly-interacting samples.

3.3 Further Directions

From the theoretical point of view, as we have already pointed out, the effect of disorder on the interacting Bose gas is an extremely interesting topic of research. In Lugan et al. (2007a) the ground state of an interacting Bose gas in a disordered potential has been deeply studied. In particular, a Lifshitz glass phase has been introduced characterizing the ground state of the system for weak interactions. Lifshitz states represent a particular class of localized states which exhibit 'weaker' localization properties than Anderson-localized states, in the sense that they show exponential decay only in the very far tails, while close to the maximum their shape mostly depends on the *local* properties of the potential. This means that they mostly resemble bound states of isolated potential wells or trapped states between barriers, differently from Anderson-localized states, whose shape

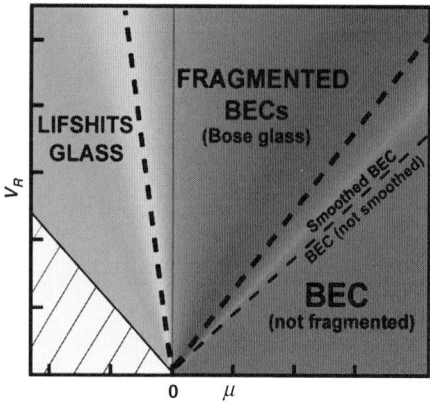

FIGURE 13 Phase diagram of an interacting disordered BEC as a function of the chemical potential μ and of the disordered potential height V_R. Reprinted figure with permission from Lugan et al. (2007a) © 2007, of the American Physical Society

is determined by *global* properties of the potential, i.e. by the combined effect of many impurities/potential wells. In Figure 13 we show the phase diagram of the interacting disordered BEC derived in Lugan et al. (2007a) as a function of the BEC chemical potential and of the speckle height. Starting from this Lifshitz glass phase and increasing interactions, a phase of fragmented interacting BECs has been proposed, which is a precursor of the Bose glass phase (see Section 4 for further discussion).

The BEC fragmented state has been previously described in Wang et al. (2004). In this work the authors studied the ground state of a BEC in the disordered potential produced by the random imperfections of a magnetic microtrap. This paper was motivated by several experimental observations (Fortágh et al., 2002; Leanhardt et al., 2003; Estève et al., 2004; Jones et al., 2004), in which fragmentation of the BEC at very close distances from the current-carrying wires of the microchip was observed. In the same work (Wang et al., 2004) the BEC dynamics in the disordered potential was also investigated. In particular, the spectral analysis of the sloshing motion after displacement of the confining potential allowed identifying different dynamical regimes: superfluid oscillations, self-trapping and an intermediate chaotic regime.

Recent theoretical works (Bilas and Pavloff, 2006; Lugan et al., 2007b) have also studied the problem of Anderson localization of excitations in a Bose–Einstein condensate. In the weakly-interacting case, BEC excitations are described by the Bogoliubov theory (Dalfovo et al., 1999). By calculating the leading-order many-body corrections to the classical BEC wavefunction, one finds that, in the absence of external potentials, excitations are described by quasi-particles with dispersion relation $\hbar\omega =$

$\sqrt{(\hbar c k)^2 + (\hbar^2 k^2/2m)^2}$, with k wavevector of the excitation and c sound velocity inside the BEC. This spectrum has two distinct regions with different k-dependences. For $k \ll \xi^{-1}$ (with ξ the healing length) the excitation spectrum is phonon-like and excitations have energy $\omega \approx ck$. For $k \gg \xi^{-1}$ the spectrum is particle-like and the energy of the excitations is $\omega \approx \omega_0 + \hbar k^2/2m$, where $\omega_0 = mc^2/\hbar$ is an energy shift due to interactions. Bilas and Pavloff (2006) have shown that, in the presence of a white-noise random potential, excitations can undergo Anderson localization, almost in the same way as the whole BEC wavefunction can undergo. Further detailed studies have been carried out in Lugan et al. (2007b) in the more realistic case of correlated disorder, showing that the localization length (and, correspondingly, the possibility to observe localization in finite-sized BECs) crucially depends on the correlation length of the disorder.

The effect of disorder on the coherent BEC dynamics can be observed also on the dephasing of Bloch oscillations which is expected to appear when a disordered or quasi-disordered potential is superimposed on a tilted optical lattice (Sanchez-Palencia and Santos, 2005; Schulte et al., 2008). Bloch oscillations are the coherent oscillations of a wavepacket in a periodic potential when a constant force is applied. This phenomenon has been observed for the first time with ultracold atoms in optical lattices (Raizen et al., 1997) because of the much longer coherence times than the ones achievable for electrons moving in real solid-state lattices, where defects and impurities strongly dephase the system in a time much shorter than the oscillation period. Also interactions lead to dephasing, as evidenced in Morsch et al. (2001) when Bloch oscillations were observed for the first time in a Bose–Einstein condensate. Later it was demonstrated that this interaction-induced dephasing can be controlled and eventually canceled by tuning the interaction strength with Feshbach resonances, as recently demonstrated in Gustavsson et al. (2008) and Fattori et al. (2008), or by using ultracold fermionic samples (Roati et al., 2004), for which interactions are forbidden by the Pauli principle. Non-interacting particles in perfectly periodic optical lattices perform undamped Bloch oscillations and, thanks to this possibility, they can be used as microscopic probes for high-precision measurements of forces at small distances (Carusotto et al., 2005). Starting from this ideal situation and adding disorder on top of the periodic lattice, one can quantitatively study the dephasing induced by disorder in a controlled way, as first theoretically studied in Schulte et al. (2008) and then recently investigated experimentally in Drenkelforth et al. (2008).

Generally speaking, disorder leads to the disruption of coherent effects. Quite interestingly, however, under certain conditions disorder can induce a spontaneous ordering of the system. This effect, known as *random-field induced order*, has been originally studied in the context of classical spin models, which in the presence of disorder may exhibit a magnetization higher than in the ordered case (Wehr et al., 2006; Sen De et al., 2007). Two-component Bose-Einstein condensates in the presence of a random Raman coupling between the two states can be used to study this class of effects, as recently proposed in Niederberger et al. (2008).

3.4 Observing Anderson Localization

Starting from the first experiments with speckle potentials (Lye et al., 2005; Fort et al., 2005; Clément et al., 2005; Schulte et al., 2005), the quest for Anderson localization in Bose–Einstein condensates has been a strongly active direction of research. On one side experimental groups have focused on producing disordered potentials on thinner length scales (Clément et al., 2006; White et al., 2008), in order to increase the amplitudes of quantum scattering and decrease the attainable localization lengths. On the other side, the challenge is to reduce atom–atom interactions in order to make localization observable. Once this is obtained, it will be even more interesting to study the effect of adding a controlled amount of interactions. For this purpose it can be strongly helpful to take advantage of Feshbach resonances to tune the scattering length a, which is the key parameter defining the strength of interactions $g = 4\pi \hbar^2 a/m$ (see Section 3). In this perspective, the choice of the element under investigation is crucial. The first experiments performed with BECs in disordered potentials (Lye et al., 2005; Fort et al., 2005; Clément et al., 2005; Schulte et al., 2005) have focused on ^{87}Rb, which is a quite convenient element for the implementation of cooling schemes, but has the disadvantage of having a quite large scattering length $a \simeq 100 a_0$ (with a_0 the Bohr radius) and no favorable Feshbach resonances at convenient magnetic fields (Marte et al., 2002). A much easier tuning of atom–atom interactions could be provided by different elements, such as ^{7}Li (which has been already studied in combination with laser speckles in Chen et al. (2008)) or ^{39}K, studied by Roati et al. (2007).

While this review was being completed (March 2008) two experiments succeeded in observing Anderson localization of coherent matter waves; by the groups of A. Aspect in France and here at LENS. In the French experiment (Billy et al., 2008) a Bose–Einstein condensate is left free to expand in a disordered waveguide produced by combining a weakly focused laser beam with a one-dimensional speckle potential. Differently from the conceptually similar experiments reported in Clément et al. (2005) and Fort et al. (2005), the analysis of the *in-situ* density profiles shows clear indication of exponentially decreasing tails, which is a signature of Anderson localization. This have been made possible by a combination of several factors: the small atomic density in the tails (necessary to reduce the counteracting effect of interactions), the small speckle autocorrelation length (necessary to have many quantum scattering events during the diffusion) and the high detection sensitivity (allowing the observation of exponential decay of the density). The observed localization is then quantitatively compared with the theory developed in Sanchez-Palencia et al. (2007, 2008).

In the experiment at LENS (Roati et al., 2008), Anderson localization has been observed for a non-interacting ^{39}K BEC in an incommensurate bichromatic lattice, similar to that used in Lye et al. (2007). Here the strategy to exclude the effect of interactions is different: instead of working with dilute samples, interactions are canceled by tuning a static magnetic field in proximity of a Feshbach resonance to set the scattering length to zero. The non-interacting condensate in the quasi-periodic potential thus realizes the non-interacting tight-binding Aubry–André model of Equation (8), which exhibits a transition from extended to localized states for increasing disorder. The crossover between extended to localized states is studied in detail by looking at the expansion of the BEC and by studying spatial and momentum distribution of the states, all of which result in agreement with the Aubry–André predictions. Experiments are now in progress in the same group to address the fundamental question on the role of interactions, that can be introduced in a controlled way by fine-tuning the scattering length via the Feshbach resonance. Preliminary studies already reveal how a weak interaction affects the observed localization transition (Roati et al., 2008).

4. STRONGLY-INTERACTING REGIME

In the previous section we have discussed the physics of disordered weakly-interacting bosonic systems. The theoretical description of this regime is provided by the semi-classical Gross–Pitaevskii equation (7), which describes the propagation of nonlinear matter waves. When interactions are strong, however, this mean-field description is not capable of fully explaining the behavior of the system. A more appropriate description is provided by a full quantum theory, taking into account quantum correlations between particles. Also in this strongly-interacting regime disorder may induce localized quantum phases: these have a different nature from Anderson localization since correlations between particles are important, whereas Anderson localization is essentially a single-particle effect.

Experimentally, a convenient way to enter the strongly-interacting regime is provided by the use of optical lattices (which we have already introduced in Section 2.2). In a deep optical lattice the system becomes effectively stronger interacting because of the combined effect of the tighter squeezing of the atom wavefunction in the potential wells (with a consequent increase of the local density) and of the increase in the effective mass due to the finite tunneling times across the potential barriers (which makes the kinetic energy less important with respect to the interaction energy).

For a system defined on a lattice, starting from the full many-body Hamiltonian, one can derive a simplified zero-temperature model, in the approximation that all the particles occupy the fundamental vibrational state of the lattice sites. In this limit the quantum state of an interacting gas of identical bosons in a lattice potential is well described by the second quantization Bose–Hubbard Hamiltonian (Fisher et al., 1989; Jaksch et al., 1998)

$$\hat{H} = -J \sum_{\langle j,j' \rangle} \hat{b}_j^\dagger \hat{b}_{j'} + \frac{U}{2} \sum_j \hat{n}_j (\hat{n}_j - 1) + \sum_j \epsilon_j \hat{n}_j \qquad (9)$$

where $\hat{b}_j$ ($\hat{b}_j^\dagger$) is the annihilation (creation) operator of one particle in the jth site, $\hat{n}_j = \hat{b}_j^\dagger \hat{b}_j$ is the number operator, and $\langle j, j' \rangle$ indicates the sum on nearest neighbors. Each of the three terms on the right-hand side of Equation (9) accounts for a different contribution to the total energy of the system: J is the *hopping energy*, proportional to the probability of quantum tunneling of a boson between neighboring sites, U is the on-site *interaction energy*, arising from atom–atom on-site short-range interactions (repulsive for ^{87}Rb, for which $U > 0$) and giving a non-zero contribution only if more than one particle occupy the same site, and $\epsilon_j \in [-\Delta/2, \Delta/2]$ is a site-dependent energy accounting for inhomogeneous external potentials superimposed on the lattice.

The quantum phase of the system depends on the interplay between these three energy scales: hopping energy J, interaction energy U and disorder Δ. We start considering the ideal case of a translationally invariant system, in which $\Delta = 0$. Assuming integer filling of the sites, when $J > U$ the system is in a superfluid (SF) state, in which the bosons are delocalized across the lattice and the tunneling ensures off-diagonal long-range coherence. Instead, when $U > J$, the system is in a localized Mott insulator (MI) state, where long-range phase coherence is lost and number Fock states are created at the lattice sites. The actual phase diagram of the system depends on the chemical potential (related to the atomic density) and shows the existence of MI lobes with integer number of atoms per site (Fisher et al., 1989). In the left graph of Figure 14 we show a qualitative sketch of the phase diagram for a three-dimensional system.

The transition from a SF to a MI for ultracold bosons in an optical lattice has been proposed in Jaksch et al. (1998) and reported for the first time in Greiner et al. (2002), where the ratio J/U was varied across the transition point by controlling the height of the lattice. The realization of a MI state does require a three-dimensional optical lattice, since, in order to enter the

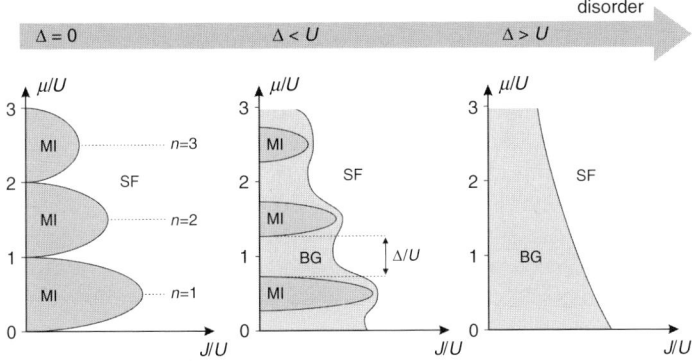

FIGURE 14 Qualitative phase diagram for a disordered system of lattice interacting bosons (as first derived in Fisher et al. (1989)). Three phases can be identified: a superfluid (SF), a Mott insulator (MI) and a Bose glass (BG).

strongly-interacting regime, the atomic wavefunction should be squeezed into tightly confining traps, with a site occupation on the order of unity. However, by using deep optical lattices effectively slicing the atomic sample into decoupled two-dimensional or one-dimensional systems, it is possible to study the SF–MI transition in lower dimensionality, as done in Stöferle et al. (2004) and Spielman et al. (2007).

In the presence of a disordered external potential the additional energy scale Δ enters the description of the system and is responsible for the existence of a new quantum phase. In the presence of weak disorder the MI lobes in the phase diagram should progressively shrink and a new *Bose glass* (BG) phase should appear (central graph of Figure 14), eventually washing away the MI region for $\Delta > U$ (right graph of Figure 14) (Fisher et al., 1989). In a simplified view, a Bose glass is half-way from a Mott insulator to a superfluid: it is an insulating state, with no long-range phase coherence, as the Mott insulator is; nevertheless, it is compressible and has no energy gap in the excitation spectrum, as a superfluid has.

The Bose glass phase was identified for the first time in Giamarchi and Schulz (1988), where a superfluid-localized transition for strongly-interacting one-dimensional bosons was predicted. The first numerical indications of the existence of this phase were given a few years later in Scalettar et al. (1991) and Krauth et al. (1991). In the 90s a great theoretical effort was made for the characterization of the phase diagram of disordered interacting bosons (Freericks and Monien, 1996; Pai et al., 1996; Prokof'ev and Svistunov, 1998; Pázmándi and Zimányi, 1998; Herbut, 1998; Rapsch et al., 1999), mostly in connection with the superfluid–insulator transition observed in many condensed-matter systems, such as ^{4}He adsorbed on porous media (Crowell et al., 1995), thin superconducting films (Goldman and Marković, 1998), arrays of

Josephson junctions (van der Zant et al., 1992) and high-temperature superconductors (Jiang et al., 1994; Budhani et al., 1994). The possible realization of a Bose glass in a system of ultracold bosons in a disordered lattice was first proposed in Damski et al. (2003) and Roth and Burnett (2003). More recently, the phase diagram of this system has been derived in other theoretical papers, considering also finite-temperature effects (Krutitsky et al., 2006; Buonsante et al., 2007a), detection schemes (Bar-Gill et al., 2006) and the possible realization of a Bose glass with incommensurate bichromatic lattices (Pugatch et al., 2006; Roscilde, 2008). Evidences for Bose glass phases have been also theoretically obtained for different classes of Bose–Hubbard models, where disorder is introduced in the hopping energy (Buonsante et al., 2007b), in the on-site interaction energy (Gimperlein et al., 2005) or by the interaction with atoms of a different species (Krutitsky et al., 2008).

The Bose glass is just the simplest disordered quantum phase that can be realized in the strongly-interacting regime. When atoms of different species, or different internal (spin) states of the same species, are considered, more complicated models can be experimentally realized and new disordered quantum phases can emerge. Atomic Bose/Fermi mixtures, in particular, represent a versatile system in which many different disordered models can be realized (Sanpera et al., 2004; Ahufinger et al., 2005). In the strong-interacting limit this system can be described in terms of composite fermionic particles corresponding to one fermion + one bosonic particle/hole in the same site. Sanpera et al. (2004) have shown that the interaction between these composite fermions can be tuned by changing the external potential: thus, a disordered potential can be used to induce an effective random interaction between the particles. This possibility allows the investigation of a variety of disordered-related models, from fermionic Ising spin glasses to models of quantum percolation (Ahufinger et al., 2005). Also different kinds of mixtures have been considered theoretically. In Roscilde and Cirac (2007) a disordered Bose / Bose mixture was investigated, evidencing a novel metastable *quantum-emulsion* state in which the two atomic species form phase-separated droplets, with properties reminiscent of a Bose glass state.

4.1 The Quest for Bose Glass

Experiments with disordered bosons in the strongly-interacting regime started at LENS in 2006. The system under investigation was a collection of one-dimensional atomic systems in a bichromatic optical lattice. A main optical lattice was used to induce the transition from a weakly-interacting superfluid to a strongly correlated Mott insulator. A secondary optical lattice was then used to add controlled quasi-disorder to the perfect crystalline structure of the MI phase. With reference to Equation (9),

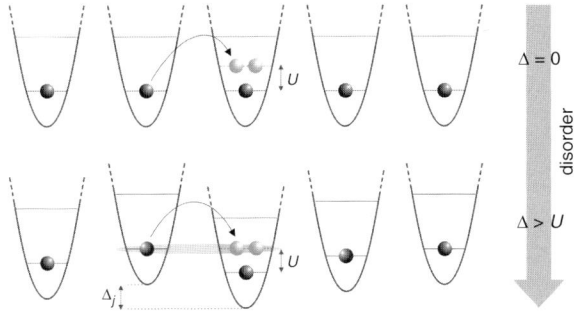

FIGURE 15 Excitations in the deep-insulating phases. (a) In a Mott insulator the tunneling of one boson from a site to a neighboring one has an energy cost $\Delta E = U$. (b) In the disordered case the excitation energy is $\Delta E = U \pm \Delta_j$, that becomes a function of the position. In the Bose glass state, in which $|\Delta_j| > U$, an infinite system could be excited at arbitrarily small energies and the energy gap would disappear

the incommensurate periodic potential superimposed on the main lattice introduces inhomogeneities of the energy landscape $\epsilon_j \in [-\Delta/2, \Delta/2]$ on the same length scale as the lattice spacing.

4.1.1 Excitation Spectrum and Coherence Properties

As introduced in the previous section, the excitation spectrum is an important observable that can be measured in order to characterize the quantum state of the system. By exploiting the possibility of time-modulating the lattice potential, as first realized in Stöferle et al. (2004), it is possible to directly measure the excitation spectrum and study how it is modified by the presence of disorder. Naively speaking, in the MI phase one realizes a crystal of atoms pinned at the lattice sites and sitting on the fundamental vibrational level, as schematically shown in the top of Figure 15. In a MI an energy gap in the excitation spectrum exists, since the elementary excitation – the hopping of a particle from a site to a neighboring one, or, in other words, the creation of a particle–hole pair – has an energy cost U, corresponding to the interaction energy of a pair of mutually repelling atoms sitting on the same site (see Figure 15).

In Figure 16(a) we show the excitation spectrum of a Mott insulator measured in the LENS experiments (Fallani et al., 2007). The plot shows a well-resolved resonance at energy U, which is distinctive of the MI state, and a second resonance at energy $2U$. While the physical origin of the excitation peak at U is the tunneling of particles between sites with the same occupancy, the second peak at $2U$ can be ascribed to several processes: it can arise from tunneling at the boundary between MI regions with different site occupancy (that are present due to the inhomogeneity of the confined sample), from higher-order processes and from nonlinear effects due to the strong modulation. A theoretical analysis of the response

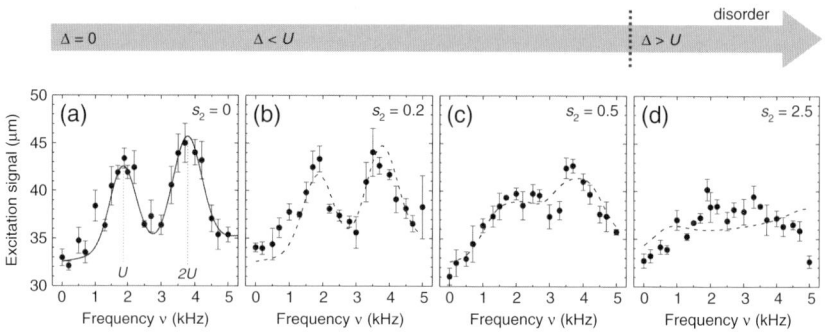

FIGURE 16 Excitation spectra of the atomic system in a Mott insulator state for increasing height of the disordering lattice. The resonances are lost and the excitation spectrum becomes flat. The data are adapted from Fallani et al. (2007)

of the bosonic system to this lattice modulation has been recently reported in Kollath et al. (2006) and Clark and Jaksch (2006).

When increasing disorder the experiment showed a broadening of the resonance peaks, which eventually become indistinguishable when $\Delta \approx U$. As a matter of fact, the presence of disorder introduces random energy differences $\Delta_j \in [-\Delta, \Delta]$ between neighboring sites (see bottom of Figure 15). As a consequence, the tunneling of a boson through a potential barrier costs $U \pm \Delta_j$, that becomes a function of the position (Guarrera et al., 2007). The excitation energy is not the same for all the bosons, differently from the pure MI case, and the resonances become inhomogeneously broadened, as can be observed in the experimental spectra at weak disorder ($\Delta < U$) shown in Figure 16(b), (c) (Fallani et al., 2007). This broadening is in agreement with a semi-classical model (Guarrera et al., 2007) and has been recently predicted in theoretical works (Hild et al., 2006; Zakrzewski, J. 2008. Private Communication), where the authors study the dynamical response of a one-dimensional bosonic gas in a superlattice potential when a periodic amplitude modulation of the lattice is applied.

Eventually, when $\Delta \gtrsim U$, one expects that an infinite system can be excited at arbitrarily small energies and that the energy gap would shrink to zero. When this happens, nearby sites become degenerate and regions of local superfluidity with short-range coherence appear in the system. This novel many-body state in which there is no gap but the system remains globally insulating is a *Bose glass*.

From the experimental point of view, additional information on the nature of the many-body ground state can be acquired by analyzing the density distribution of the atoms released from the lattice after a time-of-flight. Long-range coherence in the sample results in a density distribution with interference peaks at a distance proportional to the lattice wavevector

(Pedri et al., 2001). The visibility of these peaks provides a measurement of phase coherence. When increasing the height of the main lattice, a progressive loss of long-range coherence has been reported in Fallani et al. (2007) indicating the transition from a superfluid to an insulating state, also in the presence of disorder. The combination of the excitation spectra measurements and the time-of-flight images indicates that, with increasing disorder, the system realized in Fallani et al. (2007) goes from a MI to a state with vanishing long-range coherence and a flat density of excitations. The concurrence of these two properties cannot be found in either a SF or an ordered MI, and is consistent with the formation of a Bose glass, which is indeed expected to appear for $\Delta \gtrsim U$.

Much work has still to be done for the exhaustive characterization of such a novel disordered state. New detection schemes should be implemented, in order to have access to additional observables. This necessity is not only restricted to the study of disordered systems, being a more general issue shared by the experimental investigation of different strongly-interacting lattice systems, including e.g. systems with magnetic ordering or mixtures of different species. From the theoretical side, very recent works (Roscilde, 2008; Roux et al., 2008; Deng et al., 2008) have extensively studied the problem of one-dimensional interacting bosons in quasi-periodic lattices, working out the phase diagrams (which include the presence of Bose glass and incommensurate 'band-insulating'/'charge density wave' regions) and studying how the different phases affect experimentally detectable signals.

While this review was proof-readed, the results of a new experiment in the strongly interacting regime (White et al., 2008) were announced. In this work the disordered Bose–Hubbard model was experimentally realized by loading a Bose–Einstein condensate in the combined potential of a 3D optical lattice and a speckle pattern with short correlation length (less than 1 μm) producing disorder in a 3D geometry. The condensate fraction of the bosonic sample was measured as a function of the lattice height (across the SF–MI transition) and of the disorder amplitude, evidencing a decrease in condensate fraction as the speckle intensity was increased.

4.1.2 Noise Correlations

In the recent work of Guarrera et al. (2008) noise interferometry has been used to study interacting ^{87}Rb bosons in the bichromatic lattice. This detection technique, originally proposed in Altman et al. (2004), is based on the analysis of the spatial density–density correlations of the atomic shot noise after time-of-flight. These correlations are based on the Hanbury Brown & Twiss effect (Hanbury Brown and Twiss, 1956): if two identical particles are released from two lattice sites, the joint probability of detecting them in two separate positions (e.g. imaging them on two separate pixels of a CCD camera) depends on the distance between the detection points. These correlations, arising from quantum interference between different detection paths, were first observed for bosons in a Mott insulator state (Fölling et al., 2005) and then also for band-insulating fermions (Rom et al., 2006). The sign of the correlations depends on the quantum statistics: while bosons show positive correlations (due to their tendency to bunch, i.e. to arrive together at the detectors), fermions exhibit negative correlations (due to the antibunching, a consequence of the Pauli exclusion principle). In the case of a bosonic Mott insulator, one observes positive density–density correlation peaks at a distance proportional to the lattice wavevector k_1, as shown in the first image of the bottom row of Figure 17 for the recent experiment at LENS.

In Guarrera et al. (2008) noise correlations have been measured, starting from a Mott insulator state, for increasing heights s_2 of the secondary lattice. The absorption images after time-of-flight do not present significant differences, as shown in the top row of Figure 17, and demonstrate the absence of first-order (phase) coherence of the atomic system in the insulating state, even in the presence of the secondary lattice. However, second-order (density) correlations turn out to be significantly different with varying s_2, as illustrated in the noise correlation functions plotted in the bottom row. More precisely, with increasing s_2, one observes the appearance of additional correlation peaks at a distance proportional to the wavevector k_2 of the secondary lattice and to the beating between the two lattices $k_1 - k_2$. These peaks have to be associated with the redistribution of atoms in the lattice sites as the disordering lattice is strengthened: the MI regions characterized by uniform filling are destroyed and atoms rearrange in the lattice giving rise to a state with non-uniform site occupation, which follows the periodicity of the secondary lattice. The redistribution of atoms is then quantitatively detected by measuring the height of the additional correlation peaks.

Noise correlations thus prove to be a tool to extract important information on the lattice site occupation, which is connected to the second-order correlation function of the many-body state. The appearance of similar correlation peaks was predicted in theoretical works for hard-core bosons (Rey et al., 2006) and soft-core bosons (Roscilde, 2008) in bichromatic lattices. Future works will study the possibility of using noise interferometry to get additional insight into the nature of the disordered insulating states produced in the experiment, in particular in connection with the realization of a Bose glass phase.

4.2 Experiments with Atomic Mixtures

As we have discussed in Section 2.3, disorder can be produced by letting the atoms interact with randomly-distributed scatterers of a different atomic species. The configuration proposed in Gavish and Castin (2005) has not yet been realized experimentally. However, in 2006 the first

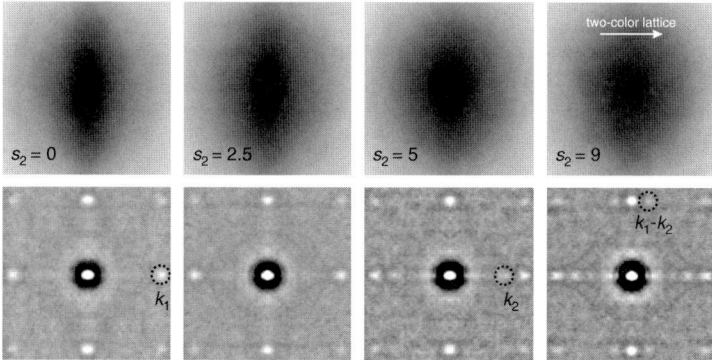

FIGURE 17 (Top) Time-of-flight absorption images of atoms in a Mott insulator state for increasing height of the secondary lattice s_2. (Bottom) Density–density correlation functions corresponding to the pictures above. The additional correlation peaks for large s_2 arise from the destruction of the Mott domains and the redistribution of the atoms in the lattice. The data are adapted from Guarrera et al. (2008)

experiments with binary mixtures in optical lattices have been realized almost at the same time by two different groups, in Zurich (Günter et al., 2006) and in Hamburg (Ospelkaus et al., 2006). In these experiments ^{87}Rb bosons and ^{40}K fermions were mixed together in a three-dimensional optical lattice. Since the two atoms have very similar resonance wavelengths, the depth of the optical lattice is almost identical for the two species, however potassium is lighter than rubidium, therefore its mobility is favored. As a result, for a range of lattice heights, rubidium can be localized in a Mott insulator state, while potassium atoms are still able to move across the lattice.

In these experiments the superfluid to Mott insulator transition of ^{87}Rb was investigated as a function of the concentration of ^{40}K impurities, typically in the range 0–20%. In particular, the visibility of the interference pattern after time-of-flight was investigated. The observation reported by the two groups was a downshift of the lattice height value at which coherence starts to be lost, when potassium atoms are introduced in the system. Different interpretations for this effect have been given, including finite-temperature effects, disorder-like induced localization, or effects connected with the strong attractive interaction between the two species.

Recently, a closely related system has been investigated at LENS (Catani et al., 2008) by using a binary bosonic mixture of ^{87}Rb and ^{41}K in a three-dimensional optical lattice. Similarly to the experiments described above, loss of coherence in the rubidium sample induced by the presence of potassium has been observed. However, this latter experiment differs from the former ones in two points: the mixture is bosonic/bosonic (instead of bosonic/fermionic) and the interspecies interaction is repulsive (instead of attractive). The observation of similar effects in systems with different quantum statistics and different interaction signs rules out some of the interpretations given so far, even if a clear explanation of the observation has not yet been found. Future advances of these experiments with mixtures in optical lattices will be accelerated by the use of Feshbach resonances for fine tuning of the interspecies interaction.

5. CONCLUSIONS

The investigation of Bose–Einstein condensates in disordered potentials is a fastly growing field of research. For decades condensed-matter physicists have theoretically studied the interplay between disorder and interactions in determining the transition from metals to insulators. Now disordered systems can be realized in cold atom laboratories and, differently from traditional solid-state systems, they allow a fine tuning of both disorder and interactions, as well as the advantage of new detection capabilities, thus extending the range of the possible experimental investigations.

The first experiments were realized only a few years ago, with the successful creation of disordered and quasi-disordered potentials and the first studies of the behavior of ensembles of ultracold bosons in different regimes of interactions. Although this field of research is quite young, it already relies on an extensive literature, mostly comprising theoretical works. The physics of disordered atomic systems is indeed extremely rich, both in the weakly-interacting regime (where Anderson localization and its disruption by interactions can be studied) and in the strongly-interacting regime (where particles are strongly correlated and new quantum localized phases can emerge, as the Bose glass). The few experiments performed until now have just opened a new direction, showing the great potentialities of ultracold atoms for investigating the physics of disorder, but still leaving open questions concerning the observed localization effects. Regarding this point, much more results are likely to emerge in the near future.

We have tried to give an overview of this newborn field of research, discussing the topics of interest and the experimental efforts made up until now. We would like to conclude by noting that any review is by necessity incomplete and cannot be exhaustive of all the work done in the field. For this reason we apologize to the authors whose works which we have unintentionally forgotten to mention.

ACKNOWLEDGMENTS

This work has been supported by European projects SCALA and EUROQUAM. We would like to thank D. Clément and M. Modugno for careful reading of the manuscript and all the other members of the Cold Quantum Gases group in Florence. We also acknowledge the authors of the works described in this review, who have kindly granted us the permission for using their figures.

REFERENCES

Abrahams, E., Anderson, P.W., Licciardello, D.C. and Ramakrishnan, T.V. (1979). Scaling Theory of Localization: Absence of Quantum Diffusion in Two Dimensions. *Phys. Rev. Lett.*, **42**, 673.
Ahufinger, V., Sanchez-Palencia, L., Kantian, A., Sanpera, A. and Lewenstein, M. (2005). Disordered Ultracold Atomic Gases in Optical Lattices: A Case Study of Fermi–Bose Mixtures. *Phys. Rev. A*, **72**, 063616.
Akkermans, E., Ghosh, S. and Musslimani, Z. (2008). Numerical Study of One-Dimensional and Interacting Bose–Einstein Condensates in a Random Potential. *J. Phys. B*, **41**, 045302.
Altman, E., Demler, E. and Lukin, M.D. (2004). Probing Many-Body States of Ultracold Atoms via Noise Correlations. *Phys. Rev. A*, **70**, 013603.
Anderson, M.H., Ensher, J.R., Matthews, M.R., Wieman, C.E. and Cornell, E.A. (1995). Observation of Bose–Einstein Condensation in a Dilute Atomic Vapor. *Science*, **269**, 198.
Anderson, P.W. (1958). Absence of Diffusion in Certain Random Lattices. *Phys. Rev.*, **109**, 1492.
Anderson, P.W. (1978). Local Moments and Localized States. *Rev. Mod. Phys.*, **50**, 191.
Aubry, S. and André, G. (1980). Analyticity Breaking and Anderson Localization in Incommensurate Lattices. *Ann. Israel Phys. Soc.*, **3**, 133.
Bar-Gill, N., Pugatch, R., Rowen, E., Katz, N. and Davidson, N. (2006). Quantum Phases of Ultra Cold Bosons in Incommensurate 1D Optical Lattices. Preprint arXiv:cond-mat/0603513.
Bilas, N. and Pavloff, N. (2006). Anderson Localization of Elementary Excitations in a One-Dimensional Bose–Einstein Condensate. *Eur. Phys. J. D*, **40**, 387.
Billy, J., Josse, V., Zuo, Z., Bernard, A., Hambrecht, B., Lugan, P., Clément, D., Sanchez-Palencia, L., Bouyer, P. and Aspect, A. (2008). Direct Observation of Anderson Localization of Matter-Waves in a Controlled Disorder. *Nature*, **453**, 891.
Bloch, I., Dalibard, J. and Zwerger, W. (2008). Many-Body Physics with Ultracold Gases. *Rev. Mod. Phys.*, **80**, 885.
Boiron, D., Mennerat-Robilliard, C., Fournier, J.-M., Guidoni, L., Salomon, C. and Grynberg, G. (1999). Trapping and Cooling Cesium Atoms in a Speckle Field. *Eur. Phys. J. D*, **7**, 373.
Budhani, R.C., Holstein, W.L. and Suenaga, M. (1994). Columnar-Defect-Induced Resistivity Minima and Bose Glass Scaling of Linear Dissipation in $Tl_2Ba_2CaCu_2O_8$ Epitaxial Films. *Phys. Rev. Lett.*, **72**, 566.
Buonsante, P., Penna, V., Vezzani, A. and Blakie, P.B. (2007a). Mean-field Phase Diagram of Cold Lattice Bosons in Disordered Potentials. *Phys. Rev. A*, **76**, 011602(R).
Buonsante, P., Massel, F., Penna, V. and Vezzani, A. (2007b). Mean-Field Phase Diagram for Bose–Hubbard Hamiltonians with Random Hopping. *Laser Phys.*, **17**, 538.
Burger, S., Bongs, K., Dettmer, S., Ertmer, W., Sengstock, K., Sanpera, A., Shlyapnikov, G.V. and Lewenstein, M. (1999). Dark Solitons in Bose–Einstein Condensates. *Phys. Rev. Lett.*, **83**, 5198.
Cataliotti, F.S., Fallani, L., Ferlaino, F., Fort, C., Maddaloni, P. and Inguscio, M. (2003). Superfluid Current Disruption in a Chain of Weakly Coupled Bose–Einstein Condensates. *New J. Phys.*, **5**, 71.
Catani, J., De Sarlo, L., Barontini, G., Minardi, F. and Inguscio, M. (2008). Degenerate Bose-Bose Mixture in a Three-Dimensional Optical Lattice. *Phys. Rev. A*, **77**, 011603(R).
Carusotto, I., Pitaevskii, L., Stringari, S., Modugno, G. and Inguscio, M. (2005). Sensitive Measurement of Forces at the Micron Scale Using Bloch Oscillations of Ultracold Atoms. *Phys. Rev. Lett.*, **95**, 093202.

Chen, Y.P., Hitchcock, J., Dries, D., Junker, M., Welford, C. and Hulet, R.G. (2008). Phase Coherence and Superfluid-Insulator Transition in a Disordered Bose–Einstein Condensate. *Phys. Rev. A*, **77**, 033632.

Clark, S.R. and Jaksch, D. (2006). Signatures of the Superfluid to Mott-Insulator Transition in the Excitation Spectrum of Ultracold Atoms. *New J. Phys.*, **8**, 160.

Clément, D., Varón, A.F., Hugbart, M., Retter, J.A., Bouyer, P., Sanchez-Palencia, L., Gangardt, D.M., Shlyapnikov, G.V. and Aspect, A. (2005). Suppression of Transport of an Interacting Elongated Bose–Einstein Condensate in a Random Potential. *Phys. Rev. Lett.*, **95**, 170409.

Clément, D., Varón, A.F., Retter, J.A., Sanchez-Palencia, L., Aspect, A. and Bouyer, P. (2006). Experimental Study of the Transport of Coherent Interacting Matter-Waves in a 1D Random Potential Induced by Laser Speckle. *New J. Phys.*, **8**, 165.

Clément, D., Bouyer, P., Aspect, A. and Sanchez-Palencia, L. (2008). Density Modulations in an Elongated Bose–Einstein Condensate Released from a Disordered Potential: What Can We Learn? *Phys. Rev. A*, **77**, 033631.

Courteille, Ph.W., Deh, B., Fortágh, J., Günther, A., Kraft, S., Marzok, C., Slama, S. and Zimmermann, C. (2006). Highly Versatile Atomic Micro Traps Generated by Multifrequency Magnetic Field Modulation. *J. Phys. B: At. Mol. Opt. Phys.*, **39**, 1055.

Cristiani, M., Morsch, O., Malossi, N., Jona-Lasinio, M., Anderlini, M., Courtade, E. and Arimondo, E. (2004). Instabilities of a Bose–Einstein Condensate in a Periodic Potential: An Experimental Investigation. *Opt. Express*, **12**, 4.

Crowell, P.A., Van Keuls, F.W. and Reppy, J.D. (1995). Superfluid-Insulator Transition in ^{4}He Films Adsorbed in Vycor Glass. *Phys. Rev. Lett.*, **75**, 1106.

Dalfovo, F., Giorgini, S., Pitaevskii, L.P. and Stringari, S. (1999). Theory of Bose–Einstein Condensation in Trapped Gases. *Rev. Mod. Phys*, **71**, 463.

Damski, B., Zakrzewski, J., Santos, L., Zoller, P. and Lewenstein, M. (2003). Atomic Bose and Anderson Glasses in Optical Lattices. *Phys. Rev. Lett.*, **91**, 080403.

Deng, L., Hagley, E.W., Wen, J., Trippenbach, M., Band, Y.B., Julienne, P.S., Simsarian, J.E., Helmerson, K., Rolston, S.L. and Phillips, W.D. (1999). Four-Wave Mixing with Matter Waves. *Nature*, **398**, 218.

Deng, X., Citro, R., Minguzzi, A. and Orignac, E. (2008). Phase Diagram and Momentum Distribution of an Interacting Bose Gas in a Bichromatic Lattice. *Phys. Rev. A*, **78**, 013625.

Diener, R.B., Georgakis, G.A., Zhong, J., Raizen, M. and Niu, Q. (2001). Transition Between Extended and Localized States in a One-Dimensional Incommensurate Optical Lattice. *Phys. Rev. A*, **64**, 033416.

Drenkelforth, S., Kleine Büning, G., Will, J., Schulte, T., Murray, N., Ertmer, W., Santos, L. and Arlt, J.J. (2008). Damped Bloch Oscillations of Bose–Einstein Condensates in Disordered Potential Gradients. *New J. Phys.*, **10**, 045027.

Estève, J., Aussibal, C., Schumm, T., Figl, C., Mailly, D., Bouchoule, I., Westbrook, C.I. and Aspect, A. (2004). Role of Wire Imperfections in Micromagnetic Traps for Atoms. *Phys. Rev. A*, **70**, 043629.

Falco, G.M., Pelster, A. and Graham, R. (2007). Collective Oscillations in Trapped Bose–Einstein-Condensed Gases in the Presence of Weak Disorder. *Phys. Rev. A*, **76**, 013624.

Fallani, L., De Sarlo, L., Lye, J.E., Modugno, M., Saers, R., Fort, C. and Inguscio, M. (2004). Observation of Dynamical Instability for a Bose–Einstein Condensate in a Moving 1D Optical Lattice. *Phys. Rev. Lett.*, **93**, 140406.

Fallani, L., Lye, J.E., Guarrera, V., Fort, C. and Inguscio, M. (2007). Ultracold Atoms in a Disordered Crystal of Light: Towards a Bose Glass. *Phys. Rev. Lett.*, **98**, 130404.

Fattori, M., D'Errico, C., Roati, G., Zaccanti, M., Jona-Lasinio, M., Modugno, M., Inguscio, M. and Modugno, G. (2008). Atom Interferometry with a Weakly Interacting Bose–Einstein Condensate. *Phys. Rev. Lett.*, **100**, 080405.

Feynman, R.P. (1982). Simulating Physics with Computers. *Int. J. Theor. Phys.*, **21**, 467.

Fisher, M.P.A., Weichman, P.B., Grinstein, G. and Fisher, D.S. (1989). Boson Localization and the Superfluid-Insulator Transition. *Phys. Rev. B*, **40**, 546.

Fölling, S., Gerbier, F., Widera, A., Mandel, O., Gericke, T. and Bloch, I. (2005). Spatial Quantum Noise Interferometry in Expanding Ultracold Atom Clouds. *Nature*, **434**, 481.

Fort, C., Fallani, L., Guarrera, V., Lye, J.E., Modugno, M., Wiersma, D. and Inguscio, M. (2005). Effect of Optical Disorder and Single Defects on the Expansion of a Bose–Einstein Condensate in a One-Dimensional Waveguide. *Phys. Rev. Lett.*, **95**, 170410.

Fortágh, J., Ott, H., Kraft, S., Günther, A. and Zimmermann, C. (2002). Surface Effects in Magnetic Microtraps. *Phys. Rev. A*, **66**, 041604(R).

Freericks, J.K. and Monien, H. (1996). Strong-coupling Expansions for the Pure and Disordered Bose–Hubbard Model. *Phys. Rev. B*, **53**, 2691.

Gavish, U. and Castin, Y. (2005). Matter-Wave Localization in Disordered Cold Atom Lattices. *Phys. Rev. Lett.*, **95**, 020401.

Giamarchi, T. and Schulz, H.J. (1988). Anderson Localization and Interactions in One-Dimensional Metals. *Phys. Rev. B*, **37**, 325.

Gimperlein, H., Wessel, S., Schmiedmayer, J. and Santos, L. (2005). Ultracold Atoms in Optical Lattices with Random On-Site Interactions. *Phys. Rev. Lett.*, **95**, 170401.

Goldman, A.M. and Marković, N. (1998). Superconductor-Insulator Transitions in the Two-Dimensional Limit. *Phys. Today*, **51**, 39.

Goodman, J.W. (2006). *Speckle Phenomena in Optics: Theory and Applications*. Roberts and Company Publishers.

Greiner, M., Mandel, O., Esslinger, T., Hänsch, T.W. and Bloch, I. (2002). Quantum Phase Transition from a Superfluid to a Mott Insulator in a Gas of Ultracold Atoms. *Nature*, **415**, 39.

Grempel, D.R., Fishman, S. and Prange, R.E. (1982). Localization in an Incommensurate Potential: An Exactly Solvable Model. *Phys. Rev. Lett.*, **49**, 833.

Grimm, R., Weidemüller, M. and Ovchinnikov, Y.B. (2000). Optical Dipole Traps for Neutral Atoms. *Adv. At. Mol. Opt. Phys.*, **42**, 95.

Guarrera, V., Fallani, L., Lye, J.E., Fort, C. and Inguscio, M. (2007). Inhomogeneous Broadening of a Mott Insulator Spectrum. *New J. Phys.*, **9**, 107.

Guarrera, V., Fabbri, N., Fallani, L., Fort, C., van der Stam, K.M.R. and Inguscio, M. (2008). Noise Correlation Spectroscopy of the Broken Order of a Mott Insulating Phase. *Phys. Rev. Lett.*, **100**, 250403.

Günter, K., Stöferle, T., Moritz, H., Köhl, M. and Esslinger, T. (2006). Bose-Fermi Mixtures in a Three-Dimensional Optical Lattice. *Phys. Rev. Lett.*, **96**, 180402.

Gustavsson, M., Haller, E., Mark, M.J., Danzl, J.G., Rojas-Kopeinig, G. and Nägerl, H.-C. (2008). Control of Interaction-Induced Dephasing of Bloch Oscillations. *Phys. Rev. Lett.*, **100**, 080404.

Hanbury Brown, R. and Twiss, R.Q. (1956). Correlation Between Photons in two Coherent Beams of Light. *Nature*, **177**, 27.

Harper, P.G. (1955). Single Band Motion of Conduction Electrons in a Uniform Magnetic Field. *Proc. Phys. Soc. A*, **68**, 674.

Herbut, I.F. (1998). Dual Theory of the Superfluid-Bose-glass Transition in the Disordered Bose–Hubbard Model in One and Two Dimensions. *Phys. Rev. B*, **57**, 13729.

Hild, M., Schmitt, F. and Roth, R. (2006). Response of Bose Gases in Time-Dependent Optical Superlattices. *J. Phys. B: At. Mol. Opt. Phys.*, **39**, 4547.

Horstmann, B., Cirac, J.I. and Roscilde, T. (2007). Dynamics of Localization Phenomena for Hard-core Bosons in Optical Lattices. *Phys. Rev. A*, **76**, 043625.

Inouye, S., Andrews, M.R., Stenger, J., Miesner, H.-J., Stamper-Kurn, D.M. and Ketterle, W. (1998). Observation of Feshbach Resonances in a Bose–Einstein Condensate. *Nature*, **392**, 151.

Ioffe, A.F. and Regel, A.R. (1960). Non-Crystalline, Amorphous, and Liquid Electronic Semiconductors. *Prog. Semicond.*, **237**, 4.

Jaksch, D., Bruder, C., Cirac, J.I., Gardiner, C.W. and Zoller, P. (1998). Cold Bosonic Atoms in Optical Lattices. *Phys. Rev. Lett.*, **81**, 3108.

Jiang, W., Yeh, N.-C., Reed, D.S., Kriplani, U., Beam, D.A., Konczykowski, M., Tombrello, T.A. and Holtzberg, F. (1994). Evidence of a Bose-Glass Transition in Superconducting $YBa_2Cu_3O_7$ Single Crystals with Columnar Defects. *Phys. Rev. Lett.*, **72**, 550.

Jones, M.P.A., Vale, C.J., Sahagun, D., Hall, B.V., Eberlein, C.C., Sauer, B.E., Furusawa, K., Richardson, D. and Hinds, E.A. (2004). Cold atoms probe the magnetic field near a wire. *J. Phys. B: At. Mol. Opt. Phys.*, **37**, L15.

Kollath, C., Iucci, A., Giamarchi, T., Hofstetter, W. and Schollwöck, U. (2006). Spectroscopy of Ultracold Atoms by Periodic Lattice Modulations. *Phys. Rev. Lett.*, **97**, 050402.

Krauth, W., Trivedi, N. and Ceperley, D. (1991). Superfluid-insulator Transition in Disordered Boson Systems. *Phys. Rev. Lett.*, **67**, 2307.

Krutitsky, K.V., Pelster, A. and Graham, R. (2006). Mean-field Phase Diagram of Disordered Bosons in a Lattice at Nonzero Temperature. *New J. Phys.*, **8**, 187.

Krutitsky, K.V., Thorwart, M., Egger, R. and Graham, R. (2008). Phase Diagram and Momentum Distribution of an Interacting Bose Gas in a Bichromatic Lattice. *Phys. Rev. A*, **78**, 013625.

Lye, J.E., Fallani, L., Modugno, M., Wiersma, D., Fort, C. and Inguscio, M. (2005). Bose–Einstein Condensate in a Random Potential. *Phys. Rev. Lett.*, **95**, 070401.

Lye, J.E., Fallani, L., Fort, C., Guarrera, V., Modugno, M., Wiersma, D.S. and Inguscio, M. (2007). Effect of Interactions on the Localization of a Bose–Einstein Condensate in a Quasiperiodic Lattice. *Phys. Rev. A*, **75**, 061603.

Leanhardt, A.E., Shin, Y., Chikkatur, A.P., Kielpinski, D., Ketterle, W. and Pritchard, D.E. (2003). Bose–Einstein Condensates near a Microfabricated Surface. *Phys. Rev. Lett.*, **90**, 100404.

Lugan, P., Clément, D., Bouyer, P., Aspect, A., Lewenstein, M. and Sanchez-Palencia, L. (2007a). Ultracold Bose Gases in 1D Disorder: From Lifshitz Glass to Bose–Einstein Condensate. *Phys. Rev. Lett.*, **98**, 170403.

Lugan, P., Clément, D., Bouyer, P., Aspect, A. and Sanchez-Palencia, L. (2007b). Anderson Localization of Bogolyubov Quasiparticles in Interacting Bose–Einstein Condensates. *Phys. Rev. Lett.*, **99**, 180402.

Marte, A., Volz, T., Schuster, J., Dürr, S., Rempe, G., van Kempen, E.G.M. and Verhaar, B.J. (2002). Feshbach Resonances in Rubidium 87: Precision Measurement and Analysis. *Phys. Rev. Lett.*, **89**, 283202.

Massignan, P. and Castin, Y. (2006). Three-Dimensional Strong Localization of Matter Waves by Scattering from Atoms in a Lattice with a Confinement-Induced Resonance. *Phys. Rev. A*, **74**, 013616.

Modugno, M. (2006). Collective Dynamics and Expansion of a Bose–Einstein Condensate in a Random Potential. *Phys. Rev. A*, 013606.

Morsch, O. and Oberthaler, M. (2006). Dynamics of Bose–Einstein Condensates in Optical Lattices. *Rev. Mod. Phys.*, **78**, 179.

Morsch, O., Müller, J.H., Cristiani, M., Ciampini, D. and Arimondo, E. (2001). Bloch Oscillations and Mean-Field Effects of Bose–Einstein Condensates in 1D Optical Lattices. *Phys. Rev. Lett.*, **87**, 140402.

Mott, N. (1978). Electrons in Glass. *Rev. Mod. Phys.*, **50**, 203.

Niederberger, A., Schulte, T., Wehr, J., Lewenstein, M., Sanchez-Palencia, L. and Sacha, K. (2008). Disorder-Induced Order in Two-Component Bose-Einstein Condensates. *Phys. Rev. Lett.*, **100**, 030403.

Ospelkaus, S., Ospelkaus, C., Wille, O., Succo, M., Ernst, P., Sengstock, K. and Bongs, K. (2006). Localization of Bosonic Atoms by Fermionic Impurities in a Three-Dimensional Optical Lattice. *Phys. Rev. Lett.*, **96**, 180403.

Pai, R.V., Pandit, R., Krishnamurthy, H.R. and Ramasesha, S. (1996). One-Dimensional Disordered Bosonic Hubbard Model: A Density-Matrix Renormalization Group Study. *Phys. Rev. Lett.*, **76**, 2937.

Paredes, B., Verstraete, F. and Cirac, J.I. (2005). Exploiting Quantum Parallelism to Simulate Quantum Random Many-Body Systems. *Phys. Rev. Lett.*, **95**, 140501.

Pázmándi, F. and Zimányi, G.T. (1998). Direct Mott Insulator-to-superfluid Transition in the Presence of Disorder. *Phys. Rev. B*, **57**, 5044.

Pedri, P., Pitaevskii, L., Stringari, S., Fort, C., Burger, S., Cataliotti, F.S., Maddaloni, P., Minardi, F. and Inguscio, M. (2001). Expansion of a Coherent Array of Bose–Einstein Condensates. *Phys. Rev. Lett.*, **87**, 220401.
Prokof'ev, N.V. and Svistunov, B.V. (1998). Comment on "One-Dimensional Disordered Bosonic Hubbard Model: A Density-Matrix Renormalization Group Study". *Phys. Rev. Lett.*, **80**, 4355.
Pugatch, R., Bar-Gill, N., Katz, N., Rowen, E. and Davidson, N. (2006). Identifying the Bose Glass Phase. Preprint arXiv:cond-mat/0603571.
Raizen, M., Salomon, C. and Niu, Q. (1997). New Light on Quantum Transport. *Phys. Today*, **50**, 30, and references therein.
Rapsch, S., Schollwöck, U. and Zwerger, W. (1999). Density Matrix Renormalization Group for Disordered Bosons in One Dimension. *Europhys. Lett*, **46**, 559.
Rey, A.M., Satija, I.I. and Clark, C.W. (2006). Quantum Coherence of Hard-Core Bosons: Extended, Glassy, and Mott Phases. *Phys. Rev. A*, **73**, 063610.
Roati, G., de Mirandes, E., Ferlaino, F., Ott, H., Modugno, G. and Inguscio, M. (2004). Atom Interferometry with Trapped Fermi Gases. *Phys. Rev. Lett.*, **92**, 230402.
Roati, G., Zaccanti, M., D'Errico, C., Catani, J., Modugno, M., Simoni, A., Inguscio, M. and Modugno, G. (2007). ^{39}K Bose–Einstein Condensate with Tunable Interactions. *Phys. Rev. Lett.*, **99**, 010403.
Roati, G., D'Errico, C., Fallani, L., Fattori, M., Fort, C., Zaccanti, M., Modugno, G., Modugno, M. and Inguscio, M. (2008). Anderson Localization of a Non-Interacting Bose–Einstein Condensate. *Nature*, **453**, 895.
Rom, T., Best, Th., van Oosten, D., Schneider, U., Fölling, S., Paredes, B. and Bloch, I. (2006). Free Fermion Antibunching in a Degenerate Atomic Fermi Gas Released from an Optical Lattice. *Nature*, **444**, 733.
Roscilde, T. (2008). Bosons in One-Dimensional Incommensurate Superlattices. *Phys. Rev. A*, **77**, 063605.
Roscilde, T. and Cirac, J.I. (2007). Quantum Emulsion: A Glassy Phase of Bosonic Mixtures in Optical Lattices. *Phys. Rev. Lett.*, **98**, 190402.
Roth, R. and Burnett, K. (2003). Phase Diagram of Bosonic Atoms in Two-Color Superlattices. *Phys. Rev. A*, **68**, 023604.
Roux, G., Barthel, T., McCulloch, I.P., Kollath, C., Schollwoeck, U. and Giamarchi, T. (2008). The Quasi-periodic Bose–Hubbard Model and Localization in One-dimensional Cold Atomic Gases. *Phys. Rev. A* (in press).
Sanchez-Palencia, L. and Santos, L. (2005). Bose–Einstein Condensates in Optical Quasicrystal Lattices. *Phys. Rev. A*, **72**, 053607.
Sanchez-Palencia, L. (2006). Smoothing Effect and Delocalization of Interacting Bose–Einstein Condensates in Random Potentials. *Phys. Rev. A*, **74**, 053625.
Sanchez-Palencia, L., Clément, D., Lugan, P., Bouyer, P., Shlyapnikov, G.V. and Aspect, A. (2007). Anderson Localization of Expanding Bose–Einstein Condensates in Random Potentials. *Phys. Rev. Lett.*, **98**, 210401.
Sanchez-Palencia, L., Clément, D., Lugan, P., Bouyer, P. and Aspect, A. (2008). Disorder Induced Trapping versus Anderson Localization in Bose-Einstein Condensates Expanding in Disordered Potentials. *New J. Phys.*, **10**, 045019.
Sanpera, A., Kantian, A., Sanchez-Palencia, L., Zakrzewski, J. and Lewenstein, M. (2004). Atomic Fermi–Bose Mixtures in Inhomogeneous and Random Lattices: From Fermi Glass to Quantum Spin Glass and Quantum Percolation. *Phys. Rev. Lett.*, **93**, 040401.
Scalettar, R.T., Batrouni, G.G. and Zimanyi, G.T. (1991). Localization in Interacting Disordered Bose Systems. *Phys. Rev. Lett.*, **66**, 3144.
Schulte, T., Drenkelforth, S., Kruse, J., Ertmer, W., Arlt, J., Sacha, K., Zakrzewski, J. and Lewenstein, M. (2005). Routes Towards Anderson-Like Localization of Bose–Einstein Condensates in Disordered Optical Lattices. *Phys. Rev. Lett.*, **95**, 170411.
Schulte, T., Drenkelforth, S., Kruse, J., Tiemeyer, R., Sacha, K., Zakrzewski, J., Lewenstein, M., Ertmer, W. and Arlt, J.J. (2006). Analysis of Localization Phenomena in Weakly Interacting Disordered Lattice Gases. *New J. Phys.*, **8**, 230.

Schulte, T., Drenkelforth, S., Kleine Büning, G., Ertmer, W., Arlt, J., Lewenstein, M. and Santos, L. (2008). Dynamics of Bloch Oscillations in Disordered Lattice Potentials. *Phys. Rev. A*, **77**, 023610.
Schwartz, T., Bartal, G., Fishman, S. and Segev, M. (2007). Transport and Anderson Localization in Disordered Two-Dimensional Photonic Lattices. *Nature*, **446**, 52.
Sen De, A., Sen, U., Wehr, J. and Lewenstein, M. (2007). Classical Spin Models with Broken Continuous Symmetry: Random Field Induced Order and Persistence of Spontaneous Magnetization. Preprint arXiv:0712.0829.
Shapiro, B. (2007). Expansion of a Bose–Einstein Condensate in the Presence of Disorder. *Phys. Rev. Lett.*, **99**, 060602.
Skipetrov, S.E., Minguzzi, A., van Tiggelen, B.A. and Shapiro, B. (2008). Anderson Localization of a Bose-Einstein Condensate in a 3D Random Potential. *Phys. Rev. Lett.*, **100**, 165301.
Spielman, I.B., Phillips, W.D. and Porto, J.V. (2007). Mott-Insulator Transition in a Two-Dimensional Atomic Bose Gas. *Phys. Rev. Lett.*, **98**, 080404.
Stöferle, T., Moritz, H., Schori, C., Köhl, M. and Esslinger, T. (2004). Transition from a Strongly Interacting 1D Superfluid to a Mott Insulator. *Phys. Rev. Lett.*, **92**, 130403.
van der Zant, H.S.J., Fritschy, F.C., Elion, W.J., Geerligs, L.J. and Mooij, J.E. (1992). Field-Induced Superconductor-to-Insulator Transitions in Josephson-Junction Arrays. *Phys. Rev. Lett.*, **69**, 2971.
Vignolo, P., Akdeniz, Z. and Tosi, M.P. (2003). The Transmittivity of a Bose-Einstein Condensate on a Lattice: Interference from Period Doubling and the Effect of Disorder. *J. Phys. B: At. Mol. Opt. Phys.*, **36**, 4535.
Wang, D.-W., Lukin, M.D. and Demler, E. (2004). Disordered Bose–Einstein Condensates in Quasi-One-Dimensional Magnetic Microtraps. *Phys. Rev. Lett.*, **92**, 076802.
Wehr, J., Niederberger, A., Sanchez-Palencia, L. and Lewenstein, M. (2006). Disorder Versus the Mermin-Wagner-Hohenberg Effect: From Classical Spin Systems to Ultracold Atomic Gases. *Phys. Rev. B*, **74**, 224448.
White, M., Pasienski, M., McKay, D., Zhou, S., Ceperley, D. and DeMarco, B. (2008). Strongly Interacting Bosons in a Disordered Optical Lattice. Preprint arXiv:0807.0446.
Wiersma, D.S., van Albada, M.P., van Tiggelen, B.A. and Lagendijk, A. (1995). Experimental Evidence for Recurrent Multiple Scattering Events of Light in Disordered Media. *Phys. Rev. Lett.*, **74**, 4193.
Wiersma, D.S., Bartolini, P., Lagendijk, A. and Righini, R. (1997). Localization of Light in a Disordered Medium. *Nature*, **390**, 671.
Wu, B. and Niu, Q. (2001). Landau and Dynamical Instabilities of the Superflow of Bose–Einstein Condensates in Optical Lattices. *Phys. Rev. A*, **64**, 061603R.

CHAPTER 4

Dipole–Dipole Interactions of Rydberg Atoms

Thomas F. Gallagher[a,b] and Pierre Pillet[b]

[a] *Department of Physics, University of Virginia, Charlottesville, VA 22903, USA*
[b] *Laboratoire Aimé Cotton, CNRS, Univ Paris-Sud, 91405 Orsay cedex, France*

Contents		
	1. Introduction	162
	2. Principles of Resonant Dipole–Dipole Collisions	166
	2.1 Dipole Transitions in Rydberg Atoms	166
	2.2 Dipole–Dipole Interactions	168
	3. Verification of the Predictions	174
	4. Manipulating Resonant Collisions	181
	5. Dipole–dipole Interactions in the Frozen Rydberg Gas	189
	6. Line Broadening and Blockades	198
	7. Mechanical Effects	204
	8. Conclusion	215
	Acknowledgments	216
	References	216

Abstract   The dipole–dipole interaction is the longest-range interaction possible between two neutral atoms or molecules, and in atoms, which have no permanent moments, it is usually the transition dipole moments which are important. There are many consequences of the dipole–dipole interaction, ranging from photo-association to energy transfer in biological molecules. Nowhere, however, is the effect of the dipole–dipole interaction displayed more clearly than in the interactions between Rydberg atoms having high principal quantum number n, whose dipole moments scale as n^2. One of the attractions of Rydberg atoms is that it is possible to

Advances in Atomic, Molecular, and Optical Physics, Volume 56 © 2008 Elsevier Inc.
ISSN 1049-250X, DOI 10.1016/S1049-250X(08)00013-X All rights reserved.

tune the Rydberg energy levels through resonance for the dipole–dipole energy transfer. For example, in binary energy transfer collisions resonances as narrow as 1 MHz have been observed. The cross-sections for these resonant collisions can be described very simply in terms of radio frequency spectroscopy and more formally as the evolution of pairs of two atom, or molecular, states. The latter description can be extended easily to the description of radiatively-assisted collisions, those in which the colliding pair of atoms emits or absorbs photons during the collision. The experimental investigations of binary resonant energy transfer collisions between Rydberg atoms, including radiatively-assisted collisions, are described. Laser cooling has opened several new avenues of research using the dipole–dipole interactions of Rydberg atoms. It is straightforward to produce an atomic sample dense enough that the cold atoms move a few per cent of the typical interatomic spacing on the 1 μs time scale of a typical experiment. Thus, on the time scale of interest, the atoms are frozen in place, and their properties more closely resemble those of an amorphous solid than a gas. Since the atoms are essentially stationary, an atom can interact simultaneously with all its neighbors, and the interactions are no longer binary, but many-body interactions. The experimental exploration of these interactions in a range of experiments is described. While the many-atom interactions are the new feature of cold atoms, binary interactions are also important. For example, the use of a dipole blockade as a quantum gate has been proposed, and several variants of partial or local blockades have been observed. Finally, the dipole–dipole interaction leads to an attractive or repulsive force between two atoms, so that even if the atoms are initially at rest, they can attract each other, collide, and ionize, initiating the evolution of a cold Rydberg gas into an ultracold plasma.

1. INTRODUCTION

One of the reasons why dipole–dipole and other interactions between atoms and molecules are important is that they provide a means for energy transfer from one atom or molecule to another. An excellent biological example of this is the Förster resonant energy transfer (FRET) process often observed in biological systems [1]. In this process the photon energy absorbed by the chromophore is passed to another molecule by resonant dipole–dipole energy transfer. Resonant energy transfers which may be more familiar to physical scientists are those which occur in lasers. For example, the well-known He–Ne laser is based on resonant collisions between excited He atoms and ground state Ne atoms [2]. The discharge in the He–Ne mixture leads to an accumulation of population in the metastable He states. In collisions of the excited He atoms with ground state Ne atoms, the electronic energy of the He atoms is resonantly transferred to the Ne atoms producing Ne atoms in the upper levels of the visible and infrared laser transitions. Roughly analogous resonant energy transfer processes are also important in the CO_2 laser [3].

Historically, dipole–dipole interactions have been of interest in atomic and molecular collision physics. However, the ability to cool and trap atoms has led to a much broader interest in dipole–dipole interactions. One of the first such examples is photo-association spectroscopy in a magneto-optical trap (MOT) [4], which is efficient because of the resonant dipole–dipole interaction between the ground state and the first excited state, for example, the Cs 6s and 6p states, which leads to a long-range $1/R^3$ attractive potential in the excited molecular state, the Cs_2 6s6p state. Here R is the internuclear separation of the two atoms. Even though the initial 6s6s molecular state is unbound, since the atoms are cold, the free–bound transitions result in narrow spectral lines, and the long series of vibrational states approaching the excited dissociation limit is the well-known signature of photo-association. An unexpected consequence of photo-association spectroscopy and the dipole–dipole interaction is that the spectrum of vibrational states in the $1/R^3$ potential allows one to obtain excellent values for the resonance line oscillator strengths [5]. Finally, the dipole–dipole interaction and the large atomic fine structure of cesium lead to the existence of a long-range molecular well, the rovibrational levels of which can be excited by photo-association of cold atoms, forming molecules which then decay to the ground state of the molecule. The formation of cold molecules via photo-association is a direct consequence of the dipole–dipole interaction [6].

Dipole–dipole interactions of cold atoms and molecules are of intrinsic interest as well. In Bose–Einstein condensates of Cr the interactions between the permanent magnetic dipoles of chromium atoms are being investigated [7]. A notion which has attracted much attention is using dipole–dipole interactions as the basis of quantum gates. Two general avenues are being explored. The first is to use polar molecules, usually mixed alkali dimers which can have dipole moments of up to a few ea_0 [8], where e is the magnitude of the electron charge and a_0 is the Bohr radius. Such a method is possible, although not trivial, since the molecules must be in low vibrational states to have large dipole moments, and creating such mixed dimers starting from a dual species MOT is a challenge. Alternatively, Stark deceleration methods may provide access to a broader class of polar molecules [9]. The second avenue is to use the dipole–dipole interaction between cold Rydberg atoms [10–13]. The large dipole matrix elements, $\sim n^2 ea_0$, lead to dipole–dipole interactions scaling as n^4, where n is the principal quantum number. Since producing atoms having $n \sim 50$ is straightforward, it is possible to see the effects of the dipole–dipole interactions clearly even at very low density.

In general it is difficult to study dipole–dipole and other energy transfer processes in a systematic way. The efficiency of energy transfer depends on how nearly coincident the energy intervals are. There is, however, no practical way to tune the energy levels of a specific pair of interacting atoms or molecules through resonance. A way of probing the resonance profile is the approach used in the study of resonant electronic to vibrational energy transfer, i.e. simply choosing molecular collision partners with a range of vibrational frequencies which straddle the electronic energy spacing of the atomic collision partner [14].

Rydberg atoms constitute an attractive system for the systematic study of resonant energy transfer, because their energy spacings are tunable. The simplest approach is based on the fact that the Rydberg levels have n-dependent energy level spacings, which provides tunability, although in discrete steps. The frequency of the CD_4 v_4 vibrational mode matches the frequency of the Na 7s to 5d interval. When the depopulation cross-sections of the Na ns states resulting from CD_4 collisions are plotted vs n, the result is a smooth curve of cross-section $\sigma \approx 100$ Å^2, with a sharp peak at 7s, where the cross-section rises to 350 Å^2 [15]. The peak is absent when CH_4, which has very different vibrational frequencies, is used as the collision partner, although there are resonant collisions with CH_4 which depopulate the 5s and 6s states. At $n = 7$ the energy level spacing is too large to know how sharp the Na–methane collisional resonances are, but experiments with polar molecules and atoms in Rydberg states of Xe provide information on the widths of the collisional resonances [16]. In a sample of NH_3 at 300 K, many rotational states are present. If a NH_3 molecule undergoes $\Delta J = -1$ transition in a resonant collision with a Xe Rydberg atom the atom is left in a higher energy state. When Xe Rydberg atoms are allowed to collide with NH_3 molecules and the final states are analyzed by selective field ionization, the resulting spectrum consists of a series of peaks corresponding to atomic transitions resonant with the molecular $\Delta J = -1$ transitions for initial rotational states of $J = 1 - 7$. Unlike the Na–methane experiments these experiments give explicitly the widths of the collisional resonances to be ~ 5 cm^{-1}.

In both the Na–methane and Xe–ammonia experiments the variation of the zero field energies of the Rydberg states was used for the tuning. The energies of the Rydberg states can also be tuned continuously with electric fields, and this approach has been used to study dipole–dipole energy transfer collisions between room temperature Rydberg atoms and the interactions among quasi-static cold Rydberg atoms obtained by exciting Rb and Cs atoms in a MOT [17–19]. What makes the Rydberg–Rydberg system so attractive is that, with their large electric dipole matrix elements, Rydberg atoms can be tuned easily over tens of GHz, making it a straightforward matter to tune the levels through a dipole–dipole resonance, even on very short time scales. Magnetic tuning of Feshbach resonances, and thus the scattering length, in cold atom collisions is an analogous technique, although the mechanism is different [20]. In addition, the large dipole moments ensure that the primary coupling between a pair of Rydberg atoms is the long-range dipole–dipole interaction in which the atoms are separated by distances large compared to their orbital radii, which makes these interactions particularly easy to understand.

Experiments on resonant collisional energy transfer between Rydberg atoms at temperatures from 300 K to 1 K have verified that the collision cross-sections are enormous, scaling as n^4, and are orders of magnitude larger than the geometric cross-sections of the Rydberg atoms [17,21]. Furthermore, the duration of the collisions is long. At 300 K the collisions of $n = 20$ atoms last longer than 1 ns, compared to 1 ps for typical 300 K atom–atom collisions, and at 1 K they have been observed to last 1 µs. The combination of the long duration of the collisions and the large dipole matrix elements of Rydberg atoms makes it straightforward to alter the collisions using radiation fields [22,23]. Using Rydberg atoms it has been possible to make quantitative measurements of radiatively-assisted collisions in which colliding pairs of Rydberg atoms absorb or emit microwave photons. A novel feature of these experiments is that they extend such experiments into the strong field regime in which the interaction with radiation field is stronger than the interaction between the atoms, a previously inaccessible regime due to the short lifetime, 1 ps, of the transient molecules formed in other radiative collisions [24]. A further attraction of radiatively-assisted collisions is that it is possible to control the strength of the interatomic interaction with the microwave field [25].

As the temperature of the Rydberg atoms is decreased the atoms move more slowly, which leads to an increased temporal duration of the collisions when the temperature is lowered from 300 K to 1 K. In a MOT the temperatures are below 1 mK, which can, in principle, lead to collisions lasting tens of microseconds. More interesting, at the densities attainable in a MOT, $> 10^9$ cm^{-3}, it is possible to enter a new regime. Specifically, on the one microsecond time scale of a typical experiment 300 µK atoms at a density of 10^9 cm^{-3} move 3% of the average interatomic spacing. On this time scale the gas of Rydberg atoms is frozen and is, in effect, an amorphous solid. Since the atoms do not move it is no longer true that binary encounters in which two atoms collide are the dominant interaction mechanism. Rather, the static interactions between an atom and all its neighbors are important [18,19]. Consequently, the frozen Rydberg gas is intrinsically a many-body system, more like a solid than a gas. Nonetheless, the frozen gas retains the attractive features of Rydberg atoms, in particular the large electric dipole matrix elements, which allow tunability of the energy levels and strong interatomic interactions. Since atoms are essentially frozen, collisions do not occur, but the exchange of energy among static atoms does occur, and diffusion of excitation through the frozen Rydberg gas is analogous to the motion of an exciton in a solid; that is the excitation travels although the atoms do not move. The strong dipole–dipole interactions of Rydberg atoms in the frozen Rydberg gas have prompted suggestions that they be used as quantum gates [10–13].

Although the most commonly observed manifestation of the dipole–dipole interactions is the energy transfer among bound atoms, resulting in redistributing population among excited states, dipole–dipole interactions can also result in ionization of the frozen Rydberg gas, initiating the evolution to a plasma [26,27]. Since the dipole–dipole potential varies as $1/R^3$, a pair of dipole coupled atoms is on either an attractive or a repulsive potential. If the atoms are on an attractive potential they are drawn together and collide, resulting in an ion and a free electron [28,29]. The dipole–dipole interaction can also lead to ionization without motion of the ions, by a process analogous to that which leads to energy transfer in the bound states. If the dipole–dipole interaction couples a pair of bound atoms to a pair of states in which one of the atomic states is at lower energy and the other lies above the ionization limit, the process of molecular auto-ionization can occur [30]. In most cases, molecular auto-ionization of pairs of atoms occurs at rates too slow to be of much practical importance, but in dense samples of Rydberg atoms ionization is observed which is too rapid to be attributed to motion of the atoms as described above and is ascribed to a many-atom variant of molecular ionization [31].

2. PRINCIPLES OF RESONANT DIPOLE–DIPOLE COLLISIONS

2.1 Dipole Transitions in Rydberg Atoms

The matrix elements associated with dipole transitions between Rydberg states can be huge. In the case of the hydrogen atom, the matrix element

$$\langle n, l \, |r| \, n, l - 1 \rangle = \frac{3}{2} n \sqrt{n^2 - l^2} a_0 \qquad (1)$$

scales as n^2, and for low l states such that $l \ll n$, this matrix element is $\frac{3}{2}n^2$. The square of this matrix element is, in many cases, larger than the sum of the squares of the other matrix elements $\langle nl \, |r| \, n', l - 1 \rangle$, with $n' \neq n$ [32]. In other words, the dipole matrix elements between states of different energies are small. In contrast, in alkali atoms the large quantum defects of the dipole coupled low l states often differ by roughly $1/2$, leading to large,

$\approx n^2$, matrix elements between states with appreciable energy differences. For this reason alkali atoms are particularly interesting for energy transfer processes. In Cs, for example, the $l < 3$ states have appreciable quantum defects, so one can expect to observe efficient transfer of substantial amounts of energy between dipole coupled low l states. For example, two Cs atoms, one in the ns state and the other in the np state can exchange energy by the process

$$ns + np \longrightarrow np + ns. \tag{2}$$

Similarly, a pair of atoms in the np and nd states can exchange energy by the process

$$np + nd \longrightarrow nd + np. \tag{3}$$

An nf, ng pair can also exchange energy, but the amount is small since both states have quantum defects <0.02 [33,34].

2.1.1 Interaction between Two Atoms

We consider the case of two-Rydberg-level atoms. The two Rydberg levels are assumed to be coupled by an electric dipole transition. By treating the coupling of the ensemble of N atoms with the vacuum of the electromagnetic field, we can demonstrate that the master equation is given by

$$\begin{aligned}
\frac{d\tilde{\sigma}(t)}{dt} = &-\frac{\Gamma}{2} \sum_i [[r_i^+ r_i^-, \tilde{\sigma}(t)]_+ - 2r_i^- \tilde{\sigma}(t) r_i^+] \\
&- \frac{1}{\hbar^2} \frac{1}{8\pi^3} \sum_{i \neq j} \frac{\hbar c}{2\varepsilon_0} |\langle 2| \vec{\mu} |1\rangle|^2 \frac{4\pi^2}{c} \\
&\times \Bigg\{ \Bigg\{ (1 - \cos^2\theta) \frac{k_0^2}{R_{ij}} \sin(k_0 R_{ij}) + (1 - 3\cos^2\theta) \frac{k_0}{R_{ij}^2} \cos(k_0 R_{ij}) \\
&- (1 - 3\cos^2\theta) \frac{1}{R_{ij}^3} \sin(k_0 R_{ij}) \Bigg\} \Big[[r_i^+ r_j^-, \tilde{\sigma}(t)]_+ - 2r_i^- \tilde{\sigma}(t) r_j^+ \Big] \\
&- i \Bigg\{ (1 - \cos^2\theta) \frac{k_0^2}{R_{ij}} \cos(k_0 R_{ij}) - (1 - 3\cos^2\theta) \frac{k_0}{R_{ij}^2} \sin(k_0 R_{ij}) \\
&- (1 - 3\cos^2\theta) \frac{1}{R_{ij}^3} \cos(k_0 R_{ij}) \Bigg\} [r_i^+ r_j^-, \tilde{\sigma}(t)] \Bigg\}
\end{aligned} \tag{4}$$

with i (or j) for the atom label. We have introduced the transition operators for the atom i

$$r_i^+ = |2, i\rangle \langle 1, i|, \qquad r_i^- = |1, i\rangle \langle 2, i|. \tag{5}$$

The Rydberg levels are labeled 1 and 2, with respective internal energies W_1 and W_2, such that $W_1 < W_2$.

The first term, $-\frac{\Gamma}{2} \sum_i \left[[r_i^+ r_i^-, \tilde{\sigma}(t)]_+ - 2 r_i^- \tilde{\sigma}(t) r_i^+\right]$, of the equation corresponds to the spontaneous emission of each atom alone. The other terms correspond to the sum of the interactions of each atom with the dipolar electromagnetic field emitted by all the other atoms [35].

2.2 Dipole–Dipole Interactions

We consider now the case of an ensemble of atoms in a volume on the order of $(\lambda_0/2\pi)^3$ or smaller, where λ_0 is the wavelength associated with the transition, $1 \longrightarrow 2$ ($\lambda_0 = 2\pi/k_0$). We can write the equation as

$$\begin{aligned}
\frac{d\tilde{\sigma}(t)}{dt} = & -\frac{\Gamma}{2} \sum_i \left[[r_i^+ r_i^-, \tilde{\sigma}(t)]_+ - 2 r_i^- \tilde{\sigma}(t) r_i^+\right] \\
& - \sum_{i \neq j} \frac{\Gamma}{2} \left[\left[r_i^+ r_j^-, \tilde{\sigma}(t)\right]_+ - 2 r_i^- \tilde{\sigma}(t) r_j^+\right] \\
& + i \sum_{i \neq j} \frac{1}{4\hbar\pi\varepsilon_0} |\langle 2| \vec{\mu} |1\rangle|^2 \left\{\frac{1}{2}\left(1 + \cos^2\theta\right) \frac{k_0^2}{R_{ij}} \right. \\
& \left. - \left(1 - 3\cos^2\theta\right) \frac{1}{R_{ij}^3}\right\} \left[r_i^+ r_j^-, \tilde{\sigma}(t)\right].
\end{aligned} \tag{6}$$

It is interesting to note that there is not only the familiar dipole–dipole interaction term scaling as R^{-3}, but a long-range term scaling as R^{-1} as well. The two terms differ by the factor $(k_0 R_{ij})^2$. Consequently, the $1/R$ term is usually negligible for a pair of close atoms, and the dipole–dipole interaction between the two atoms, labeled 1 and 2, can be written as

$$H_{dd} = \frac{1}{4\pi\varepsilon_0}\left[\frac{\vec{\mu}_1 \cdot \vec{\mu}_2}{R^3} - \frac{3\left(\vec{\mu}_1 \cdot \vec{R}\right)\left(\vec{\mu}_2 \cdot \vec{R}\right)}{R^5}\right], \tag{7}$$

where R is the distance R_{12} between the two atoms, and $\vec{\mu}_{1,2}$ the dipole transition operator associated with each atom

$$\vec{\mu}_{1,2} = e \vec{r}_{1,2}. \tag{8}$$

For a large number, N, of atoms it is less obvious that we can always ignore the term scaling as R^{-1}. The angular part of the dipole–dipole interaction depends on the angle θ between the dipole vector $\vec{\mu}$ and the vector $\vec{R}_{ij}$. The angular integral of the $1/R^3$ term vanishes, while the angular integral of the $1/R$ term gives $2/3$. We shall see that the $1/R$ term should be considered in an ensemble of cold Rydberg atoms interacting all together. In this case the total Hamiltonian for the interaction between two atoms can be written

$$H_{dd}^T = \frac{1}{4\pi\varepsilon_0}\left[\frac{\vec{\mu}_1 \cdot \vec{\mu}_2}{R^3} - \frac{3\left(\vec{\mu}_1 \cdot \vec{R}\right)\left(\vec{\mu}_2 \cdot \vec{R}\right)}{R^5} \right.$$
$$\left. - k_0^2\left[\frac{\vec{\mu}_1 \cdot \vec{\mu}_2}{2R} + \frac{\left(\vec{\mu}_1 \cdot \vec{R}\right)\left(\vec{\mu}_2 \cdot \vec{R}\right)}{2R^3}\right]\right]. \quad (9)$$

The $1/R$ term should be considered if the typical size of the atomic sample is of the same order as $k_0^{-1} = \lambda_0/2\pi$, because, unlike the $1/R^3$ term, the angular integration of the $1/R$ term does not result in a vanishing result.

2.2.1 Resonant Dipole–Dipole Collisional Energy Transfer

In all the alkali atoms but Li the quantum defects of the s and p states differ by almost exactly $1/2$. Consequently, in the presence of an electric field it is a straightforward proposition to tune the s (or p) levels midway in energy between the neighboring p (or s respectively) levels. Under these conditions the dipole–dipole collisions

$$ns + ns \longrightarrow np + (n-1)p$$

or respectively

$$np + np \longrightarrow ns + (n+1)s,$$

become resonant.

To develop a feeling for the characteristics of these resonant energy transfer collisions we begin with a simple description based on the notions of radio frequency spectroscopy. For concreteness we consider the collision of two Na 17s atoms, 1 and 2, which produces a 16p and a 17p atom. Imagine that atom 1 passes atom 2 with velocity v and impact parameter b, as shown in Figure 1. Imagine that atom 1 produces an oscillating field at atom 2 of magnitude $E_1 \approx \mu_1/R^3$, where R is the distance between the two atoms and μ_1 is the dipole moment from the initial state to the upper final state, for example the 17s–17p dipole

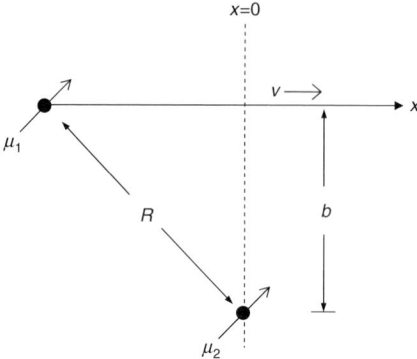

FIGURE 1 Schematic diagram of a resonant collision in which atom 1, having dipole moment $\vec{\mu}_1$ passes atom 2, having dipole moment $\vec{\mu}_2$, with velocity $\vec{v}$ and impact parameter b

moment. On resonance this oscillating field can drive atom 2 down to the lower final state if

$$\mu_2 E_1 \tau = 1 \tag{10}$$

where μ_2 is the dipole matrix element connecting the initial state to the lower final state, e.g., the 17s–16p dipole moment, and τ is the duration of the collision. E_1 is most important when $R \sim b$, so we set $E_1 \approx \mu_1/b^3$ and, correspondingly, set $\tau = b/v$. With these substitutions we find

$$\sigma \approx b^2 = \frac{\mu_1 \mu_2}{v} \tag{11}$$

and

$$\tau = \sqrt{\frac{\mu_1 \mu_2}{v^{3/2}}}. \tag{12}$$

Since $\mu_1 \approx \mu_2 \approx n^2$, in this case we find

$$\sigma = \frac{n^4}{v} \tag{13}$$

and

$$\tau = \frac{n^2}{v^{3/2}}. \tag{14}$$

Equations (13) and (14) are given in atomic units, and if we re-express them in laboratory units, for $n = 20$, we find $\sigma = 3 \times 10^{-8}$ cm^2 and $\tau = 10^{-9}$ s. Although they may be hard to believe, these values for the cross-section and duration of the collision are correct.

It is possible to construct a more rigorous description of the collisions in the following way. The coupling between the two atoms (labeled 1 and 2) is given by the dipole–dipole Hamiltonian, H_{dd}, of Equation (7). We consider again the case in which two ns Rydberg atoms exchange internal energy to give one atom in the np level and one in the $(n-1)p$ level. The system can be considered as a two-level system $|ss\rangle$ and $|pp'\rangle$, defined as the product of states

$$|ss\rangle = |ns, 1\rangle \otimes |ns, 2\rangle \tag{15}$$

and

$$|pp'\rangle = \left[|np, 1\rangle \otimes |(n-1)p, 2\rangle + |(n-1)p, 1\rangle \otimes |np, 2\rangle\right]/\sqrt{2} \tag{16}$$

where, for example, $|ns, 1\rangle$ is atom 1 in the ns state.

The dipole–dipole Hamiltonian, H_{dd}, does not couple the level $|ss\rangle$ with the antisymmetric product

$$|pp''\rangle = \left[|np, 1\rangle \otimes |(n-1)p, 2\rangle - |(n-1)p, 1\rangle \otimes |np, 2\rangle\right]/\sqrt{2}.$$

The energy of the levels ss and pp' are given by

$$W_{ss} = 2W_{ns}(E) \text{ and } W_{pp'} = W_{np}(E) + W_{(n-1)p}(E)$$

and the coupling between the two atoms is given by the matrix element, $V = \langle ss|H_{dd}|pp'\rangle$, which depends on the interatomic distance, R. The resonance occurs at the amplitude, E, of the electric field for which $W_{ss} = W_{pp'}$.

We consider two colliding Rydberg atoms. Their interatomic distance, R, varies as shown in Figure 1, in which we have assumed that the atoms follow straight line trajectories. In the reference frame of atom 2, atom 1 passes the first atom with an impact parameter, b, at $x = 0$ with a relative velocity v, parallel to the x axis. Although we ignored the orientation of the dipoles in our simple radio frequency description, it is important. Figure 2 shows two examples of the variation of the matrix element, $\langle 1|H_{dd}|2\rangle$, in the case of the dipoles, $\vec{\mu}_{1,2}$, oriented perpendicular and parallel to the x axis [36].

The total wave function for the system can be written as

$$|\Psi(t)\rangle = C_{ss}(t)|ss\rangle + C_{pp'}(t)|pp'\rangle. \tag{17}$$

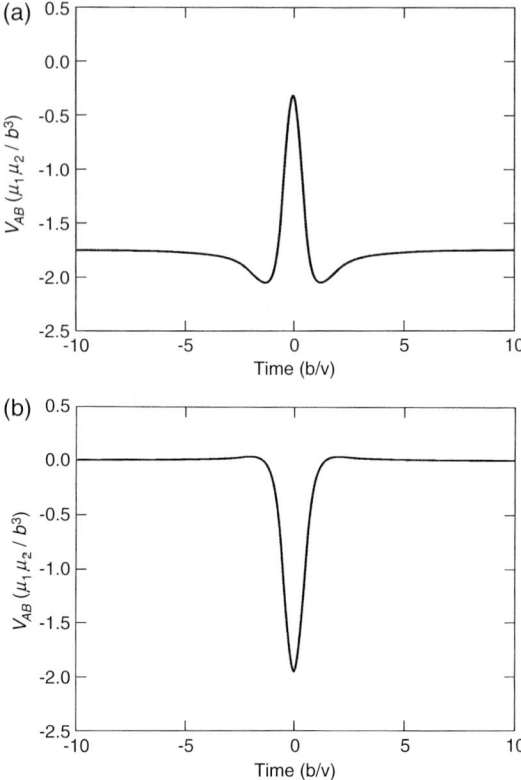

FIGURE 2 Variation in the coupling matrix element V in arbitrary units over the course of a collision for (a) the case in which $\vec{v}_c \parallel \vec{E}$ (b) the case in which $\vec{v}_c \perp \vec{E}$ and $\varphi = \pi/2$ (from Ref. [36])

The Hamiltonian of the system is given by

$$H = H_0 + V(R), \tag{18}$$

where H_0 is the Hamiltonian of the two atoms at $R \to \infty$, and can be written;

$$H_0 = \begin{pmatrix} W_{ss}(E) & 0 \\ 0 & W_{pp'}(E) \end{pmatrix}. \tag{19}$$

For the atoms initially in state $|ss\rangle$, at resonance, $W_{pp'}(E) = 0$, we obtain

$$C_{pp'}(t = \infty) = -i \sin \left[\int_{-\infty}^{+\infty} V(vt) \, dt' \right].$$

By considering the case where the dipoles are parallel to the axis x, we get

$$C_{pp'}(t=\infty) = -i\sin\left[\int_{-\infty}^{+\infty} \frac{\chi}{(b^2+v^2t^2)^{3/2}} - 3\frac{\chi v^2 t^2}{(b^2+v^2t^2)^{5/2}}\,dt\right]$$
$$= 0, \tag{20}$$

where

$$\chi = \mu_{sp}\mu_{sp}$$

and by considering the case where they are perpendicular

$$C_{pp'}(t=\infty) = -i\sin\left[\int_{-\infty}^{+\infty} \frac{\chi}{(b^2+v^2t^2)^{3/2}} - 3\frac{\chi b^2 \sin^2\varphi}{(b^2+v^2t^2)^{5/2}}\,dt\right]$$
$$C_{pp'}(t=\infty) = -i\sin\left[\frac{2(1-2\sin^2\varphi)\chi}{b^2 v}\right] \tag{21}$$

where φ is the angle between the dipole $\vec{\mu}$ and the vector $\vec{u}$ perpendicular to the plane containing the trajectories of both atoms (out of the paper in Figure 1). The calculation can be performed easily for any orientation of the dipoles. For experiments in an atomic beam where faster atoms collide with the slower ones, the cross-section is given by

$$\sigma = \int_0^{2\pi}\int_0^{+\infty} |C_p(t=\infty)|^2\, b\,db d\varphi, \tag{22}$$

which gives at resonance for the case of dipoles perpendicular to the x-axis

$$\sigma = \int_0^{2\pi}\int_0^{+\infty} \sin^2\left(\frac{2(1-2\sin^2\varphi)\chi}{vb^2}\right) b\,db d\varphi = \frac{\pi^2 \chi}{\sqrt{2}v}. \tag{23}$$

By assuming for an atomic beam that the average of the relative velocity can in a first approximation be written as $\langle\frac{1}{v}\rangle \approx \frac{1}{\langle v\rangle} \approx \sqrt{\frac{2M}{k_B T}}$, [37] we obtain

$$\sigma \approx \frac{\pi^2 \chi}{\sqrt{2}\langle v\rangle}. \tag{24}$$

We can define the impact parameter at resonance, b_0, for $\sigma = \pi b_0^2$, and

$$b_0 \approx \frac{\sqrt{\pi}}{2^{1/4}} \sqrt{\frac{\chi}{\langle v \rangle}} \sim \sqrt{\frac{\chi}{\langle v \rangle}}, \tag{25}$$

and we define the characteristic time for the collision as

$$\tau = \frac{b_0}{\langle v \rangle} \sim \frac{\sqrt{\chi}}{\langle v \rangle^{3/2}}. \tag{26}$$

Using these expressions we obtain the following orders of magnitude for room temperature resonant collisions of two Cs np atoms to give an ns and an $(n+1)s$ atom. We have $\langle np, m| \mu |ns \rangle \simeq \langle np, m| \mu |(n+1)s \rangle \simeq 0.55 n^{*2}$ a.u. For $n^* = 20$, at a temperature $T = 500$ K, we take $\langle v \rangle \sim 300$ m/s, and we obtain $b_0 \sim 1.5$ μm, $\tau \sim 5$ ns.

The width of the resonance can be obtained easily using the weak coupling approximation

$$C_{pp'}(t) = -i \int_{-\infty}^{t} V(vt') \exp\left(i\Delta(t' - t)\right) dt'.$$

The spectrum of the resonance is given by $|C_{pp'}(t = +\infty)|^2$, which depends on the detuning compared to the resonance, Δ. In the case of parallel dipoles with $\varphi = 0$, we have

$$C_{pp'}(t = \infty) = -i \int_{-\infty}^{+\infty} \frac{\chi}{(b^2 + v^2 t^2)^{3/2}} \exp\left(i\Delta(t' - t)\right) dt',$$

and the width of the resonance line will be of the order of

$$\Delta_R \sim \tau^{-1} = \frac{\langle v \rangle}{b_0} \sim \frac{\langle v \rangle^{3/2}}{\sqrt{\chi}}. \tag{27}$$

3. VERIFICATION OF THE PREDICTIONS

The initial experiments, done with Na atoms, provided verification that resonant dipole–dipole collisions between Rydberg atoms do indeed lead to the cross-sections and collision times described in the previous section [21]. The relevant energy levels are shown in Figure 3, a diagram of the energy levels of the Na 16p, 17s, and 17p levels as functions of an applied electric field. The np levels are split into $|m_l| = 0$ and 1 levels by the electric

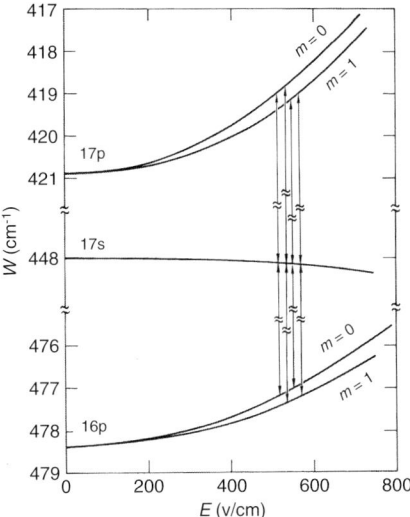

FIGURE 3 Energy level diagram for the collision Na 17s + Na 17s ⟶ Na 16p + Na 17p. There are four fields at which the resonant collisions occur (from Ref. [21])

field. For all three states the spin of the electron can be either up or down, and, for the moment, we ignore it. At four fields near 550 V/cm the 17s level is midway between the two p states, and at these fields two 17s atoms can collide to produce a 17p and a 16p atom by the process

$$\text{Na } 17s + \text{Na } 17s \rightarrow \text{Na } 17p + \text{Na } 18p. \qquad (28)$$

The experiments are done by passing a thermal Na beam between a pair of plates 1.27 cm apart where they are excited by two pulsed dye lasers from the ground 3s state to the 3p state and then to the 17s state. The atoms are allowed to collide in the presence of an electric tuning field for a time of 1 to 3 μs, after which they are exposed to a field ionization pulse which rises slowly so that atoms in the 17p state are ionized earlier than those in the 17s state, resulting in easily distinguished signals from atoms in the two states. Detecting the 17p signal, as the tuning voltage is swept over many shots of the laser leads to the collisional resonances shown in Figure 4. These resonances are labeled by the $|m_l|$ values of the upper and lower p states; for example, the resonance at 545 V/cm is due to the 17p $|m_l| = 0$ and 16p $|m_l| = 1$ states.

The fact that a large fraction of the initial 17s population is transferred to the 17p state demonstrates that the collision cross-sections are as large

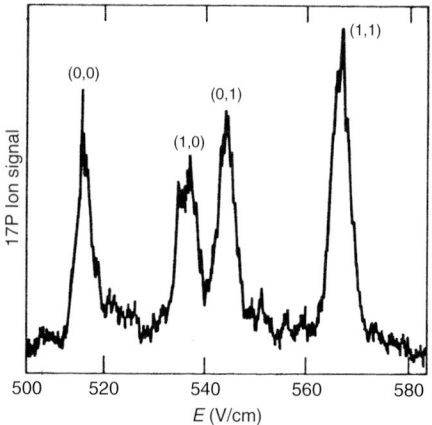

FIGURE 4 Observed resonances for the process Na 17s + Na17s → Na 17p + Na 16p. The resonances are labeled by the $|m_l|$ values of the (lower, upper) final p states, as shown in Figure 3 (from Ref. [21])

as stated above. We can express the collision rate Γ as

$$\Gamma = N\sigma v, \tag{29}$$

where N is the atomic number density and v the typical collision velocity. To observe a signal requires $\Gamma \sim 10^6$ s^{-1}. Using $N \approx 10^8$ cm^{-3} and $v = 10^5$ cm/s leads to $\sigma \sim 10^{-7}$ cm^2.

Since both the 17p and the 17s states can be detected with the same field ionization pulse, it is straightforward to measure the yield of 17p atoms as a function of the initial 17s population. By repeating similar measurements for $16 \leq n \leq 27$ it is possible to obtain the n dependence of the Na $ns + ns \longrightarrow np + (n-1)p$ cross-sections. As shown in Figure 5, they are best fit by the expression

$$\sigma = 3.3(6) \times 10^4 n^{3.7(5)} \text{ Å}^2, \tag{30}$$

in reasonably good agreement with both the predicted n^4 dependence and the magnitude suggested by the simple model of the previous section (Equation (24)). Considering the possible uncertainty of up to a factor of five in determining the number of Na atoms, the agreement is good.

Although the cross-sections are enormous compared to other atom–atom cross-sections, the widths of the collisional resonances are perhaps even more interesting. Inspection of Figure 4 reveals that the (0, 0) resonance is the narrowest, which is not surprising since in all the other resonances the spin–orbit interaction of the p states splits

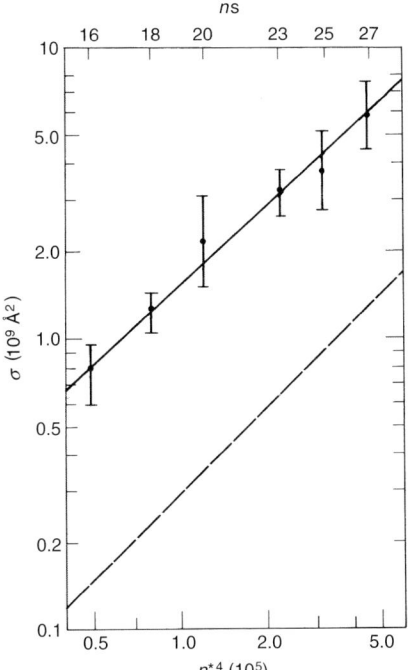

FIGURE 5 Observed cross-sections $vs\ n^{*4}$ for the process $Na\,ns + Na\,ns \rightarrow Na\,np + Na\,(n-1)p$ (•), (—) fit curve; (- - -) the curve calculated from $\sigma = 2.90 \times 10^3 n^{*4}$ Å^2, a slight variant of Equation (24) given in Ref. [21] (from Ref. [21])

the p $|m_l|$ state into two states; that is, each of the resonances other than the (0, 0) resonance contains at least two unresolved resonances. In the initial experiments the spin–orbit splittings were not resolved, in large part because the Na atoms bounced off the walls of the vacuum chamber, effectively converting the beam into a vapor cell. In subsequent, more controlled, experiments the spin–orbit splittings were resolved. Experiments with a magnetic field of 30 Gauss parallel to the electric field showed that in cases in which the spin and orbital angular momenta were uncoupled (the Na example we have been considering is such a case), the electron spins of both atoms either remained unchanged or both are flipped [38]. Never is a single spin flipped. Since it is unlikely that both spins are flipped, ignoring the electron spin is reasonable.

Due to the absence of spin–orbit splitting the (0, 0) resonance provides the most useful information about the widths, and in Figure 6 we show the n dependence of the width of the (0, 0) resonances. The field widths of the resonances, shown in Figure 4, are converted to frequency widths, using the known Na Stark shifts with the result shown in Figure 6, in which

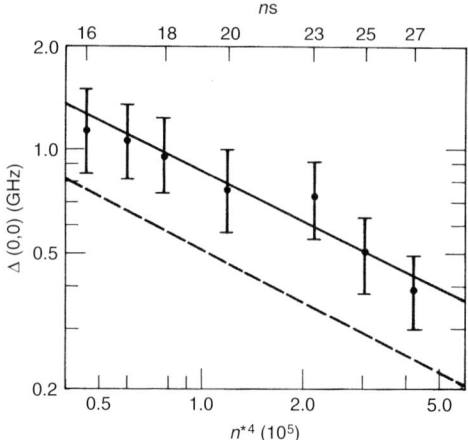

FIGURE 6 Frequency widths of the Nans + Nans → Nanp + Na$(n-1)p(0,0)$ resonances vs n^{*4}, (—) fit curve; (- - -) the curve calculated from $\Delta = 1.74 \times 10^2 n^{*-2}$ GHz, a slight variant of Equation (27) given in Ref. [21] (from Ref. [21])

the observed values are compared to the predictions of the simple model [Equation (27)] of the previous section. The fit to the experimental points gives the width (FWHM) as

$$\Delta\nu = 235(40) n^{-1.95(20)} \text{ GHz}, \tag{31}$$

which is in excellent agreement with the expected magnitude and n dependence. The frequency widths of the collisional resonances imply collisions of durations of at least 1ns, three orders of magnitude longer than the typical 300 K collision times of 1ps.

One of the more interesting aspects of resonant collisions is the velocity dependence of the cross-section and resonance width. The decrease in the width was demonstrated using resonant collisions in K, specifically the process

$$K(n+2)s + Knd \longrightarrow K(n+2)p + K(n+1)p \tag{32}$$

which is resonant at lower fields than those used in the Na experiments, as shown by the energy level diagram of Figure 7 [39]. To verify the velocity dependence of the widths of the collisional resonances a velocity selected beam of K atoms was used in which the beam passed through a chopper wheel which passed pulses of atoms having durations as short as 25 μs. The atoms were excited 20 cm away by a 5 ns laser pulse after a time delay of 250 μs. By varying the pulse length of atoms passed by the chopper

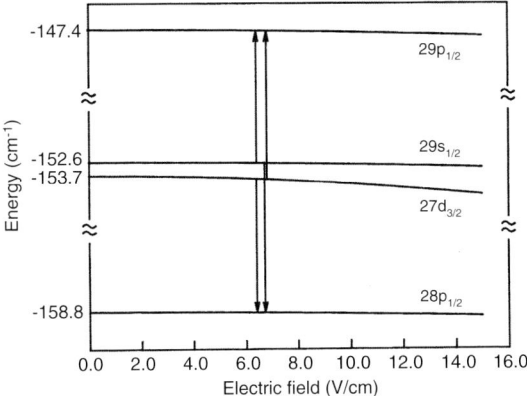

FIGURE 7 Potassium energy levels for the resonant collision K 29s + K 27d → K 29p + K 28p (from Ref. [41])

wheel, one changes the width of the velocity distribution. Aside from the velocity selection of the atoms and the fact that two Rydberg states of K were excited, the experiments were conducted in the same manner as the Na experiments described earlier. The results of these experiments are shown in Figure 8. When the K beam is used with no velocity selection and no liquid nitrogen in the cold trap the K atoms bounce from the walls, and the beam fills the entire vacuum chamber so that it becomes a cell of K atoms moving in all directions. Under these conditions the resonance of Equation (31) is 240 MHz wide, as shown in the upper frame of Figure 8. Adding liquid nitrogen to the cold trap removes the cell atoms so that all the atoms are moving in the same direction, which reduces the width of the resonance to 57 MHz. Finally, the lowest frame of Figure 8 shows the resonance obtained with a velocity selected beam, which should correspond to a temperature of 5 K. The 6 MHz width of the resonance is not quite as narrow as expected for a beam with this velocity distribution. As it turns out, the excess width is due to the magnetic splitting due to the earth's field. With the earth's field eliminated the resonances become as narrow as 1 MHz, or stated another way, the duration of the collisions has now reached 1 µs, a time comparable to the duration of the experiment [40].

In all of the experimental work discussed thus far the angular dependence of the dipole–dipole interaction has been ignored. As pointed out by Equations (17) and (18), the orientation of the dipoles and the velocity vector of the colliding atoms are important. For, example, as shown by Equation (17), for dipoles oscillating parallel to the velocity vector the collision cross-section vanishes at resonance. It is possible to observe the dependence of the collisional resonances on the angle between

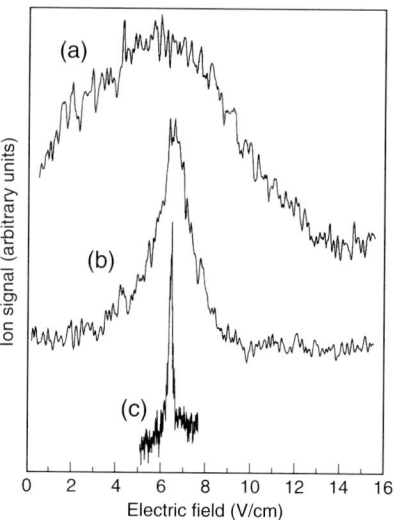

FIGURE 8 Potassium 29p signals observed in the resonant collision K 29s + K 27d → K 29p + K 28p with (a) a cell of K atoms, giving a resonance width of 240 MHz, (b) a beam of atoms, giving a width of 57 MHz, and (c) a velocity selected beam of atoms, giving a width of 5 MHz (from Ref. [39])

the applied field and the relative motion of the atoms [41]. If we consider the Na $ns + ns$ collisions shown in Figure 1 there are two natural choices of orientations, $\vec{v}_c \parallel \vec{E}$ and $\vec{v}_c \perp \vec{E}$, where $\vec{v}_c$ is the relative velocity of the colliding atoms and $\vec{E}$ is the tuning electric field. We consider the (0, 0) collisional resonance, in which the dipoles oscillate in the field direction. It is straightforward to reproduce the conditions in which Equations (17) and (18) apply by the choice of experimental geometry. Explicitly, $\vec{v}_c \parallel \vec{E}$ matches Equation (17), and $\vec{v}_c \perp \vec{E}$ matches Equation (18). In the case of $\vec{v}_c \parallel \vec{E}$ the interaction changes sign twice in the course of the collision, and, at resonance, the dipole–dipole interaction vanishes when integrated along the trajectory of the moving atom. Consequently, on resonance the cross-section vanishes, and off resonance, where the cancellation is not complete, the partial cancellation results in the requirements that the atoms come closer together than in the $\vec{v}_c \perp \vec{E}$ case for energy transfer to occur. In the case of $\vec{v}_c$ perpendicular to $\vec{E}$ the interaction does not generally vanish when integrated along the trajectory, so there is no cancellation at resonance, which allows energy transfer to occur at larger impact parameters than in the $\vec{v}_c$ parallel to $\vec{E}$ case.

The experimental test of the angular dependence is based on the fact that if the atoms are in a well-collimated beam the relative atomic velocity $\vec{v}_c$ is in the direction of the atomic beam. Accordingly, the experiments

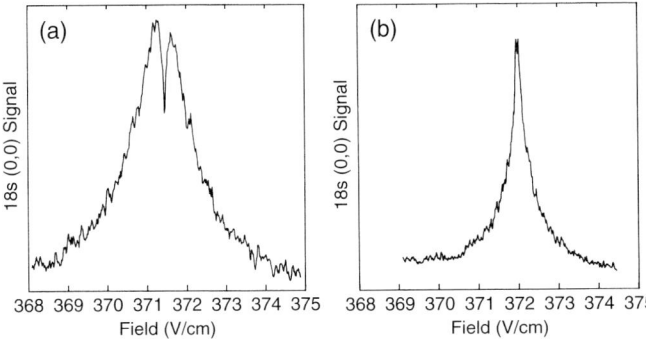

FIGURE 9 Measurements of the Na 18s + Na 18s ⟶ Na 18p + Na 17p signal, in arbitrary units, for (a) $\vec{v}_c \parallel \vec{E}$ and (b) $\vec{v}_c \perp \vec{E}$. The dip in the cross-section in (a) results from the fact that the interaction changes sign during the collision (see Figure 2), resulting in a cancellation of the interaction when integrated over the collision (from Ref. [39])

were conducted with a beam, not a cell. For $\vec{v}_c \perp \vec{E}$ the atomic beam passed midway between a pair of field plates, with the relative atomic velocity parallel to the plates. The case in which $\vec{v}_c$ is parallel to $\vec{E}$ is realized by passing the atomic beam through opposing holes in the two field plates, so that the beam travels perpendicular to the plates. The results of using these two configurations are shown in Figure 9 which displays the (0, 0) resonance for the process

$$\text{Na } 18s + \text{Na } 18s \longrightarrow \text{Na } 18p + \text{Na } 17p \qquad (33)$$

for both $\vec{v}_c \parallel \vec{E}$ and $\vec{v}_c \perp \vec{E}$ [41]. Both of the above-mentioned features are apparent. While the cross-section for $\vec{v}_c \parallel \vec{E}$ does not go to zero at resonance, presumably due to field inhomogeneities and the fact that the relative velocities are not perfectly aligned with the atomic beam direction, the sharp dip at resonance is readily apparent. Similarly, it is evident that the resonance is substantially broader for $\vec{v}_c \parallel \vec{E}$ than for $\vec{v}_c \perp \vec{E}$, a reflection of the substantial cancellation of the interaction when integrated along the trajectory, which means smaller impact parameters are required. Close inspection of Figure 9 reveals an asymmetry in the resonance line shape, which is due to the interaction of field induced permanent dipole moments of the atoms.

4. MANIPULATING RESONANT COLLISIONS

Resonant Rydberg–Rydberg collisions last for a long time compared to most atomic collisions, and, as a result, these collisions can be altered by external perturbations far more easily. An excellent example of how easily they are perturbed is a radiative collision, one in which the colliding atoms absorb or emit photons during the collision [42]. For example, the process

$$\text{Na } 17s + \text{Na } 18s + m\,h\nu \rightarrow \text{Na } 17p + \text{Na } 18p \tag{34}$$

where m is a positive or negative integer, is resonant at a slightly different field from the resonance of Equation (32) [22,23,25]. For the moment we consider the case in which the microwave frequency ω exceeds the line width of the resonant collision, or the inverse of the collision time τ, i.e. $\omega \gg 1/\tau$. We term this the high frequency regime.

The radiative collisions of Equation (33) were studied by passing a Na atomic beam through a microwave cavity made of a piece of WR 90 waveguide with closed ends. The atoms were excited by two dye laser beams counterpropagating to the atomic beam and tuned to the 3s–3p and 3p–ns transitions. The atomic beam and the laser beams pass above a septum in the cavity which allowed the application of a tuning field and a field ionization pulse. There was a 1 mm diameter hole in the center of the top of the cavity, at an antinode of the microwave field, and only the atoms beneath this hole were detected. The microwave field was present during laser excitation and turned off 3 µs after the laser excitation. Then a field ionization pulse was applied which allowed the signals from field ionization of the ns and np states to be separated. The np field ionization signal was recorded as the static tuning field was scanned over many shots of the laser. The radiative collisional resonances for the Na 18s state in several amplitudes of a 15.4 GHz field are shown in Figure 10.

In Figure 10 collisional resonances are visible in which up to three microwave photons are emitted during the collision. In spite of the fact that multiple photons are emitted, the microwave powers are low. For example, the highest field shown in Figure 10 corresponds to a microwave intensity of 340 W/cm^2, which is to be compared to the MW/cm^2 laser intensities required to observe radiative collisions with lower-lying states [24]. In Figure 10 it appears that the one-photon-assisted collision disappears at high microwave field, and the zero-photon-assisted collision disappears at 105 V/cm, only to reappear at 165 V/cm. The variation in the radiative collision resonances suggested by Figure 10 becomes obvious when the resonance signals for each of the m-photon-assisted resonances are plotted vs microwave field in Figure 11. For comparison, Figure 11 also shows the result of a model calculation of the resonant collision signals, or cross-sections, for the $m = 0$ to 3 resonances [25].

In the model the effect of the strong microwave field is taken into account by a Floquet approach. If we consider first the case in which there is no microwave field, the two states of the problem are the levels

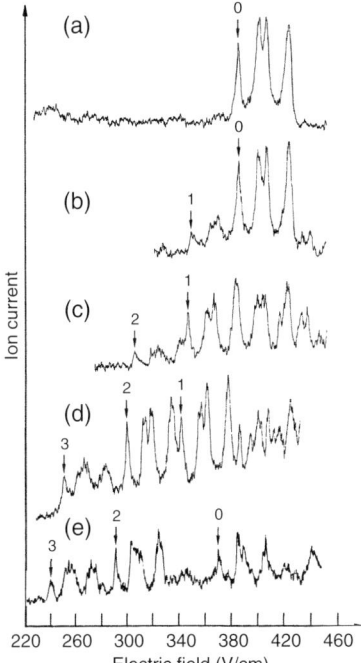

FIGURE 10 Observed signals, in arbitrary units, for Na 18s + Na 18s → Na 18p + Na 17p collisions assisted by zero to three 15.4 GHz microwave photons. Above the (0, 0) resonances are indicated how many photons have been emitted. In each trace there are obvious sets of four resonances corresponding to those of Figure 4. Microwave fields (a) 0 V/cm, (b) 13.5 V/cm, (c) 50 V/cm, (d) 105 V/cm, and (e) 165 V/cm (from Ref. [25])

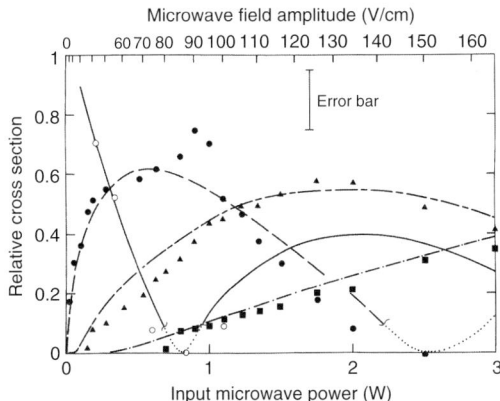

FIGURE 11 Relative cross-sections for collisions assisted by zero photons (○ and solid line), One photon (● and dashed line), two photons (▲ and long-short-dashed line), three photons (■, dot-dashed line). The lines are calculated from Eq. (36) (from Ref. [25])

defined earlier in Equations (15) and (16), $|ss\rangle$ and $|pp'\rangle$, direct products of pairs of atomic states. In isolated atoms only the np states have non-negligible Stark shifts, so in the molecular picture all the Stark tuning is due to the pp' state. At infinite internuclear separation, $R = \infty$; the ss and pp' levels cross at the resonance field. At any finite separation, R_0, there is an avoided crossing, $\Omega_0 = 2\mu\mu'/R_0^3$ due to the dipole–dipole coupling, and the eigenstates are $\psi_\pm = (\psi_{ss} \pm \psi_{pp'})/\sqrt{2}$.

Using the molecular picture we can derive the cross-section at resonance in the following way, first with no microwave field. At the beginning of the collision the pair of atoms is in the ss state, which is degenerate with the pp' state since $R = \infty$. The ss state is the superposition $(\psi_+ + \psi_-)/\sqrt{2}$ at the beginning of the collision, and both the ψ_+ and ψ_- energy levels are followed during the collision. During the collision R decreases to the impact parameter b and then increases to infinity. As R varies the energy between the ψ_+ and ψ_- states increases to $2\mu\mu'/b^3$, and then decreases to zero. If the energy, or frequency, separation between the two curves integrated over the collision is π, there is a phase shift of π between the ψ_+ and ψ_- states, with the result that the final state at the end of the collision is $(\psi_+ - \psi_-)/\sqrt{2}$, or pp'. Using the approximation used earlier in Equation (11) the phase integral is given by $(\mu\mu'/b^3)(b/v)$, and the cross-section is given by $\sigma_0 = \mu\mu'/v$. More important for our present purposes, the cross-section is proportional to $\Omega_0 = 2\mu\mu'/R_0^3$, the size of the avoided crossing of the ss and pp' levels at an arbitrary internuclear separation R_0.

We consider now the effect of adding a microwave field $E \sin\omega t$ polarized in the same direction as the static field. This field has no effect on the ss state, but since the differential Stark effect of the pp' state is linear, the effect of the added microwave field is to modulate the energy of the pp' state. Just as modulating the frequency of a laser beam produces sidebands, modulating the energy of the pp' state produces sidebands. To represent the sidebands we replace the original pp' wave function with one given by a Bessel function expansion. Explicitly,

$$\psi_{pp'} \to \psi_{pp'} \sum_m J_m\left(\frac{kE}{\omega}\right) e^{-im\omega t}, \tag{35}$$

where $k = dW_{pp'}/dE$ is the permanent dipole moment of the pp' state. Now there are avoided crossings of the ss state with all the sidebands of the pp' state. Since we have made the replacement of Equation (34), at $R = R_0$ the mth sideband of the pp' state has an avoided crossing with the ss state of magnitude

$$\Omega(E)_m = \Omega_0 \left| J_m\left(\frac{kE}{\omega}\right) \right|. \tag{36}$$

Since the cross-section is proportional to the magnitude of the avoided crossing separation at $R = R_0$, we immediately can relate the cross-section for the m-photon-assisted collision in the microwave field of amplitude E, $\sigma(E)_m$, to the resonant collision cross-section with no microwave field, σ_0. Explicitly,

$$\sigma(E)_m = \sigma_0 \left| J_m\left(\frac{kE}{\omega}\right) \right|. \tag{37}$$

Evaluating Equation (36) leads to the lines in Figure 11, which are in good agreement with the experimental cross-sections. Similar experiments were done with the K system of Equation (31) using a velocity selected beam [43]. In spite of the fact that in the K case the frequency of the oscillating field was much smaller, $\approx$ 4 MHz, since the line widths of the collisional resonances were yet smaller, the radiative collisions were still in the high frequency regime in which the colliding atoms are exposed to multiple field cycles during a collision.

At the other extreme is the low frequency regime, $\omega \ll 1/\tau$, in which an oscillating field varies slowly compared to the time of a collision. We consider adding a slowly varying field $E_{rf} \cos(\omega t + \phi)$ which is parallel to the static field E_S. If the collisional resonance is observed at the field E_R in the absence of the rf field, at time $t = 0$ it is observed at the static field for which the total field equals E_R, i.e.

$$E_S = E_R - E_{rf} \cos\phi. \tag{38}$$

Unlike the high frequency regime, the phase ϕ now matters.

As is often true, the two limiting cases, $\omega \gg 1/\tau$ and $\omega \ll 1/\tau$, are easily understood, but it is not at all obvious that they can be connected in a simple way. The low frequency regime suggests that as $\omega \sim 1/\tau$ is approached from the low frequency regime, the phase of the field at which the collision occurs will be important. To control the phase, the transform limited collisions of K atoms in a velocity selected beam were studied [44], in particular, the system of Equation (31). In a transform limited collision the time during which the atoms are allowed to collide is shortened to the point that the collisional resonances exhibit transform broadening. In such a collision the beginning and end of a collision occur at known times, and if the collisions occur in the presence of an rf field phase locked to the collision time, it is possible to study the $\omega \sim 1/\tau$ regime. In the experiments the allowed collision time was 0.7 μs, the range -0.35 μs $< t < 0.35$ μs, and the collisions took place in the presence of the rf field $E_{rf} \cos(\omega t + \phi)$. In Figure 12 we show the population transfer to the K 29p state as a function of the static field with no rf field and with an rf field of amplitude 0.21 V/cm, frequency $\omega/2\pi = 0.75$ MHz, and phases

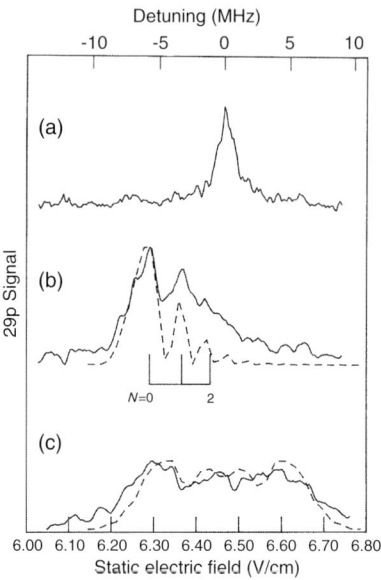

FIGURE 12 Observed signals, in arbitrary units, for K 29s + K 27d → K 29p + K 28p resonance observed in the presence of a 0.21 V/cm 0.75 MHz field $E\cos(\omega t + \phi)$. The allowed collision time of 0.7 μs is symmetric about $t = 0$. (a) $E = 0$; (b) $\phi = 0$; (c) $\phi = \pi/2$. The dotted lines are the result of numerical integration of the Schrödinger equation, and the stick figures are the resonances due to integral numbers of 2π phase shifts (from Ref. [44])

$\phi = 0$ and $\pi/2$. With no rf field the resonance is observed at a static field of 6.5 V/cm. The $\phi = 0$ trace has its main peak at 6.3 V/cm and subsidiary peaks at higher fields. The $\phi = \pi$ trace is a mirror image of the $\phi = 0$ trace, reflected through $E_S = 6.5$ V/cm. The $\phi = \pi/2$ trace, on the other hand, exhibits a broad, almost rectangular peak, which looks nothing like the $\phi = 0$ and $\phi = \pi$ traces.

The broken lines of Figure 12 were calculated by numerical integration of the Schrodinger equation, and they are reasonable representations of the experimental data. We can develop a physical picture of why different phases lead to such different results by examining Figure 13, which shows the energy levels, W_i and W_f, of the initial, 29s + 27p, and final, 29p + 28p, states as functions of time for $\phi = 0$ and $\phi = \pi/2$. For the $\phi = 0$ case of Figure 13(a), when the static field is less than 6.3 V/cm, the two levels never come into resonance. At 6.3 V/cm the levels come into resonance once, leading to the large peak in Figure 12(b). As the static field is raised above 6.3 V/cm the levels come into resonance twice, on the rising and falling edges of the rf field cycle. There are thus two avoided level crossings, and at the first one the amplitude is split and follows two coherent paths. Depending on the difference in the phase accumulation along

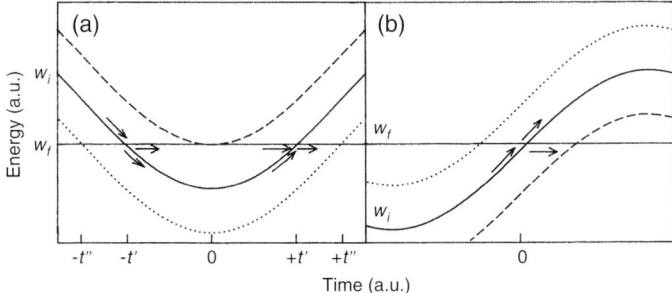

FIGURE 13 The initial- and final-state energies, W_i and W_f in a 0.75 MHz rf field with phases (a) $\phi = 0$ and (b) $\phi = \pi/2$. In both cases, the allowed collision time is 0.7 µs. As the static electric field is scanned, the single interaction period at $t = 0$, when $E_S = E_R - E_{rf}$ (dashed line), evolves into two interaction periods at $t = \pm t'$ (solid line) and $\pm t''$ (dotted line) when $E_S > E_R - E_{rf}$, as shown in (a). The transition amplitudes interfere constructively when the total phase shift of Φ between $-t'$ and t' is $2\pi N$. In (b) only one interaction period occurs irrespective of E_S (from Ref. [49])

the two paths, when the two amplitudes are recombined at the second avoided crossing there is an increase or decrease in the transition probability to the final state. The subsidiary peaks at 6.4 and 6.5 V/cm are cases in which the phase accumulations between the two avoided crossings are 2π and 4π. Related phenomena are Stuckelburg oscillations observed in differential scattering, multi-photon transitions power shifted back and forth through resonance by strong radiation pulses, and cold molecular states passed back and forth across a Feshbach resonance [45–48]. Unlike the $\phi = 0$ trace of Figure 12(b), the $\phi = \pi/2$ trace of Figure 12(c) is essentially structureless, and the reason is evident in Figure 13(b). The levels come into resonance only once, so there is no possibility of interference.

At this point, we have a description of the $\omega \sim 1/\tau$ regime in the terms used to describe the low frequency regime, but it is not apparent how this description passes to the Bessel function description of the high frequency regime, given by Equation (36). The connection can be made rigorously, but it is useful to consider a pictorial approach, which is similar to Figure 13 [49]. In the high frequency regime there are many cycles of the field during the collision, and in Figure 14 we show the time dependence of the W_i and W_f energy levels over two cycles of the microwave field. As we have already discussed, the transition amplitudes are generated at the avoided crossings. For constructive addition of two transition amplitudes one cycle apart, such as A and B in Figure 14, the accumulated phase difference between the two curves must be a multiple of 2π, a requirement which we write as

$$\Phi_- - \Phi_+ = \int_{t_0}^{t_0+2\pi/\omega} (W_f - W_i) dt = 2\pi m \qquad (39)$$

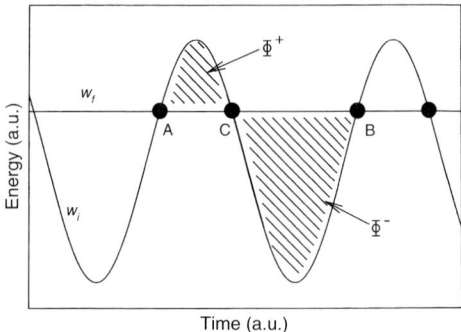

FIGURE 14 The initial- and final-state energies, W_i and W_f in a high frequency rf field. Only two cycles of the rf field are shown. Each time $W_i = W_f$ a collisional interaction occurs and the relative amplitudes of ψ_i and ψ_f change. If $\Phi_- - \Phi_+ = 2\pi N$, then the transition amplitudes over successive cycles add constructively and a resonance in the cross-section is observed. For the maxima in the cross-section Φ_- and Φ_+ must separately be equal to integral multiples of 2π (from Ref. [49])

where m is an integer. Since W_i has a sinusoidal time dependence and we are integrating over one cycle, Equation (38) reduces to

$$W_f - W_i = m\omega \tag{40}$$

which is simply the resonance condition for the m-photon-assisted collision. Meeting the requirement of Equation (38), or equivalently Equation (39), ensures that the transition amplitudes from points A and B add constructively, but it tells us nothing about how the amplitude from point C adds. To have the amplitudes of points A, B, and C, all add constructively, we impose the additional requirement that Φ_+, and therefore Φ_- as well, be integral multiples of 2π. Why this requirement leads to the Bessel function dependence in the high frequency regime can be appreciated in the following way. Imagine that the static field is tuned to resonance for the three-photon-assisted resonance, $m = 3$, which occurs at a static field removed from the resonance field by E_Δ. When the microwave field amplitude is increased from zero to E_Δ the transition probability monotonically increases, reaching a maximum when the field amplitude equals E_Δ, at which point the levels come into resonance once on each field cycle. At this point $\Phi_+ = 0$ and $\Phi_- = 6\pi$. As the field is further increased both Φ_+ and Φ_- increase. When $\Phi_+ = \pi$ and $\Phi_- = 7\pi$, the amplitudes from successive avoided crossings exactly cancel, and the cross-section vanishes. When the field is further increased, so that $\Phi_+ = 2\pi$ and $\Phi_- = 8\pi$ the amplitudes all add constructively again. Extending this reasoning to higher microwave fields we can see how the oscillatory Bessel function behavior of the radiatively-assisted collision cross-sections comes about.

5. DIPOLE–DIPOLE INTERACTIONS IN THE FROZEN RYDBERG GAS

As the temperature of a sample of Rydberg atoms is lowered from 300 K to 1 K the velocity of the colliding atoms decreases by a factor of seventeen. Correspondingly, the cross-section for resonant energy transfer increases by the same factor, and the duration of the collision, the inverse of the resonance line width, increases by a factor of almost seventy. Since it is possible to produce samples of atoms at temperatures below 1 mK, one might expect to observe even larger changes as the temperature is lowered from 1 K to 1 mK. In fact, something more interesting occurs; we obtain a frozen Rydberg gas, a system which behaves in many ways more like a solid than a gas. Specifically, on the time scale of interest, the atoms do not move, and many-body interactions among the static atoms become important. In contrast, at 300 K only the binary encounters of collisions are important.

Resonant energy transfer in the frozen Rydberg gas analogous to collisional resonant energy transfer at higher temperatures provides a clear illustration of the qualitative difference between 300 µK and 300 K. The initial experiments were done with both Rb and Cs, and we consider the Cs case as a concrete example [18,19,50,51]. In particular, we consider the process

$$\text{Cs } 24p_{3/2} + \text{Cs } 24p_{3/2} \rightarrow \text{Cs } 24s + \text{Cs } 24s, \tag{41}$$

which is resonant at 80 V/cm for the $|m| = 1/2$ levels of the 25p state, as shown in Figure 15. In the experiment Cs atoms at a temperature of 150 µK were held in a MOT at densities of up to 10^{10} cm^{-3}. Using a 7 ns long 515 nm dye laser pulse, one excites the atoms from the 6p state to the 24p state in a static field of ~80 V/cm. The electric field admixes 23d character into the 24p state, making the excitation possible. The atoms are allowed to interact with each other for 3 µs, after which a field ionization pulse is applied. During the time between the laser pulse and the field ionization pulse some of the atoms are resonantly transferred to the 24s and 25s states by the dipole–dipole interaction. The field pulse rises slowly enough so that the field ionization of atoms in the more weakly bound 25s state is easily separated from field ionization of the 24p atoms. The 25s signal is recorded as the static field is slowly swept through the resonance over many shots of the laser.

Typical recordings of the 25s signal after a 3 µs interaction time for several densities are shown in Figure 16. Inspection of Figure 16 reveals

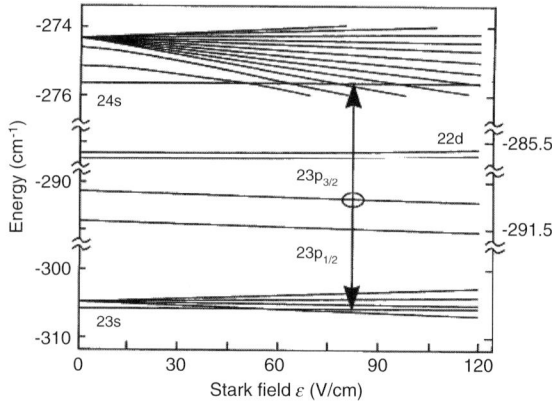

FIGURE 15 Stark diagram of a Cs atom in the vicinity of the 23p state. The arrows indicate the energy resonance for the internal energy exchange Cs 23p$_{3/2}$ + Cs 23p$_{3/2}$ → Cs 23p$_{3/2}$ + Cs 23p$_{3/2}$ (from Ref. [19])

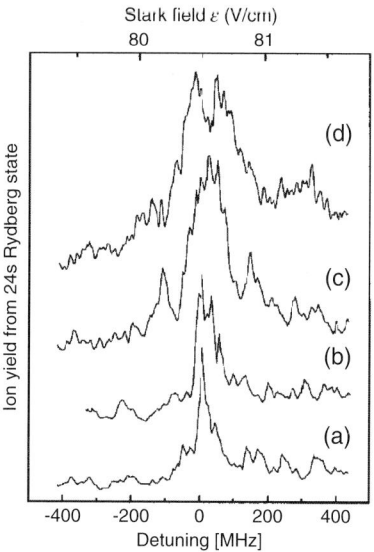

FIGURE 16 Observed signals, in arbitrary units, for the energy transfer resonance Cs 23p$_{3/2}$ + Cs 23p$_{3/2}$ → Cs 23p$_{3/2}$ + Cs 23p$_{3/2}$ for 23p$_{3/2}$, $|m| = 1/2$. Estimated densities: (a) 4, (b) 7, (c) 40, (d) 100 × 10^8 cm^{-3} (from Ref. [19])

that the width increases with density and is approximately 80 MHz wide at a density of 10^9 cm^{-3}. The contributions of inhomogeneous E and B fields to the widths of the resonances shown in Figure 16 are estimated to be 20 MHz. The maximum fractional population transfer from the 24p state is approximately 15%.

Although the resonance shown in Figure 16 appears qualitatively similar to those seen at 300 K, if we carefully consider the conditions of the experiment, it is evident that the observed resonances are not the effect of transient binary encounters between atoms. In 3 μs, atoms at a density of 10^9 cm^{-3} move only 7% of their average spacing, i.e. they are effectively frozen in place so that collisions simply do not occur. Furthermore, the characteristics of the observed resonances of Figure 16 do not match those of binary collisions of atoms at a temperature of 150 μK. At this temperature the cross-section for a binary resonant energy transfer collision, from Equation (11), is 10^{-4} cm^2, implying an impact parameter of 60 μm. At a density of 10^9 cm^{-3} there are 1000 other atoms closer to the target atom than the projectile atom at its point of closest approach. Using Equation (14) we estimate the width of the binary collision resonance to be 2 kHz. Finally, in binary resonant energy transfer collisions the rate increases with density, but the resonance only broadens when the energy transfer process begins to saturate the number of Rydberg atoms. Since the maximum population transfer in this experiment is only 15%, the process is not close to saturation.

To a first approximation, the Rydberg gas is frozen, and it is a reasonable approximation to assume the atoms to be fixed in place. With the laser we excite many 24p atoms, but due to the interatomic interactions, a collection of 24p atoms is not an eigenstate of the system, but a superposition which evolves in time. If there were only two atoms we would expect to see quantum beats, but in a random many-atom system such simple oscillatory behavior is less likely. The obvious interatomic interaction is the dipole–dipole interaction of Equation (40) between an atom and its nearest neighbor. Evaluating this interaction for the process of Equation (40) gives a resonance width proportional to the density, but at a density of 10^9 cm^{-3} it is ∼1.5 MHz, two orders of magnitude lower than observed. If we assume that atoms which are farther away are uniformly distributed, their interaction integrates to zero. To explain the much larger widths, other always resonant interactions were introduced. Consider a sample composed initially of 24p atoms. After a few of them undergo the transfer process of Equation (40), the resulting 24s and 25s atoms find themselves in a sea of 24p atoms, and they interact by the processes

$$\text{Cs 25s} + \text{Cs 24p}_{3/2} \rightarrow \text{Cs 24p}_{3/2} + \text{Cs 25s} \tag{42}$$

and

$$\text{Cs 24s} + \text{Cs 24p}_{3/2} \rightarrow \text{Cs 24p}_{3/2} + \text{Cs 24s}. \tag{43}$$

These final-state interactions broaden the final state into a band of levels, leading to increased widths of the resonances. In the descriptions of the resonant transfer which have appeared to date only the static dipole–dipole interaction has been considered. The retarded interaction of Equation (9) between the dipoles was disregarded because it differs from the static interaction by a factor of $(\lambda_0/2\pi R)^2$. Since $\lambda_0/2\pi$ is much larger than the average spacing of the atoms this term appears negligible. However, it has an angular dependence such that the interaction does not vanish when integrated over the sample. Furthermore, it has a $1/R$ dependence, so atoms which are farther apart are relatively more important than they are for the static dipole–dipole interaction. Even if it is only integrated to $R = \lambda/2\pi$, its effect is equivalent to that of the static interaction. An interesting feature of this interaction is that it is a true collective, many-atom interaction rather than many binary interactions.

Later experiments were carried out to probe different aspects of the picture of the resonant transfer given above. Carroll et al. verified that the interactions display the expected angular dependence for a dipole–dipole interaction [52]. Specifically, they used an essentially one-dimensional cylindrical sample of atoms, the axis of which could be rotated with respect to the tuning electric field, which defines the quantization axis. The system they studied was

$$\text{Rb } 32d + \text{Rb } 32d \rightarrow \text{Rb } 34p + \text{Rb } 30k \qquad (44)$$

where $30k$ is a Stark state of $n = 30$. The energy transfer of Eq. (43) is resonant for different $30k$ Stark states at fields from 0 to 4 V/cm. Using a one-dimensional sample minimizes the many-body effects, so the analogs of Equations (41), (42) are not important. They focused on two resonances, at 2.75 and 3.75 V/cm, and they observed this pair of resonances for different angles between the cylinder axis and the field, with the result shown in Figure 17. Both of the observed resonances have several m components, and when they are all taken into account the predicted dependence is shown by the calculated line, which has a minimum at 45°, in reasonable accord with the measurements, and not far from the magic angle of 54° naively expected for a dipole–dipole resonance. This experiment is roughly analogous to the collision experiment in which the atoms moved perpendicular to or parallel to the tuning field [41].

The many-body aspect of the initial explanation has been probed by several experiments. An attractive approach is to start with a one-dimensional cylindrical sample, in which each atom interacts essentially only with its neighbor on either side, and increase the radius of the cylinder to introduce many-body effects. By exciting Rb atoms in a MOT at densities of up to approximately 6×10^8 cm^{-3} with a blue dye laser beam focused to radii from 6.5 to 20 μm. Carroll et al. were able to

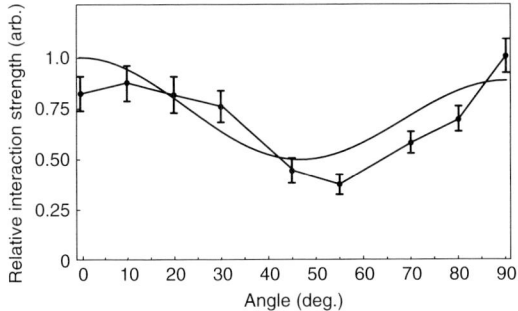

FIGURE 17 Interaction strength, or population transfer, in arbitrary units, for the Process Rb32d + Rb32d ⟶ Rb34p + Rb30k vs angle between the static field, which orients the dipoles, and the sample axis. the smooth curve is the calculated angular dependence for the dipole–dipole interaction, and the experimental points have error bars (from Ref. [52])

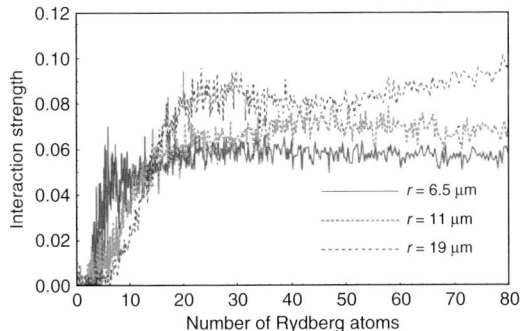

FIGURE 18 (Color online) Interaction strength, or population transfer, in arbitrary units, for the process Rb31d + Rb31d ⟶ Rb33p + Rb29k as a function of the number of atoms in sample tubes of three different radii. As the number of atoms is raised, the larger diameter tube exhibits a stronger interaction strength, in spite of the lower density of atoms, showing the importance of many-body effects (from Ref. [53])

change their sample from being one dimensional to approximately three dimensional [53]. At the highest density the spacing between the atoms was approximately 6.5 μm, so at any lower density the sample of 6.5 μm radius was one dimensional. They studied the system

$$\text{Rb}\,31d + \text{Rb}\,31d \rightarrow \text{Rb}\,33p + \text{Rb}\,29k, \tag{45}$$

which is resonant at 3 V/cm. They measured the fraction of atoms transferred to the 33p state in 3 μs after laser excitation for three different radii of the cylinder of Rydberg atoms, as shown in Figure 18. Note that the horizontal axis is the total number of Rydberg atoms, so the densities of the 6.5

and 19 μm radii samples are different by a factor of ten. For a small number of atoms the fraction of atoms transferred rises much more rapidly for the small radius sample than for the larger ones, which is not surprising; the density is higher, and the many-body effects are presumably less important at low densities. However, when the number of atoms exceeds 30 the fraction of atoms transferred to the 33p state is larger in the larger cylinders than in the 6.5 μm radius cylinder, in spite of the lower density. When the atoms have interactions not only along the cylinder axis but also perpendicular to it, the energy transfer increases, indicating the importance of many-body interactions. Raising the dimensionality also increases the likelihood of percolation, which may play a role.

Using the same system originally examined by Anderson et al. [18], Mourachko et al. have examined the effect of the always resonant interactions [55]. To be specific, they examined the system

$$Rb\,25s + Rb\,33s \rightarrow Rb\,24p + Rb\,34p, \qquad (46)$$

which is resonant at fields of 3.1 and 3.4 V/cm. This system possesses the always resonant processes

$$Rb\,25s + Rb\,24p \rightarrow Rb\,24p + Rb\,25s \qquad (47)$$

and

$$Rb\,33s + Rb\,34p \rightarrow Rb\,34p + Rb\,33s, \qquad (48)$$

which broaden the observed resonances. Their approach was to add another resonant interaction which had nothing to do with the tuned resonance of Eq. (45). Specifically, they converted up to 20% of the 33s atoms to 34s atoms, introducing the always resonant interaction

$$Rb\,34s + Rb\,34p \rightarrow Rb\,34p + Rb\,34s. \qquad (49)$$

The dipole matrix elements of the 25s–24p, 33s–34p, and 25s–24p transitions are 490, 126, and 930 ea_0. In the initial experiment of Anderson et al. the 25s and 33s populations were equal, so the strengths of the interactions of Equations (45), (46) and (47) are in the ratio 4:1:16. Consequently, a substantial amount of the broadening of the resonances is due to the interaction of Equation (46). To ensure that the interaction of Equation (46) is not the dominant interaction Mourachko et al. used a 33s density of 5×10^8 cm^{-3} and a 25s density 40 times lower. With only 25s and 33s atoms present, when the 34p state population was observed by scanning the tuning field across the resonances, the energy transfer

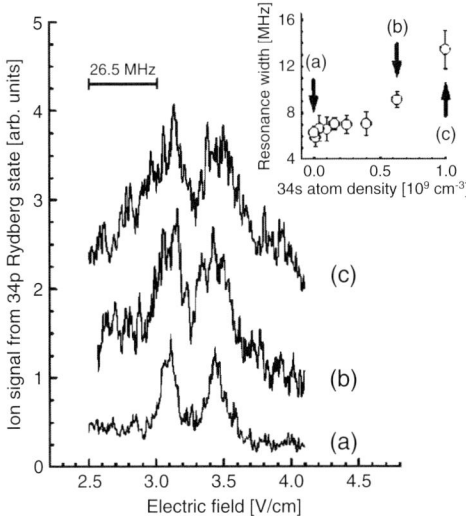

FIGURE 19 Energy transfer resonances for the process Rb25s + Rb33s ⟶ Rb24p + Rb34p Observed with (a) no 34s atoms, (b) 34s atoms at a density of 6.3×10^8 cm^{-3} (c) 34s atoms at a density of 10^9 cm^{-3}. The traces are offset vertically for clarity. The resonances become both broader and higher as 34s atoms are added, in spite of the decreasing number of 33s atoms. The inset shows the width of the observed resonances vs 34s density (from Ref. [55])

resonances were 4 to 5 MHz wide, as shown by Figure 19(a), the width presumably reflecting the B field inhomogeneity of the trap. As shown by Figure 19, converting some of the 33s atoms to 34s atoms makes the resonances both broader and stronger, in spite of the fact that the number of 33s atoms decreased. As shown by the inset of Figure 19, the width begins to increase linearly with the density of 34s atoms once the strength of the interaction of Equation (48) exceeds 5 MHz. While the increase in the width is expected if many-body interactions are present, the increase in the magnitude of the peak signal may be surprising. This is attributed to the fact that the interaction of a 34p atom with a 34s atom coherently splits the final state of the energy transfer by much more than the strength of the interaction of Eq. (45). In contrast, the interaction of a 34p atom with a 33s atom leads to negligible splitting. As a result, when integrated over the resonance the 34p signal increases when 34s atoms are present. Since the resonances at different sites are unresolved, the peak signal increases as well. The increase in the integrated signal occurs for the same reason that the integrated radiative collision resonances contain more signal than the resonant collision resonances on which they are based [49]. Irrespective of the details of the explanation, it is evident that the always resonant interactions broaden the energy transfer resonances.

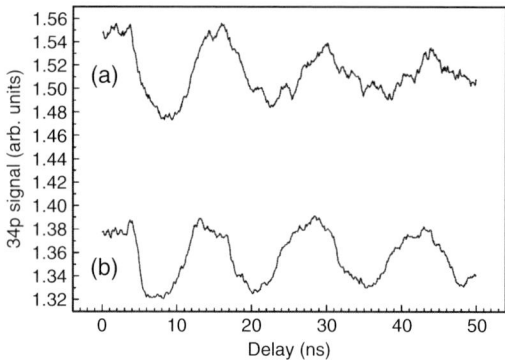

FIGURE 20 Ramsey fringes observed for the process Rb25s + Rb33s ⟶ Rb24p + Rb34p at two densities, (a) 7.6×10^9 cm^{-3}, (b) 1.9×10^9 cm^{-3} (from Ref. [56])

A final demonstration showing that the always resonant interactions are present is a Ramsey interference experiment using the Rb system of Eqs. (45)–(47) [56]. The mixture of cold Rb 25s and 33s atoms was exposed to a field of 4.5–8 V/cm, and it was brought into resonance at 3.4 V/cm with two 100 ns long electric field pulses separated by a variable time T. The population transfer to the 34p state was monitored as the time T was varied for different total densities of the atoms. The signals observed for densities of 7.6 and 1.9×10^9 cm^{-3} and an off resonant static field of 4.7 V/cm are shown in Figure 20. The frequency of the oscillations is the frequency separation of the 25s33s and 24p34p states in the off resonant field and does not depend on the density. If there were no interaction between the atoms, and they did not move, the oscillatory signal would not decay, but it is evident that at high density it does decay, and the decay rate increases with density. The decay rates for the two traces of Figure 20 are 33.9 and 11.4×10^9 s^{-1}, which are roughly comparable to the widths of the energy transfer resonances at these two densities, so it is reasonable to attribute the damping to the always resonant interactions of Equations (7) and (8), which are present even when the static field is detuned from the resonance at 3.4 V/cm.

In all the cold atom Förster resonant exchange measurements described thus far the two kinds of participating atoms have been intermingled. For example, in the work of Anderson et al. [18], the 25s and 33s atoms occupied the same spatial volume. Van Ditzhuijzen et al. have recently observed the process

$$\text{Rb}\,41d_{3/2} + \text{Rb}\,49s_{1/2} \longleftrightarrow \text{Rb}\,42p_{1/2} + \text{Rb}\,49p_{3/2}$$

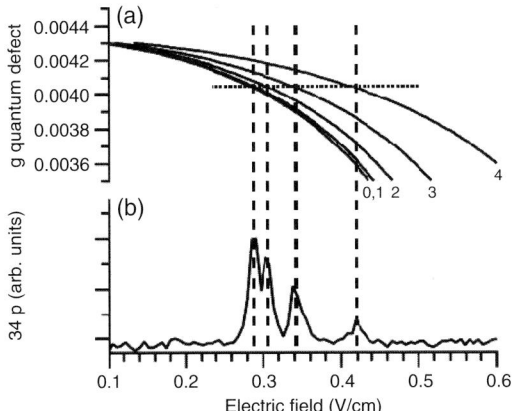

FIGURE 21 Measurement of the Rb ng quantum defect using the dipole–dipole energy transfer Rb 32d + Rb 32d ⟶ Rb 34p + Rb 30g. (a) Resonance fields for different values of m calculated as a function of the Rb ng quantum defect (b) observed resonances, corresponding to different values of m, as a function of the tuning field (from Ref. [33])

using spatially separated linear samples of $41d_{3/2}$ and $49s_{1/2}$ atoms, formed by two focused laser beams as much as 41 μm apart [57].

Due to the omnipresent interatomic interactions the frozen Rydberg gas does not appear to be a promising system for precision spectroscopy. However, at low densities, the width of an energy transfer resonance is in principle very narrow, and this feature has been used by Afrousheh et al. to obtain the best value to date of the Rb ng quantum defect, 0.004 09(6) at $n = 30$ [33]. They used the system

$$\text{Rb } 32d_{5/2} + \text{Rb } 32d_{5/2} \to \text{Rb } 34p_{3/2} + \text{Rb}30g \tag{50}$$

which is resonant at fields from 0.3 to 0.4 V/cm for different m values of the 30g state. This system is the same as that used by Carroll et al. [52], but the tuning field is so low that the lowest $n = 30$ Stark state is still predominantly 30g. To extract useful spectroscopic information they nulled stray electric fields perpendicular to the tuning field and turned off the trap gradient fields, reducing the magnetic field to 17 mG. The atoms were allowed to interact for 21 μs after which the population in the 34p state was detected by field ionization. They observed four resonances corresponding to the $m = 0$ and 1, 2, 3, and 4 levels of the 30g state, as shown in Figure 21. They were able to observe these resonances in spite of the fact that the dipole–dipole coupling of Equation (49) is small, occurring only due to the small admixture of 30f into the 30g state, leading to dipole matrix elements of less than 20 ea_0.

Most resonant energy transfer experiments have been done using static fields to tune the levels into resonance. While static field tuning is straightforward, the tuning is unidirectional. For example a Rb $(n + 2)\text{p}(n-2)f$ pair can only be tuned down in energy relative to an ndnd pair. Consequently, using static field tuning this resonant energy transfer can only be observed for $n < 42$. For $n > 42$ these pairs cannot be brought into resonance using a static field. However, using an off resonant microwave field it is a straightforward matter to shift levels in either direction by judicious choice of microwave frequency. Since the shift is proportional to the microwave power, tuning is accomplished by attenuation of the power. This approach has been used by Bohlouli-Zanjani et al., who used a 28.5 GHz microwave field to bring the otherwise inaccessible Rb 43d + Rb 43d → Rb 45p + Rb 41f transition into resonance [58]. It could be used, for example to study, for any n, the process Cs $n\text{p}_{3/2}$ + Cs $n\text{p}_{3/2}$ → Cs ns + Cs $(n+1)$s, which can only be tuned into resonance with a static field for $n < 42$.

6. LINE BROADENING AND BLOCKADES

The long-range interactions and long interaction times of the frozen Rydberg gas have stimulated great interest, due to the possibility of studying and controlling the coherent evolution of a small number of atoms [18,19,54]. Of particular interest are potential applications in quantum information processing [10–12].

The dipole–dipole interactions between cold Rydberg atoms inside a volume with a typical dimension, L, leads to the existence of a band of levels. Such a band of levels corresponds to a density-dependent broadening of the atomic levels. This broadening can be observed spectroscopically by microwave or high resolution laser spectroscopy [59, 60]. The line width of the probe source, and any power broadening, must be substantially less than any expected density-dependent broadening. One difficulty of these experiments is the elimination of spurious effects leading to a broadening of the lines, such as the presence of ions. The electric field created by one ion at a distance of 10 μm is 0.14 V/cm, which is not at all negligible in many experiments involving high n Rydberg states. For example, this field can lead to a Stark shift of 300 MHz for $n = 50$ atoms. Several mechanisms lead to the formation of ions. The first is blackbody radiation, with a typical photo-ionization rate of one thousand per second for n ranging from 20 to 40. For higher n, the ionization rate by blackbody radiation decreases and can be ignored. A more likely source of ions at high n is dipole–dipole induced cold Rydberg–Rydberg ionization which will be discussed in detail in the following section. It has been impossible to prevent completely the formation of ions, at least in the

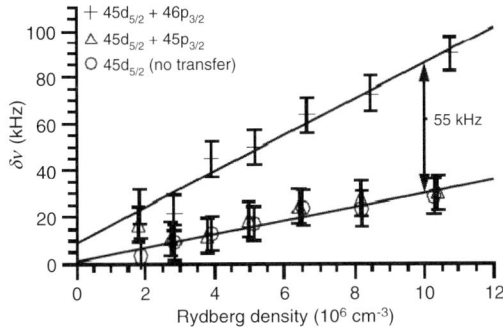

FIGURE 22 Broadening of the Rb $45d_{5/2} - 46d_{5/2}$ probe transition as a function of Rydberg atom density with and without microwave transfer pulses to populate neighboring p states with large dipole couplings, 46p (+), and small dipole couplings, 45p (△) (from Ref. [59]).

samples examined to date. Consequently, to limit the presence of ions to an acceptably low level, only low Rydberg atomic densities can be used.

Afrousheh et al. observed dipole–dipole line broadening using microwave spectroscopy. The $45d_{5/2}$ Rydberg state of rubidium was excited then probed with the two-photon microwave transition $45d_{5/2} \rightarrow 46d_{5/2}$ [59]. The measured line widths are sensitive to the Rydberg atomic density, suggesting the influence of interatomic interactions. They could convert half of the $45d_{5/2}$ initially excited atoms into $46p_{3/2}$ ones using a microwave transfer pulse tuned to the $45d_{5/2} \rightarrow 46p_{3/2}$ transition. When $46p_{3/2}$ atoms were present the line width of the two-photon probe transition is broadened by a quantity corresponding to the characteristic dipole–dipole coupling at the average interatomic distance. Figure 22 shows the evolution of the broadening of the $45d_{5/2} \rightarrow 46d_{5/2}$ probe transition as a function of the average density with and without the transfer pulse. To verify that the broadening was not an artifact introduced by the transfer pulse, they used a transfer pulse tuned to the $45d_{5/2} \rightarrow 45p_{3/2}$ transition. It did not lead to any broadening because, while states $45d_{5/2}$ and $46p_{3/2}$ are strongly coupled, $45d_{5/2}$ and $45p_{3/2}$ are not. In contrast to the Förster processes of Equations (40)–(42) and (45)–(47), these results indicate that most of the physics of the dipole–dipole interactions in this Rydberg sample can be understood within the framework of two-body interactions between close Rydberg atoms. The problem is, however, different from the case of the Förster resonances where the very broad resonances cannot be explained simply within the framework of only two-body interactions [55]. In the Förster case the pairs of very close atoms play a particular role in the observation of broad resonances, but a complete model has to take into account both the Förster reaction and the migration of the products of the reaction by exchange of Rydberg excitation between

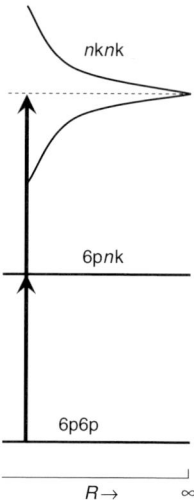

FIGURE 23 An illustration of the essential idea behind the dipole blockade in the excitation of a pair of Cs atoms from the 6p6p state to the $nknk$ State (both atoms in the Rydberg nk Stark state). The first photon, at the atomic 6p − nk frequency, can be absorbed, but the second cannot be since the dipole–dipole interaction of two nk atoms shifts the energy from the energy of two isolated atoms, shown by the broken line

the ensemble of the atoms. The problem of Förster resonance is therefore more complex. Finally, the back and forth transfer in the reaction Cs np + Cs np ⟷ Cs ns + Cs $(n-1)$s has been examined using narrow-band de-excitation laser spectroscopy by Mudrich et al. [60]. After excitation at the Förster resonance they could observe both np and $(n+1)$s field ionization signals. Using a delayed narrow-band cw laser they pumped atoms down to the 7s state, via their np parts, and they observed that the depletion of the $(n+1)$s field ionization signal occurred only with the laser tuned to the blue of the $R = \infty$ transition frequency, suggesting that only on the repulsive curve do pairs of atoms retain their coherence.

In the above experiments, the band of levels is excited using a broadband pulsed dye laser beam. Using a narrow bandwidth laser offers novel interesting possibilities. One exciting process is the possibility of using the dipole blockade of the excitation of a second Rydberg atom as a very efficient realization of a scalable quantum logic gate [11]. The principle of the dipole blockade is easily understood with the aid of Figure 23, which shows the excitation of a pair of atoms both initially in the Cs 6p state. One of the atoms can be excited by the absorption of a photon from a laser tuned to the Cs 6s-Rydberg transition, producing the singly excited pair shown in Figure 23. If both atoms are excited to the Rydberg state there exists a dipole–dipole interaction between them which shifts the levels out of resonance with the laser. Consequently, the excitation of the second

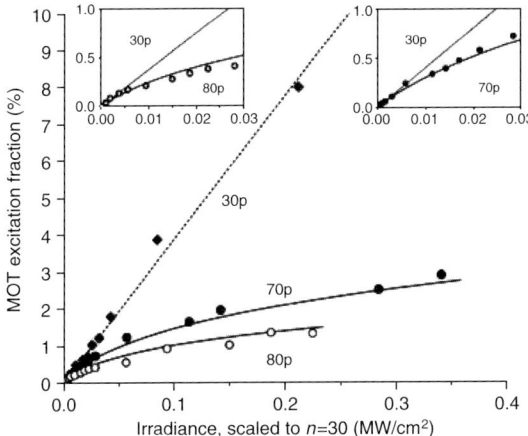

FIGURE 24 Van der Waals blockade observed in the excitation of the Rb 30p, 70p, and 80p states from the ground state. The laser irradiance is scaled by $1/n^3$ to account for the variation in oscillator strength. The excitation of the 70p and 80p states is suppressed below the linear dependence of the 30p state by the van der Waals blockade of the excitation (from Ref. [61])

atom is blocked, hence the term dipole blockade. To be effective as a gate, the dipole–dipole interaction must preclude the excitation of a second atom anywhere in the sample, which imposes the requirement

$$\frac{\mu^2}{L^3} > \Delta\omega_{\text{laser}}. \tag{51}$$

This requirement has not yet been met in any experiment, largely because the samples of atoms have been of large spatial extent. However, local, or partial blockades have been observed in which only one Rydberg atom within a blockade radius R_b is excited.

The first evidence for a blockade was the observation of the limitation of the laser excitation due to the van der Waals, or second order dipole–dipole, coupling between Rydberg atoms [61]. Specifically, a suppression of the excitation corresponding to a local blockade has been reported in excitation of high Rb np states ($n \sim 70$–80) using a pulse amplified single mode laser [61]. Figure 24 shows clearly the evidence of the blockade for $n = 70$ and $n = 80$ compared to $n = 30$. The limitation of the excitation is not a laser power saturation effect but is due to the van der Waals interactions between the atoms providing the blockade. Excitation with cw lasers has also been used to show the suppression of the excitation of high-lying nd states of Rb [62] and the excitation of Rb Stark states [63]. In the latter case a novel photon statistics approach was employed. A clear demonstration of the utility of narrow-band excitation

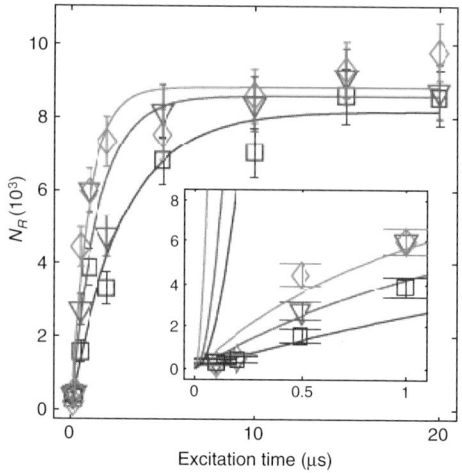

FIGURE 25 (Color online) Strongly blocked excitation to the Rydberg states. The number N_R of Rydberg atoms excited from the Rb ground state to the 43s state as a function of exposure time to the exciting lasers for high intensity (two-photon Rabi frequency 210 kHz) and three ground state densities; 3.2×10^{13} (◊), 66.6×10^{12} (∇), 2.8×10^{12} (□) cm^{-3}. The inset shows the comparison at short times to a collective excitation model (from Ref. [64])

was provided by Heidemann et al., who observed a very strong blockade in the two-photon excitation of the Rb 43s state from the ground 5s state [64]. With a two-photon Rabi frequency of 240 kHz they observed the saturation shown in Figure 25, in which they compare their observations to saturation curves for several densities of Rb. In addition, they compare their observations to a collective excitation model in the inset of the figure. One of the potential problems with cw excitation is that if the exciting laser light is present for a long time the appearance of ions due to cold collisions of atoms excited early in the excitation can produce Stark shifts which mimic the blockade later in the laser excitation. The duration of the exciting laser light should be short enough to avoid any formation of ions during the excitation. Another limitation of the blockade is laser power broadening. For a broad-band high-intensity excitation, the suppression of the excitation is no longer expected, since pairs of close atoms can be excited. It is interesting to note that for the same density of Rydberg atoms, the blockade will be more difficult to observe with low initial density of cold ground state atoms, simply because higher power will be required to produce the same number of Rydberg atoms.

The dipole blockade of laser excitation to a Rydberg state can be observed in many configurations in which there is a dipole–dipole coupling between cold atoms. A very simple one consists of creating in each Rydberg atom a significant permanent dipole by adding a small static

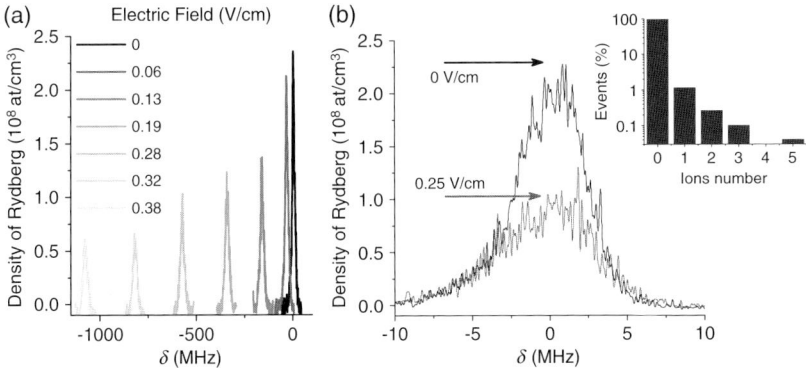

FIGURE 26 (Color online) (a) Excitation spectra of the Cs $75p_{3/2}$ level in small electric fields showing the Stark shift. (b) Excitation spectra for zero field and 0.25 V/cm showing the suppression of the excitation in the latter case due to the dipole blockade. The frequency scale is relative to the center of the unshifted and Stark shifted lines, respectively. The inset shows the probability of the presence of an ion during the excitation (from Ref. [65])

electric field. An electric field induced dipole blockade has been observed with the cesium atom in laser excitation of high-lying Stark states adiabatically connected to the np ($60 < n < 85$) states [65]. Figure 26 shows the efficiency of the dipole blockade in the case of the $75p_{3/2}$ excitation using a three-step process $6s_{1/2} \longrightarrow 6p_{3/2} \longrightarrow 7s_{1/2} \longrightarrow np_{3/2}$, with three cw laser beams applied as pulses of duration 300 ns at an 80 Hz repetition rate. Two single mode diode lasers, tuned to the wavelengths 852 nm and 1.47 µm respectively, are used for the transitions $6s_{1/2} \longrightarrow 6p_{3/2}$ and $6p_{3/2} \longrightarrow 7s_{1/2}$. A Ti:Sa laser (wavelength: 770–800 nm) is used to excite the atoms in the transition $7s_{1/2} \longrightarrow np_{3/2}$. The resolution, $\Delta\omega_L \sim 5$ MHz, of the laser excitation is mostly due to the lifetime of the $7s_{1/2}$ level. The excitation spectra of the $75p_{3/2}$ level are recorded for different static electric fields by scanning the frequency of the Ti:Sa laser. The origin of the strong decrease (up to 70 %) of the maximum intensity of the lines cannot be explained simply as a consequence of the decreasing p character of the Stark level with increasing field but is the evidence of the dipole blockade. The ensemble of the data can be understood by considering the interaction of each atom with its nearest neighbor. The onset of the local blockade occurs when the average interaction for a pair of close atoms is comparable to the resolution of the laser excitation

$$\mu^2 n_{Ry} \sim \hbar \Delta \omega_L, \tag{52}$$

where n_{Ry} is the Rydberg atomic density.

The first evidence for the dipole blockade using a Förster configuration has been demonstrated in the case of the cesium atom, specifically, the reaction $np + np \longrightarrow ns + (n-1)s$ [66]. As previously mentioned, the levels can be tuned to the Förster resonance only for n smaller than 42, which limits the efficiency of the dipole–dipole coupling. Nevertheless, Figure 27 shows a limitation (30%) of the excitation at the resonance for the $38p_{3/2}$ level. While a complete understanding of the dipole blockade at the Förster resonance is more complex than the electric field induced interaction of permanent electric dipoles, the central feature is the same. In the Stark energy diagram of pairs of atoms, the avoided crossing between the curves $(np_{3/2}np_{3/2})$ and $(ns(n-1)s)$ prevents, at resonance, the excitation of a second atom.

A clear indication of the destruction of the coherence of excitation has been reported by Johnson et al., who exposed ground state Rb atoms in a far off resonance trap to two-photon excitation to the nd states. With one atom in the trap they saw clear Rabi oscillations at a frequency of 2 MHz in the excitation probability, but when more than one atom was in the trap they observed loss of contrast of the Rabi oscillations due to the van der Waals interaction of the excited atoms, a variant of the dipole blockade mechanism that is operational in Figure 23. The sensitivity to the presence of more than one atom is due to the small size of the trap [67]. This approach is clearly an excellent one for the realization of a dipole blockade.

7. MECHANICAL EFFECTS

The dependence of the strength of the dipole–dipole interaction on the separation between the interacting atoms has several consequences, which may be readily appreciated by looking at Figure 28(a), a generic picture of the dipole–dipole μ^2/R^3 potentials for a pair of dipole coupled atomic states. If the two atomic states at $R \longrightarrow \infty$ are, for example ns and np, at any finite value of R the eigenstates are $(|nsnp\rangle + |npns\rangle)/\sqrt{2}$ and $(|nsnp\rangle - |npns\rangle)/\sqrt{2}$, which are repulsive and attractive. A pair of atoms on the repulsive curve moves apart, and a pair on the attractive curve moves together. If the two atoms are initially at rest, both of these motions are along the line joining the two atomic nuclei.

Pure dipole–dipole curves of the type shown in Figure 28(a) exist only when the two atomic states are dipole coupled states; we would not expect two atoms in the same state to lead to $1/R^3$ potential curves like this, but rather to van der Waals $1/R^6$ potentials. If the two atomic states are Rydberg states, there are many other levels energetically nearby, which has an interesting consequence. Due to the large dipole matrix elements and small level spacings the van der Waals description breaks down at surprisingly large distances, in many cases distances comparable to the

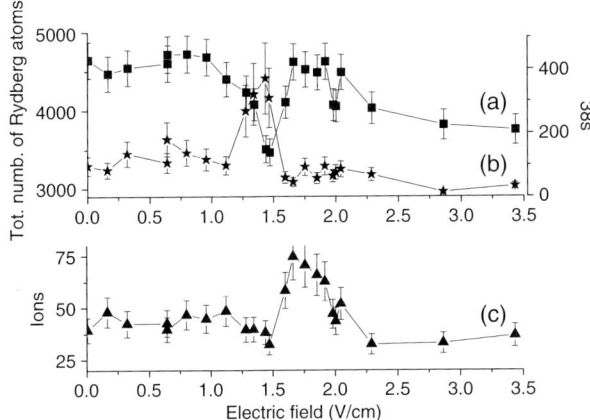

FIGURE 27 Dipole blockade in the excitation of the Cs $38p_{3/2}$ state from the 7s state observed at the Förster resonance Cs $38p_{3/2}$ + Cs $38p_{3/2}$ → Cs $38s$ + Cs$39s$. (a) Total number of Rydberg atoms excited (b) number of 39s atoms observed (c) number of ions produced (from Ref. [66])

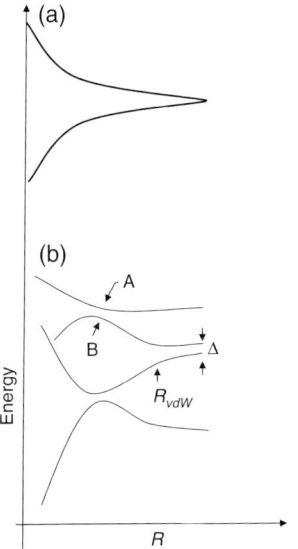

FIGURE 28 (a) Attractive and repulsive potentials, in arbitrary units, of a pair of dipole coupled atomic states. (b) Generic potential curves in arbitrary units. Two states separated by Δ at $R = \infty$ exhibit a van der Waals interaction at large R, but it becomes a dipole–dipole interaction for $R < R_{vdW}$. There are also avoided level crossings between levels of different n. Flat spots in the potentials, such as A and B, generate molecular spectral features, which may be bound (A), or dissociative (B)

separation of atoms in a MOT. For example, in Figure 28(b) we show levels composed of pairs of even parity atomic states, for example, $ns(n+1)s$ and $npnp$, which are spaced by the energy Δ. It is useful to define a van der Waals radius, R_{vdW}, given by [28]

$$R_{vdW} = (\mu\mu'/\Delta)^{1/3} \tag{53}$$

where μ and μ' are the atomic dipole matrix elements coupling the pair of states. For $R > R_{vdW}$ the dipole–dipole interaction between the two atoms has a $1/R^6$ dependence, but for $R < R_{vdW}$ the potential (labeled by R_{vdW} in Figure 28(b)) approaches a $1/R^3$ dependence, in spite of the fact that it is connected to a pair of atomic states of the same parity at $R = \infty$. Since the dipole matrix elements between nearby states vary as $\approx n^2$, and $\Delta = \delta/n^3$, with δ often taking values of 0.03 or less, R_{vdW} scales as $n^{7/3}$, and for $n = 50$ and $\delta = 0.01$, $R_{vdW} = 1$ μm, a distance comparable to the interatomic distances commonly encountered in traps. In Figure 27(b), R_{vdW} is indicated by the arrow. It is important to note that while Equation (53) is a good rule of thumb, there are resonances at which $R_{vdW} \longrightarrow \infty$.

The second consequence of the presence of many nearly degenerate pairs of states is that the potential curves exhibit avoided crossings as a function of R, which introduces the possibility of molecular states, which can be bound or dissociative. For example, in Figure 28(b) at point A there is a flat spot in the potential, and assuming the potential for a pair of ground state atoms to be flat in this region, pairs of atoms over a range of separations can be excited to this energy, providing an observable spectral feature. In the case of point A the shallow potential well supports bound molecules. Excitation to point B should provide a similar spectral feature, but in this case the molecules would have at least a 50% chance of dissociating. Irrespective of what subsequently happens to the molecules, the avoided crossings of the molecular potential curves are likely to lead to observable molecular spectral features.

Several manifestations of the molecular potential curves have been observed. One of the first was the existence of spectral features due to the curve crossings of the molecular potentials. Farooqi et al. observed small spectral peaks to the red of the Rb np states when they excited ground state Rb atoms in a MOT to Rydberg np states by single-photon UV excitation using a pulse amplified cw laser with 100 MHz resolution [68]. They attributed these spectral features to two-photon excitation of $ns(n-1)d$ pairs coupled to the $npnp$ pairs. While the excitation was attributed to flat portions of the potentials, it was not possible to tell if the molecular state produced was bound or dissociative.

Using a resolution approaching 1 MHz Overstreet et al. excited Cs atoms from the 6p state and observed the spectrum shown in Figure 29 [69]. They observed a spectral feature, labeled 'collision resonance' in

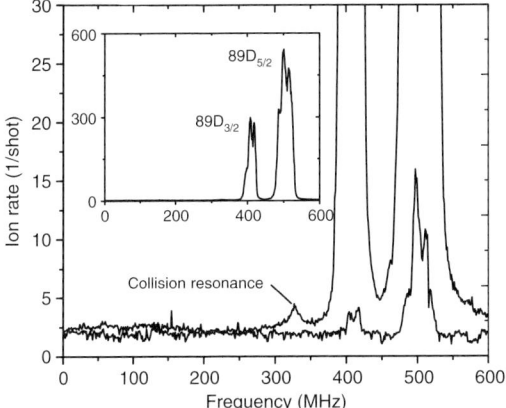

FIGURE 29 Cs+ Ions observed in excitation from the Cs 6p state. The inset shows the expected 89d fine structure states, and the full figure, on an expanded scale, shows the extra molecular feature, labeled 'collision resonance,' produced by excitation to a flat spot on a potential curve (from Ref. [69])

Figure 29, slightly to the red of the frequency required to excite the atomic Cs 89d state from the 6p state. The excitation frequency is close to that required to produce $R = \infty$, 88d90d pairs and is a two-photon excitation. By observing the expansion of the cloud of Rydberg atoms subsequent to laser excitation they measured the velocities of the atoms excited at the atomic transitions and at the 'collision resonance'. When the laser is tuned to an atomic resonance the velocity is 6 cm/s, which corresponds to the 79 μK temperature of the trapped atoms, but when tuned to the 'collision resonance', it is 17(3) cm/s. The 'collision resonance' is apparently a dissociating molecular state. The velocity of the atoms corresponds to an energy in the pair of molecules of approximately 10 MHz, which implies that the excitation of the 'collision resonance' occurs at very large values of R, 7 μm according to their model potential calculations.

Earlier we described dipole–dipole energy transfer with the assumption that the atoms were frozen in place, but from Figure 28 it is evident that such an assumption must break down at some point. Fioretti et al. observed this breakdown using the energy transfer

$$\text{Cs } 24p + \text{Cs } 24p \rightarrow \text{Cs } 25s + \text{Cs } 24s, \tag{54}$$

which is resonant at 84.7 V/cm [54]. The energy levels are the same as those used by Mourachko et al. [19], except that the m_j values of the initial p states are in this case 1/2 and 3/2 instead of both being 1/2. Fioretti et al. excited the atoms to the 24p state with a broad-band laser and detected that the energy transfer had occurred by detecting atoms in the 25s state.

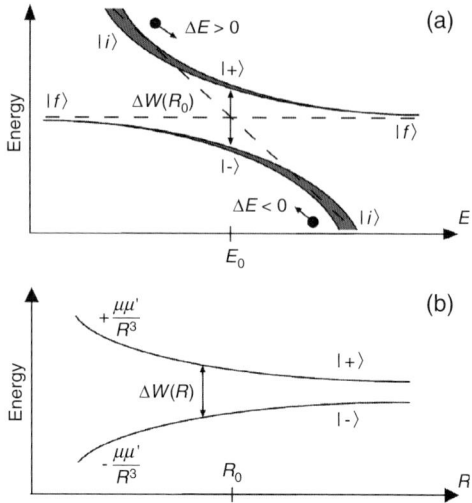

FIGURE 30 Schematic energy level diagram, in arbitrary units, for the interacting pair of states $|i\rangle$ (Cs 24p + Cs 24p) and $|f\rangle$ (Cs 24s + Cs 25s). (a) The $|i\rangle$ and $|f\rangle$ energies are plotted vs tuning field at infinite internuclear spacing (- - - -). At a finite internuclear spacing, R_0, the $|+\rangle$ and $|-\rangle$ eigenstates are linear superpositions of $|i\rangle$ and $|f\rangle$. The thickness of the lines shows the probability of excitation by the laser. (b) The $|i\rangle$ and $|f\rangle$ levels are plotted vs internuclear separation at a field slightly off resonance for a given field, E_0 (from Ref. [54])

Instead of sweeping a purely static tuning field over many shots of the laser they added a 2.4 V/cm field step rising or falling with a 1 μs time constant. The effect of the field step is easily understood with the aid of Figure 30, a schematic diagram of the levels involved. At $R = \infty$ the initial state i is the 24p24p state, and the final state f is the 25s24s state. The $R = \infty$ energies are shown by the broken lines in Figure 30(a). If there were no dipole–dipole interaction, these would be the energies for all R, and the energy levels would cross. However, the dipole–dipole interaction couples the i and f states, resulting in the new eigenstates $+$ and $-$, which exhibit an avoided crossing of $2\mu\mu'/R^3$ at the resonance. As shown by Figure 30(b), $+$ is repulsive, and $-$ is attractive. The width of the lines indicating the $+$ and $-$ eigenstates are proportional to the 24p character and are thus proportional to their probabilities of laser excitation. The probability of observing a 25s atom is simply the probability of adiabatic rapid passage. If there were no motion of the atoms, the probability of population transfer from the 24p to the 25s state should be the same for excitation in a field below the resonance followed by a positive field sweep through the resonance and excitation in a field above the resonance followed by a negative field sweep through the resonance, assuming the sweeps to be symmetric. In fact, they observed more population transfer to the 25s state when the laser excitation occurred in a field above the resonance field, and the field step was negative. As shown by Figure 30(a), excitation above the resonance is to the attractive, $|-\rangle$, eigenstate. Due to the attractive potential, the atoms move closer together as the field is swept through the resonance, increasing the size of the avoided crossing and the probability of an adiabatic passage, leading to more 25s atoms. On the other hand, excitation at a field below the resonance leads to the population of the repulsive, $|+\rangle$, state. During the sweep through the resonance the atoms move apart, reducing the coupling and the probability of transfer to the 25s state.

One of the more dramatic demonstrations of the effects of the motion along the potential curves is that it can lead to ionization, leading to the spontaneous conversion of a sample of cold Rydberg atoms into an ultracold plasma [26,27]. In terms of Figure 28(a), ionization occurs by motion along the attractive curve until the atoms collide and one of them ionizes. This picture is based on the assumption that the curve crossings, such as those shown in Figure 28(b), are traversed diabatically. If we assume that the atoms remain in their initial n state and follow a diabatic path, ionization of one of the atoms will occur when the two atoms are separated by a distance $R \approx 4n^2$. The other atom is driven to a lower-lying state. If we further assume that the dipole–dipole interaction is still a good description at this small spacing, we can compute the amount of electronic dipole–dipole energy which has been converted into kinetic energy of the atoms. This energy is

$$W_K = \mu^2/(4n^2)^3. \tag{55}$$

Assuming $\mu = n^2$ leads to an energy of $1/64n^2$. This energy is substantially smaller than the initial binding energy of the atoms. If we ignore it and assume the liberated electron to have no energy we can calculate the binding energy of the atom which is left behind to be $1/n^2$, i.e. it has a principal quantum number of $n/\sqrt{2}$. This is the upper limit for the principal quantum number of the remaining Rydberg atom. It is straightforward to calculate the time required for the atoms to move along the attractive path and ionize. For $n = 50$ and an initial separation of 5 μm the time is 4 μs [70].

Ionization by motion along pure dipole–dipole curves such as those shown in Figure 28(a) has been observed by Li et al., who populated Rb ns states with a pulsed blue laser and then drove microwave transitions from the molecular $nsns$ state to the $nsnp$ state [28]. The ionization process was modeled by Nasciemento et al. using the molecular potentials, and they found good agreement between the experiment and the calculation [71]. As shown in Figure 31(a), the $nsnp$ states can be either attractive or repulsive. The 500 ns long microwave pulse came immediately after the

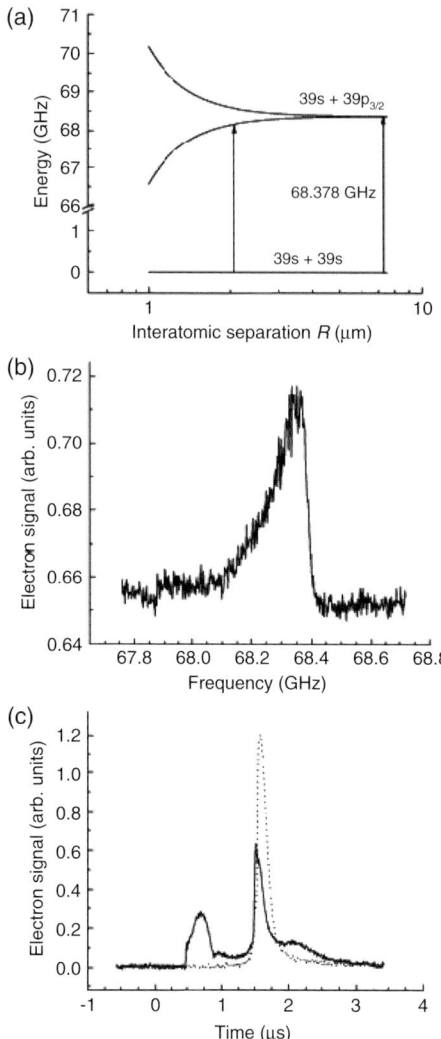

FIGURE 31 (a) Energy levels for the Rb pair states 39s39s and 39s39p as a function of internuclear separation, showing the repulsive and attractive potentials of the latter and the $R \sim \infty$ behavior. (b) Plasma electron signal observed as a function of microwave frequency. The electrons are detected after a 3.5 μs delay following a 500 ns long microwave pulse (c) Time resolved field ionization signal after 3.5 μs delay with and without the microwave pulse tuned to the attractive side of the atomic line, showing the redistribution of population and plasma formation (from Ref. [28])

laser excitation and the atoms were allowed to evolve for 3.5 μs before the ions produced were detected with an electric field pulse. They monitored the ion signal as the microwave frequency was scanned in the vicinity of the ns–np atomic transition. Figure 31(b) shows the typical result of such an experiment starting from the 39s state. Microwave excitation to the attractive curve results in ionization, but when it is to the repulsive curve it does not. On the other hand, if, under the same conditions, the 39p atoms are detected instead of the ions, a signal is observed only on the repulsive, high frequency, side of the atomic transition.

Laser excitation of the Rydberg atoms in zero field results in all the atoms being in the same state, so a dipole–dipole interaction is not possible. However, the van der Waals interaction, which is simply an off resonant dipole–dipole interaction, becomes effectively a dipole–dipole interaction for $R < R_{vdW}$, and ionization occurs for the same reason, and on a similar time scale, since R_{vdW} can be comparable to the average spacing of atoms in the trap. Li et al. observed more rapid ionization when the Rb nd states were excited than when the ns states were excited at the same density, and they attributed the difference to the fact that R_{vdW} is roughly ten times larger for the nd states than for the ns states [28].

With narrow-band laser excitation it is possible to excite pairs of atoms unambiguously to potential curves which are either attractive or repulsive and to observe the difference. Amthor et al. have examined the Rb ns and nd states [29]. In the former case it is possible to excite only the repulsive state and in the latter only the attractive state. They observe that the nd states ionize three times as fast as do the ns states, although why the ns states ionize is not completely clear. More interesting, they are able to observe the effect of the attractive $ndnd$ potential in the excitation spectra, as shown by Figure 32, the spectrum of the Rb $60d_{5/2}$ state observed using several detection methods. When the Rydberg atoms are detected a symmetric line centered on the atomic frequency is observed. When ions are observed with different delay times the signals shown in Figure 32(b) are observed. At the shortest delay time shown, 4 µs, the signal is most shifted to the red, because only those atoms lying relatively close to each other can ionize so quickly. As the delay is increased to 6 and then 11 µs the observed peak moves closer to the atomic frequency, since it is no longer necessary that the atoms be so close together to ionize in these longer times.

Cesium presents a case in which the difference between excitation to attractive and repulsive potentials can be observed in a more direct way. Specifically, the dipole–dipole energy exchange

$$\text{Cs } np + \text{Cs } np \leftrightarrow \text{Cs } ns + \text{Cs } (n+1)s \qquad (56)$$

is resonant at $n = 42$. For $n < 42$ the $npnp$ state is repulsive, and for $n > 42$ it state is attractive. Viteau et al. used narrow-band, 7 MHz, excitation of the Cs np states specifically, 300 ns long pulses of cw laser light to

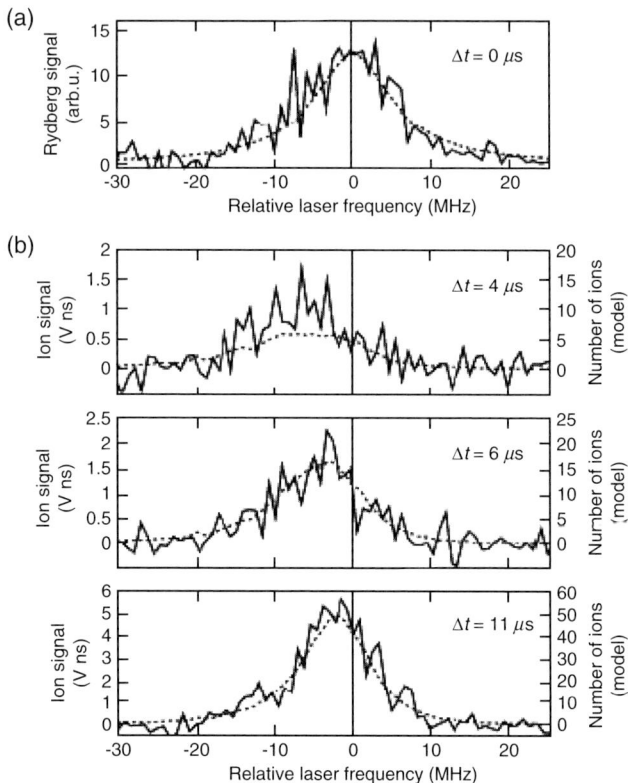

FIGURE 32 Excitation spectra of the Rb $60d_{5/2}$ state from the 5p state showing motion along the attractive potential curve, resulting in ionization. The frequency is relative to the unperturbed atomic frequency. (a) The total number of Rydberg atom signals is peaked at the atomic frequency. (b) Detection of the ions formed after several delay times. For short delays, only close pairs of atoms, excited far to the red of the atomic frequency ionize and are detected. For longer delays the atoms can be farther apart, resulting in smaller frequency displacements from the atomic line (from Ref. [29])

excite trapped Cs atoms from the 7s state to the np state [M., Viteau, Private Communication]. They observed the ions and remaining Rydberg atoms as a function of the delay after the excitation and the power of the exciting laser. Immediately, 300 ns, after laser excitation there was minimal ionization and no difference in the fractional ionization of states of $n < 42$ and $n > 42$. However, for delays of 5 μs and longer there were clear differences. First, for $n < 42$ the optical spectrum of ion production is peaked to the red of the Rydberg atom signal, while for $n > 42$ it peaked at the position of the atomic line. This result is essentially the same as that of Amthor et al. [29]. In addition, about a factor of two more ionization

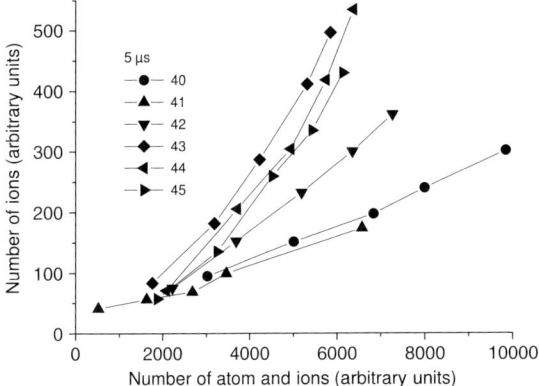

FIGURE 33 Number of ions detected after 5 μs delay vs the total number of ions and Rydberg atoms detected (the initial Rydberg population), when the Cs $np_{3/2}$ states are excited for $40 \leqslant n \leqslant 45$. Above $n = 42$ a pair of p atoms is on an attractive potential, and for n below 42 the potential is repulsive. The difference in the number of ions is apparent (from Ref. [M., Viteau Private communication])

is observed for the $n > 42$ states than for $n < 42$ states. In Figure 33 the difference between excitation to attractive, $n > 42$, and repulsive, $n < 42$, potentials is obvious, but again, it is not completely evident why there is so much ionization of the $n < 42$ states.

If the atoms are drawn together along an attractive curve and ionize, dipole–dipole electronic energy is converted into kinetic energy of the resulting atom and ion. Knuffman and Raithel have made observations consistent with this notion [72]. Under the conditions for which they began to see spontaneous ionization of the Rydberg atoms to form a plasma they observed fast Rydberg atoms. They observed field ionization of Rydberg atoms 2 cm from the MOT in which they had been excited. The velocity distribution was inferred from the time-of-flight distribution of the Rydberg atoms, which ranged from 1 to 15 ms. The most probable velocity they observed when exciting Rb $n = 90$ atoms was 5 m/s, a velocity roughly 100 times higher then the velocities observed by Overstreet et al. [69]. This velocity corresponds to an energy of 0.2 cm^{-1}, and is roughly 2% of the initial binding energy of the Rydberg atoms, in rough accord with our expectation based on Equation (54).

Ionization by motion along the attractive potential curve of Figure 28 typically occurs on a microsecond time scale, but ionization can also be observed on the time scale of tens of ns. Li et al. observed significant ionization in 250 ns, and Tanner et al. observed significant, almost complete ionization in 80 ns [31]. Although Li et al. attributed the ionization to motion of atoms which are very close together, it seems unlikely that there are enough close pairs of atoms to generate the size of the signal they saw, and it is virtually impossible to reconcile the later observations of Tanner et al. with this explanation. Tanner et al. recorded the excitation spectra in essentially the same way as did Amthor et al. [29] and Viteau et al. [M., Viteau Private Communication]. The significant differences were the use of a pulse amplified laser instead of a cw laser and a very short time delay before the field pulse. Pulse amplification increases the line width, to 100 MHz, but it also increases the intensity, allowing nonlinear excitation. The excitation spectra for the Rb 51d state exposed to an ionization pulse 80 ns after the laser pulse are shown in Figure 34 for different laser intensities. The free ions were detected as the laser frequency was scanned over the 5p–51d transition. As shown, at low intensity the observed signal approximately matches the laser line width of 100 MHz, but as the intensity is raised the signal broadens and new features appear. Figure 32 was recorded at a Rb 5p density of 5×10^{10} cm^{-3}. The laser power dependence of the spectra shows that the production of ions exhibits an I^2 intensity dependence at low power when the laser is tuned slightly off resonance, and when the laser is tuned far from resonance it exhibits a higher order, at least I^4, dependence. The nonlinear dependence on the laser power implies that more than one atom is being excited simultaneously by two-photon, and in some cases multi-photon, excitation [73]. The requirement of nonlinear excitation is probably why such rapid ionization was not observed by Amthor et al. or Viteau et al. [29], [M., Viteau, Private Communication].

Tanner et al. [31] attributed the ionization they observed to a many-atom form of molecular auto-ionization, a process suggested by Hahn.[30] The interaction in Hahn's diatomic molecular auto-ionization is the same dipole–dipole interaction responsible for resonant dipole–dipole energy transfer. The only difference is that one of the final states is a continuum state. Thus the process can be written as

$$nl + nl \longrightarrow n_f l_f + \varepsilon l''. \tag{57}$$

Due to energy conservation the state of the n_f atom must be one with principal quantum number less than $n/\sqrt{2}$. The difference from the dipole–dipole ionization discussed above is that there is no necessity for the atoms to move, so the ionization can in principle occur more quickly. In fact, however, diatomic molecular auto-ionization is simply too slow to explain the observed ionization. For example, $n = 50$ atoms at a density of 10^{10} cm^{-3} should ionize at a rate of 10^3 s^{-1}, far too slow to account for ionization occurring on a time scale of tens of ns. Many-atom molecular auto-ionization occurs by transitions through near resonant virtual intermediate states to the ionization continuum. The important differences from diatomic molecular auto-ionization are as follows. Many transitions of small Δn are required, which have much larger matrix

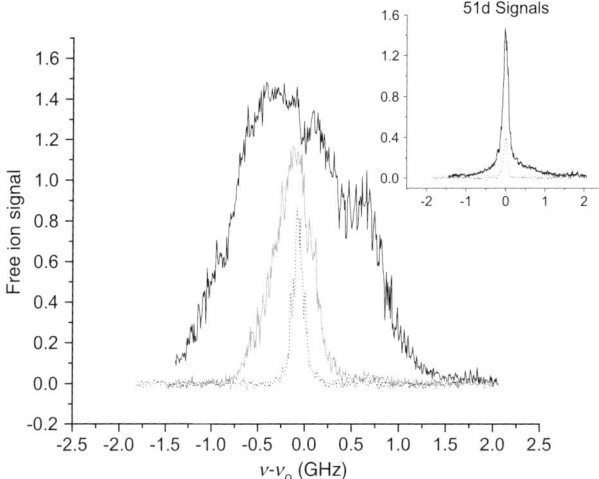

FIGURE 34 Excitation spectra of the Rb 51d State from the 5p state in which ions are detected after 100 ns at several laser intensities; 0.03 (dotted), 0.2 (gray), 1.0 (black). The frequency is relative to the atomic transition frequency. At low intensity, ions are detected only at the atomic line center, but at higher intensity ions are detected far from the atomic line center, with a nonlinear power dependence due to the nonlinear excitation of more than one atom at once. The inset shows the spectra taken under the same conditions when the Rb 51d state is detected. The atoms are presumed to be on the edge of the atom cloud (from Ref. [31])

elements than transitions of large Δn. When many atoms are taken into account the density of states becomes enormous, and the resonance denominators very small, so that a transition through multiple virtual intermediate states is not an impediment to ionization. Unlike diatomic molecular auto-ionization in which the residual atoms are deeply bound, a range of binding energies is expected for the remaining bound state atoms, and this was observed. This process is essentially the same as the process termed superexchange in polyatomic molecules and related to microwave ionization at high scaled frequency [74,75].

8. CONCLUSION

At 300 K the dipole–dipole interaction between Rydberg atoms is manifested in resonant dipole–dipole energy transfer collisions which are tuned into resonance with a small electric field. These collisions are unique among atomic collisions in that they have enormous cross-sections, scaling as n^4, and long durations, scaling as n^2. While these properties of the resonant collisions are interesting, what is more interesting is that they have allowed novel collisions to be studied. For example, it is difficult to imagine observing transform limited collisions or radiatively-assisted collisions in the strong field regime in any other system.

Rydberg atoms in a MOT constitute a physical system with qualitatively different properties. The differences stem from the fact that, on an experimentally interesting time scale, the atoms move only a few per cent of their typical internuclear spacing. The atoms are, to a first approximation, frozen in place, so the transient binary interactions of collisions in a 300 K gas are replaced by static interactions in the frozen Rydberg gas. Since an atom can have several neighbors comparably far away, an atom can now interact with more than one other atom at the same time, so that the system becomes more like an amorphous solid. In fact, the diffusion of excitation, similar to that observed in amorphous solids is observed in the frozen Rydberg gas. The fact that the frozen Rydberg gas is like a solid has led to several proposals to use the dipole–dipole interactions of Rydberg atoms as the basis of quantum gates.

Even if the Rydberg atoms are initially at rest, the dipole–dipole interaction leads to a force on the atoms, which is not negligible for pairs of atoms which are particularly close together. The force can be either repulsive or attractive, and if it is attractive the atoms can be drawn together, collide, and ionize. This process is one which helps initiating the spontaneous evolution of the cold Rydberg gas into a plasma [26, 27]. While the evolution into a plasma is probably undesirable in some contexts, the combined cold-Rydberg-gas–plasma system is one in which atoms can go both ways, from bound atoms to the plasma and vice versa. Finally, the cold Rydberg gas has characteristics of isolated atoms, an amorphous solid, and a plasma, and the dipole–dipole interactions of the Rydberg atoms play an important role in this system.

ACKNOWLEDGMENTS

The authors acknowledge many fruitful discussions with Daniel Comparat, Vladimir Akulin, Amondsen Chotia, Mattieu Viteau, Jianing Han, Paul Tanner, and Edward Shuman. This work has been supported by the Institut Francilien de Recherche sur les Atomes Froids (IFRAF), which made possible the stay of TFG at Laboratoire Aime Cotton, and the Air Force Office of Scientific Research.

REFERENCES

[1] J.R. Lakowicz (1994), in: J.R. Lakowicz (Ed.), Topics in Fluorescence Spectroscopy, Vol. 4, Plenum, New York.
[2] A. Javan, W.R. Bennet Jr. and D.R. Herriot (1962), *Phys. Rev. Lett.*, **8**, 470.
[3] C.K.N. Patel (1964), *Phys. Rev. Lett.*, **13**, 617.
[4] J.D. Miller, R.A. Cline and D.J. Heinzen (1993), *Phys. Rev. Lett.*, **71**, 2204.
[5] W.I. McAlexander, E.R.I. Abraham and R.G. Hulet (1996), *Phys. Rev. A*, **54**, 5R.

[6] A. Fioretti, D. Comparat, A. Crubellier, F. Masnou-Seeuws and P. Pillet (1998), *Phys. Rev. Lett.*, **80**, 4402.
[7] J. Stuhler, A. Griesmaier, T. Koch, M. Fattori, T. Pfau, S. Giovanazzi, P. Pedri and L. Santos (2005), *Phys. Rev. Lett.*, **95**, 150406.
[8] D. DeMille (2002), *Phys. Rev. Lett.*, **88**, 067901.
[9] H.L. Bethlem, G. Bierden and G. Meijer (1999), *Phys. Rev. Lett.*, **83**, 1558.
[10] D. Jaksch, J.I. Cirac, P. Zoller, S.L. Rolston, R. Cote and M. Lukin (2000), *Phys. Rev. Lett.*, **85**, 2208.
[11] M.D. Lukin, M. Fleischhauer, R. Côté, L.M. Duan, D. Jaksch, J.I. Cirac and P. Zoller (2001), *Phys. Rev. Lett.*, **87**, 037901.
[12] E. Brion, A.S. Mouritzen and K. Molmer (2007), *Phys. Rev. A*, **76**, 022334.
[13] M. Saffman and T.G. Walker (2005), *Phys. Rev. A*, **72**, 022347.
[14] P.L. Houston (1981), in: I. Prigogine and S.R. Rice (Eds.), *Advances in Chemical Physics*, Wiley, New York.
[15] T.F. Gallagher, G.A. Ruff and K.A. Safinya (1980), *Phys. Rev. A.*, **22**, 843.
[16] K.A. Smith, F.G. Kellert, R.D. Rundel, F.B. Dunning and R.F. Stebbings (1978), *Phys. Rev. Lett.*, **40**, 1362.
[17] K.A. Safinya, J.F. Delpech, F. Gounand, W. Sandner and T.F. Gallagher (1981), *Phys. Rev. Lett.*, **47**, 405.
[18] W.R. Andersen, J.R. Veale and T.F. Gallagher (1998), *Phys. Rev. Lett.*, **80**, 249.
[19] I. Mourachko, D. Comparat, F de Tomasi, A. Fioretti, P. Nosbaum, V.M. AKulin and P. Pillet (1998), *Phys. Rev. Lett.*, **80**, 253.
[20] S. Inoyue, M.R. Andrews, J. Steger, H.-J. Meisner, D.M. Stamper-Kurn and W. Ketterle (1998), *Nature*, **392**, 151.
[21] T.F. Gallagher, K.A. Safinya, F. Gounand, J.F. Delpech, W. Sandner and R. Kachru (1982), *Phys. Rev. A*, **25**, 1905.
[22] R. Kachru, N.H. Tran and T.F. Gallagher (1982), *Phys. Rev. Lett.*, **49**, 191.
[23] P. Pillet, R. Kachru, N.H. Tran, W.W. Smith and T.F. Gallagher (1983), *Phys. Rev. Lett.*, **50**, 1763.
[24] P. Cahuzac and P.E. Toschek (1977), *Phys. Rev. Lett.*, **40**, 1087.
[25] P. Pillet, R. Kachru, N.H. Tran, W.W. Smith and T.F. Gallagher (1987), *Phys. Rev. A*, **36**, 1132.
[26] M.P. Robinson, B. Laburthe Tolra, M.V. Noel, T.F. Gallagher and P. Pillet (2000), *Phys. Rev. Lett.*, **85**, 4466.
[27] S.K. Dutta, D. Feldbaum, A. Walz-Flannigan, J.R. Guest and G. Raithel (2001), *Phys. Rev. Lett.*, **86**, 3993.
[28] W. Li, P.J. Tanner and T.F. Gallagher (2005), *Phys. Rev. Lett.*, **94**, 173001.
[29] T. Amthor, M. Reetz-Lamour, S. Westermann, J. Denskat and M. Weidemuller (2007), *Phys. Rev. Lett.*, **98**, 023004.
[30] Y. Hahn (2000), *J. Phys. B*, **33**, L655.
[31] P.J. Tanner, J. Han, E.S. Shuman and T.F. Gallagher (2008), *Phys. Rev. Lett.*, **100**, 043002.
[32] H.A. Bethe and E.A. Salpeter (1977), *Quantum Mechanics of One and Two Electron Atoms*, Plenum, New York.
[33] K. Afrousheh, P. Bohlouli-Zanjani, J.A. Petrus and J.D.D. Martin (2006), *Phys. Rev. A*, **74**, 062712.
[34] J. Han and T.F. Gallagher (2006), *Phys. Rev. A*, **74**, 054502.
[35] J.D. Jackson (1998), *Classical Electrodynamics*, Wiley, New York.
[36] T.F. Gallagher (1994), *Rydberg Atoms*, Cambridge University Press, Cambridge.
[37] N.F. Ramsey (1954), *Molecular Beams*, Oxford University Press, Oxford.
[38] M.J. Renn, Q.S. Sun and T.F. Gallagher (1994), *Phys. Rev. A*, **50**, 4077.
[39] R.C. Stoneman, D.S. Thomson and T.F. Gallagher (1987), *Phys. Rev. Lett.*, **58**, 1324.
[40] D.S. Thomson, M.J. Renn and T.F. Gallagher (1990), *Phys. Rev. Lett.*, **65**, 3273.
[41] D.S. Thomson, R.C. Stoneman and T.F. Gallagher (1989), *Phys. Rev. A*, **39**, 2914.
[42] S.E. Harris and D.B. Lidow (1999), *Phys. Rev. Lett.*, **33**, 674.
[43] D.S. Thomson, M.J. Renn and T.F. Gallagher (1992), *Phys. Rev. A*, **45**, 358.

[44] M.J. Renn and T.F. Gallagher (1991), *Phys. Rev. Lett.*, **67**, 2287.
[45] D. Coffey Jr., D.C. Lorents and F.T. Smith (1962), *Phys. Rev.*, **187**, 201.
[46] M.C. Baruch and T.F. Gallagher (1992), *Phys. Rev. Lett.*, **68**, 3515.
[47] R.R. Jones (1995), *Phys. Rev. Lett.*, **74**, 1091.
[48] M. Mark, T. Kraemer, P. Waldburger, J. Herbig, C. Chin, H.-C. Nagerl and R. Grimm (2007), *Phys. Rev. Lett.*, **99**, 113201.
[49] M.J. Renn, D.S. Thomson and T.F. Gallagher (1994), *Phys. Rev. A*, **49**, 409.
[50] V. Akulin, I. Mourachko, F. de Tomasi and P. Pillet (1999), *Physica D*, **131**, 125.
[51] J.S. Frasier, V. Celli and T. Blum (1999), *Phys. Rev. A*, **59**, 4358.
[52] T.J. Carroll, K. Claringbould, A. Goodsell, M.J. Lim and M.W. Noel (2004), *Phys. Rev. Lett.*, **93**, 153001.
[53] T.J. Carroll, S. Sunder and M.W. Noel (2006), *Phys. Rev. A*, **73**, 032725.
[54] A. Fioretti, D. Comparat, C. Drag, T.F. Gallagher and P. Pillet (1999), *Phys. Rev. Lett.*, **82**, 1839.
[55] I. Mourachko, W. Li and T.F. Gallagher (2004), *Phys. Rev. A*, **70**, 031401(R).
[56] W.R. Anderson, M.P. Robinson, J.D.D. Martin and T.F. Gallagher (2002), *Phys. Rev. A*, **65**, 063404.
[57] C.S.E. van Ditzhuijzen, A.F. Koenderink, J.V. Hernandez, F. Robichaeux, L.D. Noordam and H.B. van Linden van den Heuvell, (2008). arXiv:0706.0110v3.
[58] P. Bohlouli-Zanjani, J.A. Petrus and J.D. Martin (2007), *Phys. Rev. Lett.*, **98**, 203005.
[59] K. Afrousheh, P. Bohlouli-Zanjani, D. Vagale, A. Mugford, M. Fedorov and J.D.D. Martin (2004), *Phys. Rev. Lett.*, **93**, 233001.
[60] M. Mudrich, N. Zahzam, T. Vogt, D. Comparat and P. Pillet (2005), *Phys. Rev. Lett.*, **95**, 233002.
[61] D. Tong, S.M. Farooqi, J. Stanojevic, S. Krishnan, Y.P. Zhang, R. Coté, E.E. Eyler and P.L. Gould (2004), *Phys. Rev. Lett.*, **93**, 063001.
[62] K. Singer, M. Reetz-Lamour, T. Amthor, L.G. Marcassa and M. Weidemuller (2004), *Phys. Rev. Lett.*, **93**, 163001.
[63] T. Cubel Liebisch, A. Reinhard, P.R. Berman and G. Raithel (2005), *Phys. Rev. Lett.*, **95**, 023405.
[64] R. Heidemann, U. Raitzsch, V. Bendkowsky, B. Butscher, R. Low, L. Santos and T. Pfau (2007), *Phys. Rev. Lett.*, **99**, 163601.
[65] T. Vogt, M Viteau, A. Chotia, J. Zhao, D. Comparat and P. Pillet (2007), *Phys. Rev. Lett.*, **99**, 073002.
[66] T. Vogt, M Viteau, J. Zhao, A. Chotia, D. Comparat and P. Pillet (2006), *Phys. Rev. Lett.*, **97**, 083003.
[67] T.A. Johnson, E. Urban, T. Henage, L. Isenhower, D.D. Yavuz, T.G. Walker and M. Saffman (2008), *Phys. Rev. Lett.*, **100**, 113003.
[68] S.M. Farooqi, D. Tong, S. Krishnan, J. Stanojevic, Y.P. Zhang, J.R. Enscher, A.S. Astrin, C. Boisseau, R. Cote, E.E. Eyler and P.L. Gould (2003), *Phys. Rev. Lett.*, **91**, 183002.
[69] K.R. Overstreet, A. Schwettmann, J. Tallant and J.P. Shaffer (2007), *Phys. Rev. A*, **76**, 011403(R).
[70] F. Robichaeux (2005), *J. Phys. B*, **38**, S333.
[71] V.A. Nasciemento, M. Reetz-Lamour, L.L. Caliri, A.L. De Oliviera and L.G. Marcassa (2006), *Phys. Rev. A*, **73**, 034703.
[72] B. Knuffman and G. Raithel (2006), *Phys. Rev. A*, **73**, 020704(R).
[73] J.M. Raimond, G. Vitrant and S. Haroche (1981), *J. Phys. B*, **14**, L655.
[74] A.A. Stuchebrukhov, H. Mehta and R.A. Markus (1993), *J. Phys. Chem.*, **97**, 12491.
[75] R.V. Jensen (1999), *Phys. Rev. Lett.*, **62**, 1476.

CHAPTER 5

Strong-Field Control of X-Ray Processes

Robin Santra, Robert W. Dunford, Elliot P. Kanter, Bertold Krässig, Stephen H. Southworth and **Linda Young**

Argonne National Laboratory, Argonne, Illinois 60439, USA

Contents		
	1. Motivation	220
	2. Basic concepts	222
	2.1 Basic X-Ray Processes	222
	2.2 Strong-Field Laser Physics	224
	3. Theoretical aspects	227
	3.1 Strong-Field Ionization	227
	3.2 Laser Dressing	233
	3.3 Laser-Induced Molecular Alignment	236
	4. General experimental considerations	240
	5. Case studies	242
	5.1 Orbital Alignment in Strong-Field Ionization	242
	5.2 Electromagnetically-Induced Transparency	247
	5.3 Laser-Induced Molecular Alignment	251
	6. Summary and outlook	253
	Acknowledgments	255
	References	255

| Abstract | In x-ray science, cross-sections for x-ray absorption in matter are typically assumed to be invariants, depending only on the atomic and spatial composition of the material. However, by utilizing strong electromagnetic fields, it is possible to modify x-ray absorption cross-sections in a controlled way. This allows one to use x-rays to gain deeper insight into strong-field processes. As an application, strong electromagnetic fields may be utilized to control the propagation of x-rays through a medium. Gas-phase |

systems are particularly suitable for illustrating the basic principles underlying combined x-ray and laser interactions. We discuss three different scenarios for modifying resonant x-ray absorption near an inner-shell edge: (1) Strong-field ionization of the target particles at laser intensities in the range 10^{14}–10^{15} W/cm^2; (2) modification of electronic structure of inner-shell-excited systems by laser dressing at 10^{12}–10^{13} W/cm^2; (3) control of resonant x-ray absorption by molecules through laser-induced spatial alignment at 10^{11}–10^{12} W/cm^2.

1. MOTIVATION

Control of x-ray processes using intense optical lasers represents an emerging scientific frontier – one which combines x-ray physics with strong-field laser control. The past decade has been one in which intense lasers at optical wavelengths have markedly expanded our ability to control the behavior of isolated atoms and molecules [1–4]. It is now routine to produce focused intensities of 10^{15} W/cm^2 [5,6]. An intensity of 10^{15} W/cm^2 corresponds to a field of $\sim$10 V/Å, comparable to the atomic field binding outer-shell electrons. By exposing an atom to an external field of this magnitude it becomes possible to substantially modify the probability of resonant x-ray absorption, i.e. the probability for promotion of an inner-shell electron to a valence or Rydberg state. Earlier studies focused on strong-field modifications to the final-state continuum, as manifested by photo-electron satellites caused by free–free transitions [7–9]. Laser-induced photo-electron satellites are widely used to characterize ultrafast radiation fields, see for example Ref. [10]. In contrast, here we focus on modification and control of *x-ray absorption* in free atoms and molecules by the application of strong laser fields. Three intensity regimes will be discussed: ionization (10^{14}–10^{15} W/cm^2), laser-dressing (10^{12}–10^{13} W/cm^2), and molecular alignment (10^{11}–10^{12} W/cm^2). Wuilleumier and Meyer [11] reviewed combined laser/x-ray experiments performed at lower laser intensities.

At high intensities, $\sim 10^{14}$ W/cm^2, atoms are rapidly ionized when an outer-shell electron tunnels through a suppressed Coulomb barrier [12], and the atom's x-ray absorption spectrum is modified irreversibly [13,14]. Probing the resulting ion with tunable, polarized x-rays reveals orbital alignment [13,15] and provides access to the ion quantum state distribution [14,16]. These studies represent the first spectroscopic probes of the residual ion; previous experiments measured fragmentation products such as ions and electrons generated by the strong laser field [17,18].

At non-ionizing intensities the x-ray absorption probability in an atom can be controlled reversibly by application of a laser-dressing field [19].

Our work focuses on modifications to x-ray absorption spectra, while earlier work addressed the manifestations of laser dressing in electron energy spectra [7–9,20]. A dramatic example of strong-field control of x-ray absorption is electromagnetically-induced transparency (EIT) for x-rays [21], predicted for the Ne 1s → 3p transition in the presence of 800-nm radiation at 10^{13} W/cm^2. EIT at optical wavelengths is reviewed in Refs. [22–24]. In EIT for x-rays, the high laser intensity is essential to couple core-excited Rydberg levels and overcome their rapid auto-ionization decay, which occurs on the femtosecond time scale. EIT control of x-ray absorption can be used to imprint laser pulse trains of femtosecond duration onto 100 picosecond x-ray pulses from existing synchrotron sources, thus producing a femtosecond-capable x-ray waveform generator [21].

An entirely different mechanism provides strong-field control of resonant x-ray absorption in molecules at even lower intensity, 10^{12} W/cm^2. A non-resonant, linearly polarized laser field will align a molecule by interaction with the molecule's anisotropic polarizability; the most polarizable axis within the molecule will align parallel to the laser polarization axis [25]. Since the laser polarization direction is under simple control with a waveplate, so is the direction of the molecule's most polarizable axis with respect to the x-ray polarization axis, which is fixed in the laboratory frame. It is well established that x-ray absorption resonances in the near edge region, resulting from the promotion of a 1s electron to an empty σ^* or π^* orbital, are sensitive to the angle between the molecular axis and the x-ray polarization axis [26]. Thus, laser control of molecular alignment implies laser control of resonant x-ray absorption.

Laser-controlled molecular alignment also enables control over x-ray diffraction; scattering from an ensemble of aligned molecules produces Bragg-like diffraction spots rather than the concentric rings observed in an isotropic gas. An important potential application is single biomolecule structure determination using coherent diffractive imaging with x-ray free-electron lasers [27]. The original concept [27] did not suggest aligned molecules, but rather proposed to scatter 10^{12} x-rays from a single biomolecule within 10 fs and collect a diffraction pattern with sufficient information to determine the molecular orientation in a single shot. Having pre-aligned molecules will vastly simplify the data collection and analysis – as pointed out by Spence and Doak in the context of electron scattering [28]. While these proposals [27,28] focus on x-ray scattering from a *single* large molecule, and multiple repetition to build up statistics, i.e. 'serial crystallography', our work focuses on x-ray probing of an *ensemble* of 10^8 small molecules in the gas phase which have been aligned with laser techniques [4]. This strategy will allow one to acquire x-ray diffraction patterns of aligned, non-interacting molecules and thus obtain Ångstrom-level molecular images using existing synchrotron sources.

Studies of strong-field control of x-ray absorption are enabled by access to tunable, polarized x-ray pulses from, e.g. Argonne's Advanced Photon Source and soon-to-be-available short pulse x-ray sources [29–32]. Strong optical fields are created by focusing ultrashort laser pulses to ultrasmall (10–100 μm) spot sizes. Probing atoms and molecules in a strong-field environment with ultrafast x-rays is an exciting experimental and theoretical challenge, and opens a new frontier merging x-ray physics and strong-field control.

2. BASIC CONCEPTS

2.1 Basic X-Ray Processes

The main photon–atom processes in the ~1–100-keV energy range are photo-electric absorption and elastic (Rayleigh) and inelastic (Compton) scattering from bound electrons [33]. Atomic K-shell ionization energies increase as Z^2, where Z is the atomic number. The ~1–100-keV range encompasses the K-shell ionization energies of Ne (0.87 keV, $Z = 10$) to Rn (98.4 keV, $Z = 86$). Photo-absorption cross-sections greatly exceed scattering cross-sections over energy ranges from below to far above the respective K edges of each atom [34,35]. Elastic (coherent) scattering is the process involved in atomic structure determinations by x-ray diffraction [36], and Compton (incoherent) scattering is used to determine electron density distributions of materials [37]. Experiments on Rayleigh and Compton scattering from free atoms have provided detailed tests of theoretical treatments of non-local exchange, electron correlation, and x-ray scattering factors [38–40]. Time-resolved x-ray diffraction can be used as a structural probe of dynamical processes, for example, laser-induced molecular dissociation [41] and is expected to be a major tool used to image atomic structures using x-ray free-electron lasers [42]. Inversion of diffracted intensities to determine molecular structures can be simplified by spatially aligning the target molecules in polarized laser fields [28]. Scattering processes are of general interest in time-resolved x-ray science, but we limit the following discussion to photo-absorption by inner-shell electrons and consequent decay processes that are relevant to the experiments discussed in this review.

X-rays have been studied using anode sources since their discovery by Röntgen in 1895, but modern research is greatly enhanced using intense, tunable, collimated, polarized, narrow-band x-ray pulses provided by synchrotron radiation sources [43]. Synchrotron radiation enables detailed studies of x-ray and inner-shell processes by employing combinations of high-resolution, differential, coincidence, or time-resolved measurements. The absorbed x-ray energy can be tuned below, through, and above specific inner-shell ionization thresholds [44–49]. The present discussion of x-ray processes assumes that x-ray interactions with matter are in the weak-field regime appropriate to synchrotron radiation experiments. Nonlinear x-ray interactions are expected when using free-electron lasers [50].

Photo-absorption by an inner-shell electron of a many-electron atom produces an excited intermediate state that rapidly relaxes either radiatively or non-radiatively by emission of a fluorescent x-ray or an Auger electron. Decay rates for radiative and non-radiative processes depend on atomic number and subshell, and have been calculated, measured, and tabulated [51–55]. Here we use the example of K-shell photo-absorption of Kr ($Z = 36$). There are several important consequences of decay processes. Rapid decay results in significant lifetime broadening, e.g., 2.7 eV in the case of Kr K vacancies, and this broadening affects the spectral resolution of photo-absorption, photo-electron, x-ray fluorescence, and Auger-electron spectra. The 1s → np ($n = 5, 6, 7...$) Rydberg excitations are completely obscured by lifetime broadening in the K-shell photo-absorption spectrum of Kr [56]. Lifetime broadening of the edge jump at the 1s ionization threshold is modeled by an arctangent shape that extends through the Rydberg states [56,57].

The short lifetimes of inner-shell vacancy states, 0.2 fs in the case of Kr K vacancies, can be exploited for time-resolved experiments. Fluorescent x-rays and Auger electrons are emitted promptly and can be detected with single-particle counters to measure x-ray absorption by transient, laser-excited targets. As described in Section 4, fluorescent x-rays are particularly useful, because they readily pass through dense gas-phase targets and low-Z window materials and can be detected with small, energy-dispersive detectors. For example, 2p → 1s (Kα) fluorescent x-rays generally pass through target atoms with low attenuation, because their energies are well below the K edge and far above the L edges. In the case of Kr, the L_3, L_2, L_1, and K-edge energies are 1.679 keV, 1.731 keV, 1.920 keV, and 14.327 keV, respectively, while the Kα fluorescence energy is 12.648 keV. Energies for atomic ionization thresholds and x-ray transitions are taken from the comprehensive evaluation of Ref. [58]. The branching ratios for radiative and non-radiative decay processes depend on atomic number and subshell [51]. Radiative processes increase in probability with increasing Z and are more likely for K vacancies than L vacancies. The radiative yield for a Kr K vacancy is 0.621 [55]. np → 1s radiative decay rates strongly decrease as n increases [59]. Among the radiative decay channels for a Kr K vacancy, the calculated branching ratio for 2p → 1s is 0.866 [59]. Combining this branching ratio with the radiative yield, the probability of a Kr K vacancy to decay by a Kα transition is 0.538. The Kα x-rays provide a prompt, efficient, atom-specific signature of x-ray absorption. These properties were exploited for the experiments on Kr and CF$_3$Br in strong laser fields discussed in Section 5.

The previous discussion implicitly assumed a two-step model in which x-ray absorption and vacancy decay are treated independently. This picture is valid for the studies presented in this review. We note that the two-step model is unsuitable for describing radiative and radiationless resonant Raman scattering [44,47,60,61], in which the vacancy creation and decay steps are coupled and described as a one-step, resonant inelastic scattering process. With the availability of intense, tunable, high-resolution synchrotron x-ray sources and high-resolution x-ray emission and electron spectrometers, several experimental studies have been reported on resonance and threshold phenomena in inner-shell photo-excitation and decay processes [45–48]. Our current research is investigating how strong laser fields can exploit, modify, and control x-ray processes.

2.2 Strong-Field Laser Physics

Developments in strong-field laser physics have tracked those in ultrafast laser technology. The development of chirped pulse amplification [62] has greatly increased the output energy of ultrashort pulsed lasers – such that table-top sources now routinely produce focused intensities of 10^{15} W/cm^2 [5]. Higher intensities of up to 10^{20} W/cm^2 can be produced at large scale facilities [63], but the regime above 10^{15} W/cm^2 is beyond the scope of the present review. The electric-field amplitude $\mathcal{E}_0$ is related to the cycle-averaged intensity I_L via $I_L = (1/2)c\varepsilon_0\mathcal{E}_0^2$, where c is the speed of light and ε_0 the electric constant, so that $\mathcal{E}_0[\text{V/Å}] = 2.745 \times 10^{-7}\sqrt{I_L[\text{W/cm}^2]}$. For 10^{15} W/cm^2, $\mathcal{E}_0 = 8.7$ V/Å – a magnitude that is comparable to the atomic Coulomb field which binds an outer-shell electron.

At these field strengths, an outer-shell electron can escape with substantial probability from its bound state, via tunneling, before the laser field reverses its sign. The Keldysh parameter γ represents the ratio of the laser frequency to the tunneling rate and is given by $\gamma = \sqrt{I_p/2U_p}$, where I_p is the energy required to ionize the electron and $U_p = e^2\mathcal{E}_0^2/4m_e\omega_L^2$ is the classical wiggle energy of an electron in an optical field with amplitude $\mathcal{E}_0$ and frequency ω_L and is referred to as the *ponderomotive energy* (e is the elementary charge and m_e is the electron mass). For $\gamma \ll 1$, the quasi-static or adiabatic approximation for the ionization step is well justified. Strictly speaking, only in this regime the simple tunneling (or over-the-barrier) ionization picture is valid. However, the adiabatic approximation sometimes proves quite useful even for $\gamma \approx 1$.

Classical mechanics describes the motion of the freed electron in the field with a wiggle amplitude $a_w = e\mathcal{E}_0/m_e\omega_L^2$ [64,65]. For 10^{15} W/cm^2 and $\lambda = 800$ nm, $U_p = 60$ eV and $a_w = 2.7$ nm. Because the freed electron is driven by the laser field, an energetic recollision with the residual ion can occur. This recollision process gives rise to highly interesting phenomena including high-harmonic generation [66] and attosecond pulse generation [10,67]. The studies described in this review are the first to provide information on the quantum state distribution within the residual ion using x-ray microprobe techniques.

Atoms are also affected at lower intensities where the external laser field is much weaker than the Coulomb field; the electronic structure of an atom is perturbed by the AC stark shift [20,68]. Precision measurements are often affected by AC stark shifts where Hz-level changes in transition frequencies can be a significant problem [69], but strong-field laser experiments are influenced as well. Because strong-field experiments have focused on electron energy distributions, the effects of non-resonant laser fields have been most widely investigated in this setting. When an intense laser, whose photon energy is below the ionization potential, ionizes an atom via a multiphoton process, strong-field effects are observed as above-threshold ionization (ATI) photo-electron structures [17,70]. In ATI, freed electrons absorb energy from the laser field, in so-called free–free transitions, after the initial ionization step. ATI produces photo-electrons, not only at the minimum energy ($E_{min} = n\hbar\omega_L - I_p$), but also at higher energies at separations given by the photon energy (e.g. at $E_{min} + \hbar\omega_L$). These effects are also observed as photo-electron satellite structures in two-color experiments involving single photon ionization in the presence of a laser-dressing field [7–9]. In these two-color experiments, the intensities of the laser-dressing fields are significantly smaller, 10^{10} W/cm^2 to 10^{12} W/cm^2 because the electron has already been liberated by absorption of a single high energy photon [7]. In this review we discuss a different effect of a laser-dressing field, namely the consequences of perturbed electronic structure on the x-ray absorption spectrum of an atom. For x-ray absorption spectra, significant modifications are observed only for laser-dressing field intensities of $\sim 10^{13}$ W/cm^2.

Molecules subjected to strong non-resonant laser fields will align and distort their molecular framework with respect to the polarization axes of the laser field. The past few years have witnessed a veritable explosion of work on the development of laser-based techniques to align and orient individual molecules in space [4,25,71–97]. By allowing the study of 'fixed-in-space' molecules and the control of the relative orientations of colliding molecules, such techniques are expanding our understanding of fundamental physical and chemical processes. Furthermore, the ability to orient individual biomolecules is extremely important for simplifying techniques for single-shot imaging and structural determinations at the Linac Coherent Light Source (LCLS) [27].

Two classes of laser-based alignment have been developed to produce 'fixed-in-space' molecules, and are referred to as adiabatic and impulsive alignment. Adiabatic laser alignment [25,71–76] is an extension of static DC field alignment [98–100], where the molecule is aligned preferentially along the laser polarization axis due to interaction with its anisotropic polarizability; the laser field is necessarily present to keep the molecule aligned. By contrast, in the impulsive case [77–97], molecules are aligned subsequent to the passage of a short laser pulse. This impulse coherently excites a superposition of rotational states, creating an aligned ensemble that then decays and revives at predictable intervals. The two approaches offer complementary sets of advantages and disadvantages. With adiabatic methods, the alignment is maintained for the duration of the alignment pulse, allowing the use of relatively long (~100-ps) x-ray-probe pulses, such as those produced at the Advanced Photon Source. However, the laser field is present when the molecules are being probed, which may affect the measurements. With the impulsive method, the recurrence of the alignment occurs long after the laser pulse is gone, allowing field-free measurements. However, this advantage is countered by the short duration of the aligned sample: it typically dephases within a few ps. Also, a higher degree of alignment is generally produced in the adiabatic regime.

Friedrich and Herschbach [25,71,72] have presented detailed theoretical discussions of the alignment of molecules by non-resonant laser fields. They have shown that for non-resonant frequencies much greater than the inverse of the pulse duration, the cycle-averaged laser–molecule interaction Hamiltonian is [25,71,72]

$$H_L = -\frac{1}{4}\mathcal{E}_0^2(\alpha_\| - \alpha_\perp)\left\{\cos^2\vartheta - \frac{1}{3}\right\}, \quad (1)$$

where $\mathcal{E}_0$ is the field amplitude, $\alpha_\|$ and $\alpha_\perp$ are the polarizability components parallel and perpendicular, respectively, to the molecular principal axis, and ϑ is the angle between the molecular principal axis and the laser polarization. The way we have written Equation (1), H_L is proportional to the spherical harmonic $Y_{2,0}$, i.e., H_L is a spherical tensor operator of rank two. This is exploited in Section 3.3 [Equation (36)].

When $\alpha_\| > \alpha_\perp$, as in the case of linear molecules, the molecule aligns parallel to the laser polarization axis. The degree of alignment is then usually characterized by $\langle\cos^2\vartheta\rangle$; $\langle\cos^2\vartheta\rangle = 1/3$ for randomly oriented molecules and $\langle\cos^2\vartheta\rangle = 1$ for completely aligned molecules. For molecules with typical values of polarizabilities, strong alignment is possible at laser intensities of 10^{11}–10^{12} W/cm². Stapelfeldt and co-workers [73–75] have used linearly polarized light to demonstrate strong alignment of I_2, and achieved a value of $\langle\cos^2\vartheta\rangle = 0.8$. Comparable values of $\langle\cos^2\vartheta\rangle$ were reported [75] for a number of larger molecules such as CS_2, CH_3I, and C_6H_5I. More recently, alignments as high as $\langle\cos^2\vartheta\rangle \geq 0.92$ have been achieved by adiabatic alignment at $\sim 10^{12}$ W/cm² of molecules near 1 K rotational temperature [101]. It has also been shown that under suitable conditions, an elliptically polarized alignment laser can be used to restrict rotation about all three axes of the molecule [76], making this an extremely powerful technique to control the motion of isolated molecules. The work described in this article uses the adiabatic alignment technique to produce an ensemble of aligned molecules which is then used to control resonant x-ray absorption.

3. THEORETICAL ASPECTS

Two modes (or two colors) of the radiation field are considered: The laser beam with photon energy ω_L and the x-ray beam with photon energy ω_X. They are assumed to be (essentially) monochromatic, linearly polarized, and copropagating. The polarization vector e_L of the laser defines the quantization axis, which is chosen to coincide with the z axis of the space-fixed coordinate system. Further, e_X denotes the polarization vector of the x-ray beam, and ϑ_{LX} is the angle between e_L and e_X. We assume throughout that x-ray absorption is a one-photon process (cf. Section 2.1). Atomic units are employed in this section.

3.1 Strong-Field Ionization

3.1.1 Adiabatic Ionization Model

Relativistic corrections to the electron dynamics driven by the laser field (see, for instance, Ref. [102]) are not important at intensities of the order of 10^{14} W/cm². The intra-atomic dynamics may, however, be affected by relativistic effects. This is particularly important for the heavier noble gases. For instance, the spin–orbit splitting between the $4p_{3/2}^{-1}$ and $4p_{1/2}^{-1}$ states of Kr$^+$ is relatively large (0.7 eV), so that the orbital angular momentum quantum numbers m_l may not be sufficiently well conserved during strong-field ionization. We have, therefore, developed an effective one-particle description of strong-field ionization with the inclusion of spin–orbit interaction [15]. Even though the methodology is quite general, we focus on applications to noble-gas atoms with an $l = 1$ outer-valence shell. Basic assumptions made are that the electronic response can be captured in a one-electron picture and that the electronic response is fast on the time scale of an optical cycle, i.e., the electronic motion follows the oscillating laser electric field adiabatically. The assumption of adiabaticity is consistent with both tunneling and over-the-barrier ionization mechanisms.

The basis of this model, which was used to calculate the ionic quantum state populations presented in Section 5.1, is the Hamiltonian

$$H = -\frac{1}{2}\nabla^2 + V_{\text{HS}}(r) + \frac{1}{2}\alpha^2 \frac{1}{r}\frac{dV_{\text{HS}}}{dr} l \cdot s - z\mathcal{E} - i\eta W(r). \tag{2}$$

The effective one-electron potential V_{HS} is calculated using the Hartree–Fock–Slater [103,104] code written by Herman and Skillman [105]; V_{HS} describes the interaction of an electron with the nucleus and with the mean field generated by the other electrons. Besides the Hartree–Fock–Slater approach itself, a key approximation made here is that the external electric field $\mathcal{E}$ is too weak to substantially modify the intra-atomic potential represented by V_{HS}. At field strengths of 0.1 a.u. or less, this approximation should be reasonable. Note that $\mathcal{E}$ in Equation (2) represents a *static* field along the z axis (the quantization axis). The strategy in the following is to use Equation (2) to calculate, as a function of $\mathcal{E}$, the strong-field ionization rates $\Gamma(\mathcal{E})$ of the outer-valence orbitals. Intensity-dependent ionization rates are then obtained by averaging $\Gamma(\mathcal{E})$ over a laser period. The operator in Equation (2) depending on the scalar product of the orbital angular momentum l and the spin s of the active electron describes the spin–orbit interaction (α is the fine-structure constant). The last operator in Equation (2), $-i\eta W$, is a complex absorbing potential (CAP) [106–108]. The real, non-negative parameter η is the CAP strength; the local one-electron potential $W(r)$ is defined in Ref. [15]. The CAP absorbs the strong-field-ionized electron and renders the associated continuum wave function square-integrable.

A matrix representation of the Hamiltonian H [Equation (2)] is constructed with respect to the ls-coupled state vectors

$$|n, l, j, m\rangle = \sum_{m_l, m_s} C(lsj; m_l m_s m) |n, l, m_l, m_s\rangle, \quad (3)$$

where $C(lsj; m_l m_s m)$ is a Clebsch–Gordan coefficient [109], and the $|n, l, m_l, m_s\rangle$ are eigenstates of

$$H_0 = -\frac{1}{2}\nabla^2 + V_{\text{HS}}(r) \quad (4)$$

subject to the boundary condition that the radial eigenfunctions vanish at both ends of the radial grid. The atomic model (H_0 plus spin–orbit term) is adjusted so that the outer-valence binding energies coincide with experimental values. The matrix elements of the dipole operator z and of the CAP are determined using standard angular momentum algebra [110]. The complex Siegert energy E [111–113] of each resonance state of interest is calculated by diagonalizing the matrix of H (exploiting the conservation of the projection quantum number m) and optimizing [106] the parameter η characterizing the strength of the CAP. The imaginary part of the Siegert energy is a direct measure of the ionization rate: $\Gamma = -2 \times \text{Im}(E)$. Note that Γ depends parametrically on the external electric field $\mathcal{E}$.

In the case of the heavier noble-gas atoms, the outer-valence orbitals are hardly mixed by the external electric field, even at the electric-field strength corresponding to the saturation intensity for ionization [15,16]. This means that, in addition to m, the total angular momentum j is (approximately) conserved for the outer-valence orbitals. Thus, we obtain, for the outer-valence states, Siegert energies that may be classified in terms of j, m. This allows us to assign an ionization rate $\Gamma_{j,m}$ to each j, m outer-valence subshell. In the experiments discussed in Section 5.1, electron dynamics during an optical cycle was not resolved, so that the relevant quantity is the ionization rate averaged over a laser period (in fact, a suitably chosen quarter period is sufficient):

$$\bar{\Gamma}_{j,m}[I_L] = \frac{2}{\pi} \int_0^{\pi/2} \Gamma_{j,m}(\mathcal{E}_0 \cos\varphi) d\varphi. \tag{5}$$

The relation between the field amplitude $\mathcal{E}_0$ and the intensity I_L in the case of a linearly polarized laser field is $\mathcal{E}_0 = \sqrt{8\pi\alpha I_L}$, where the intensity is measured in units of $E_h/(t_0 a_0^2) = 6.43641 \times 10^{15}$ W/cm² (E_h: Hartree energy; t_0: atomic unit of time; a_0: Bohr radius). To determine the time evolution of the atomic state distribution generated by a laser pulse, we solve the following rate equations:

$$\dot{\rho}_0(t) = -\sum_{j,m} \bar{\Gamma}_{j,m}[I_L(t)]\rho_0(t), \tag{6}$$

$$\dot{\rho}_{j,|m|}(t) = \left\{\bar{\Gamma}_{j,m}[I_L(t)] + \bar{\Gamma}_{j,-m}[I_L(t)]\right\} \rho_0(t). \tag{7}$$

Here, ρ_0 is the probability of finding a neutral atom ($\rho_0 = 1$ before the pulse), and $\rho_{j,|m|}$ is the probability of finding an ion with a hole in either the j, m or the $j, -m$ outer-valence orbital ($\rho_{j,|m|} = 0$ before the pulse). Production of higher charge states is not considered here.

3.1.2 X-Ray Probe of Valence Hole

Let $|j, m\rangle$ denote the spin–orbit-coupled eigenstates of the laser-produced ion with a hole in the outer-valence shell, which is assumed to be a p shell. The ion quantum state distribution generated by the strong laser field can be measured via resonant x-ray absorption, by exciting a 1s electron into the outer-valence hole. In the following, we briefly discuss the theory [14, 15] used to analyze the x-ray absorption data presented in Section 5.1.

We consider initial ion density matrices (right after the laser pulse is over) that are diagonal with respect to the conserved quantum number m:

$$\rho = \sum_m \sum_{j',j''} |j',m\rangle \rho^{(m)}_{j',j''} \langle j'',m|. \tag{8}$$

The x-ray absorption cross-section may be calculated using time-dependent perturbation theory, treating the atomic Hamiltonian (including spin–orbit interaction) as the unperturbed part and the interaction with the probe x-rays as the perturbation. We make the electric dipole approximation and assume that the decay of the core hole is much faster than the duration of the probe pulse. In addition, for the conditions we are interested in here, the laser-pump and x-ray-probe pulses are long in comparison to the valence spin–orbit period $T^{(SO)}$. Coherences in the ion density matrix remain unobserved. Therefore, the x-ray absorption signal depends only on the diagonal elements $\rho_{j,j}^{(m)}$. The x-ray absorption cross-section in the spectral vicinity of the transition between the valence-hole states and the K-shell-hole states is thus obtained by the usual procedure of averaging over the initial states of the ion and summing over the final states:

$$\sigma(\omega_X, \vartheta_{LX}) = \sum_{j,m} \rho_{j,j}^{(m)} \frac{2\pi\alpha\omega_X \Gamma_K}{(E_K - E_j - \omega_X)^2 + \Gamma_K^2/4}$$
$$\times \sum_{m'} |\langle j, m | d \cdot e_X | K, m' \rangle|^2. \qquad (9)$$

In this expression, E_K and Γ_K are the energy and decay width, respectively, of the state $|K, m'\rangle$ with a hole in the K-shell (filled valence shell), E_j is the energy of the valence state $|j, m\rangle$, and d is the electric dipole operator in length form.

The x-ray absorption cross-section of an unaligned ensemble of ions in the state $|j, m\rangle$ (fixed j, uniform population of m) is

$$\sigma_j(\omega_X) = \frac{2\pi\alpha\omega_X \Gamma_K}{(E_K - E_j - \omega_X)^2 + \Gamma_K^2/4} \times \frac{|\langle j||d||K\rangle|^2}{3(2j+1)}, \qquad (10)$$

where $\langle j||d||K\rangle$ is a reduced matrix element [110]. Thus, with the help of the Wigner–Eckart theorem [110], it follows from Equation (9) that

$$\sigma(\omega_X, 0°) = 2\rho_{3/2,1/2}\sigma_{3/2}(\omega_X) + \rho_{1/2,1/2}\sigma_{1/2}(\omega_X) \qquad (11)$$

and

$$\sigma(\omega_X, 90°) = \tfrac{1}{2}\{\rho_{3/2,1/2} + 3\rho_{3/2,3/2}\}\sigma_{3/2}(\omega_X) + \rho_{1/2,1/2}\sigma_{1/2}(\omega_X), \qquad (12)$$

where $\rho_{j,|m|} \equiv \rho_{j,j}^{(m)} + \rho_{j,j}^{(-m)}$. The quantum state populations $\rho_{j,|m|}$ may be estimated using the rate equations (6) and (7). Alternatively, the $\rho_{j,|m|}$ may be extracted from experimental data by employing Equations (11) and (12) in combination with accurate *ab initio* data [114] for $\sigma_{3/2}(\omega_X)$ and $\sigma_{1/2}(\omega_X)$. Both approaches are demonstrated in Section 5.1.

For some applications [13,115], we replaced Equations (11) and (12) by simplified expressions. These were obtained by identifying each valence-hole or core-hole state of the ion with a one-hole configuration derived from a single closed-shell determinant for the neutral atom. This is a one-particle model consistent with Koopmans' theorem. The total angular momentum j of the ion is then given by that of the hole with orbital angular momentum $l = 1$ and spin $s = 1/2$. This allows one to show that

$$\frac{|\langle j||d||K\rangle|^2}{2j+1} = \frac{1}{3}|\langle np||d||1s\rangle|^2, \qquad (13)$$

which is independent of j [cf. Equation (10)]. In addition, because of the large decay width of a K-shell hole, the two spin–orbit components ($j = 3/2$ and $j = 1/2$) are typically unresolved. Making these approximations, $\sigma_{3/2}(\omega_X) = \sigma_{1/2}(\omega_X)$. Hence,

$$\sigma(\omega_X, 0°) = A(\omega_X)\{4\rho_{3/2,1/2} + 2\rho_{1/2,1/2}\} \qquad (14)$$

and

$$\sigma(\omega_X, 90°) = A(\omega_X)\{\rho_{3/2,1/2} + 3\rho_{3/2,3/2} + 2\rho_{1/2,1/2}\}, \qquad (15)$$

where $A(\omega_X)$ is a common factor that contains the dependence on the x-ray photon energy.

3.1.3 Plasma Collisions

In the absence of collisions, the laser-produced quantum state populations $\rho_{j,|m|}$ are stationary after the laser pulse. The plasma environment the ions are in eventually destroys the alignment the ions initially have. The time-resolved x-ray absorption measurements described in Section 5.1 may be understood assuming that relaxation among the j, m sublevels is caused by electron–ion collisions. This leads to the following rate equations to describe the time evolution of the j, m sublevels after the laser pulse

$$\dot{\rho}_i(t) = \sum_{i' \neq i} \{k_{i' \to i}\rho_{i'}(t) - k_{i \to i'}\rho_i(t)\} n_{\text{el}}(t). \qquad (16)$$

Here, $i \equiv \{j, |m|\}$ and $n_{\text{el}}(t)$ is the electron density in the plasma. We assume that the plasma remains quasi-neutral throughout so that $n_{\text{ion}}(t) = n_{\text{el}}(t)$, where $n_{\text{ion}}(t)$ is the ion density. The initial plasma density is set equal to the target gas density, since we are interested in conditions under which ionization is saturated in the laser focal region probed by the x-rays. The plasma expansion is modeled assuming an exponential decay of the plasma density with a density-independent time constant.

To calculate the inelastic rate constants $k_{i \to i'}$ in Equation (16), we use the following physically transparent strategy. An electron scattering on an ion in the $|j = 3/2, m\rangle$ state sees a non-spherical charge distribution, with a leading quadrupole contribution in addition to the $-1/r$ potential. Magnetic interactions, e.g., spin–spin, are negligible in comparison to the electrostatic interactions considered here. The Hamiltonian of this simplified one-electron model is

$$H = -\frac{1}{2}\nabla^2 - \frac{1}{r} - \frac{1}{r^3}\sqrt{\frac{4\pi}{5}}(Q_2 \cdot Y_2), \tag{17}$$

where Q_2 is the atomic quadrupole tensor operator [116], Y_2 is a spherical harmonic referring to the angular motion of the colliding electron, and $(Q_2 \cdot Y_2)$ denotes the scalar product of the two spherical tensor operators [110]. Utilizing a Herman–Skillman [105] potential, we estimate the matrix elements of Q_2 with respect to the $|j, m\rangle$ states. The transition cross-sections are calculated employing a distorted wave Born approximation. That is, before and after the collision with the ionic quadrupole the electron is represented by a Coulomb wave [117]. We average over the electron impact direction with respect to the ion alignment axis. This is motivated by the fact that Coulomb collisions with the ions and other electrons lead to rapid randomization of the electron motion after strong-field ionization.

The energy distribution of the photo-electrons is obtained from a classical model [64], which is in reasonable agreement with observations [118]. For a linearly polarized laser, the distribution is quasi-exponential. The computed electron energy distribution and collision cross-sections are used to calculate the rate constants $k_{i \to i'} = \langle v \sigma_{i \to i'} \rangle$ ($\langle \cdots \rangle$ indicates an average over the electron speed v). Knowing the time evolution of the $\rho_{j,|m|}$, the time-dependent x-ray absorption in the parallel ($\vartheta_{LX} = 0°$) and perpendicular ($\vartheta_{LX} = 90°$) configurations can be calculated. Using Equations (14) and (15), and taking into consideration the time evolution of the plasma density, the x-ray absorption rates for the two configurations (apart from an overall time- and $j, |m|$-independent factor) are given by

$$R_\parallel(t) = \left[4\rho_{3/2,1/2}(t) + 2\rho_{1/2,1/2}(t)\right] n_{\text{ion}}(t), \tag{18}$$
$$R_\perp(t) = \left[\rho_{3/2,1/2}(t) + 3\rho_{3/2,3/2}(t) + 2\rho_{1/2,1/2}(t)\right] n_{\text{ion}}(t). \tag{19}$$

As demonstrated in Section 5.1, this model gives good agreement with experimental data.

3.2 Laser Dressing

In the previous subsection, the laser was assumed to be strong enough to ionize noble-gas atoms, so that subsequent x-ray absorption probes ions as well as neutral atoms. A related scheme is the following. If one overlaps the laser and x-ray fields in both space and time, but keeps the laser intensity just low enough to avoid excitation of the closed-shell atoms in their ground state, then the effect of the laser field is to modify only the final states that a core electron can reach via x-ray absorption. This scenario – x-ray absorption by laser-dressed noble-gas atoms – is discussed in the following [19,21]. The approach presented differs from earlier work [119,120] in that it describes the element-specific properties of the x-ray absorption cross-section in the immediate vicinity of an inner-shell edge.

3.2.1 Hamiltonian for Laser-Dressed Atoms

We are concerned here with the case in which ω_X is large enough to drive the excitation of an electron in the K-shell. At such high photon energies, electrons in higher-lying shells are rather insensitive to the x-ray field. On the other hand, inner-shell electrons are unaffected by the laser. As long as the laser intensity is small in comparison to an atomic unit, even the valence shell is only weakly modified, and this modification is expected to be similar before and after the absorption of an x-ray photon by a K-shell electron. These assumptions justify the use of a single-active-electron picture. Spin–orbit coupling in the Rydberg manifold of the active electron is neglected.

Our model is thus based on the Hamiltonian

$$H = H_{AT} + H_{EM} + H_{I,L} + H_{I,X} + W, \tag{20}$$

where

$$H_{AT} = -\frac{1}{2}\nabla^2 + V_{HS}(r) \tag{21}$$

is the atomic Hamiltonian [cf. Equation (4)],

$$H_{EM} = \omega_L a_L^\dagger a_L + \omega_X a_X^\dagger a_X, \tag{22}$$

is the Hamiltonian for the two electromagnetic-field modes (the operators $a^\dagger$ and a, respectively, create and annihilate a photon [121]),

$$H_{I,L} = \chi^T i\sqrt{\frac{2\pi}{V}\omega_L}\left[e_L a_L - e_L^* a_L^\dagger\right] \tag{23}$$

and

$$H_{\text{I,X}} = \chi^{\text{T}} i \sqrt{\frac{2\pi}{V}\omega_X} \left[e_X a_X - e_X^* a_X^\dagger \right] \qquad (24)$$

describe the atom's interaction with the laser and x-ray modes (V denotes the normalization volume of the electromagnetic field). The absorbing potential W used here is derived from smooth exterior complex scaling [122–124]. Details may be found in Ref. [19]. The operator χ in Equations (23) and (24) is the electron position vector obtained by complex scaling of the radial coordinate of the electron. In place of a Hermitian scalar product, a symmetric inner product is used in Equations (23) and (24) [107,125].

In comparison to all other interactions, the influence of the x-ray field may be considered as weak. We therefore separate the total Hamiltonian H in Equation (20) into a weakly interacting part, $H_{\text{I,X}}$, and a strongly interacting part, $H - H_{\text{I,X}} = H_{\text{AT}} + H_{\text{EM}} + H_{\text{I,L}} + W$.

3.2.2 Non-perturbative Treatment of Laser Interaction

Let the photon numbers in the absence of interaction with the atom be N_L for the laser mode and N_X for the x-ray mode, respectively. Hence, before x-ray absorption, the state of the active electron plus photon modes reads

$$|I\rangle = |1,0,0\rangle |N_L\rangle |N_X\rangle, \qquad (25)$$

where $|n,l,m_l\rangle$ is an eigenstate of H_{AT} and $|N_L\rangle |N_X\rangle$ is an eigenstate of H_{EM}. The eigenvalue corresponding to $|1,0,0\rangle$ is E_K, which is the negative of the K-shell binding energy. Since $H_{\text{I,X}}$ couples only the electronic and x-ray degrees of freedom, the accessible final states (the states reached via x-ray absorption) must have non-zero components with respect to $|n,l,m_l\rangle |N_L\rangle |N_X - 1\rangle$, where $l = 1$. To calculate the laser-dressed final-state manifold, we represent $H - H_{\text{I,X}}$ in the basis formed by the direct-product states

$$|\Phi_{n,l,m_l,\mu}\rangle = |n,l,m_l\rangle |N_L - \mu\rangle |N_X - 1\rangle. \qquad (26)$$

Here, the quantum numbers n, l, and m_l correspond to orbitals that are unoccupied in the atomic ground state. The number of laser photons that are absorbed (emitted) by the core-excited electron is denoted by $\mu = 0, \pm 1, \pm 2, \ldots$. All K-shell-excited states undergo rapid relaxation via Auger decay or x-ray emission. To take these effects into consideration in our effective one-electron description, we note that the decay of a K-shell hole involves primarily other inner-shell electrons; the excited electron is a spectator. We, therefore, assign a width Γ_K to each $|n, l, m_l\rangle$ accompanied by a core hole in the many-particle wave function. To a good approximation, Γ_K may be assumed to be independent of the laser field and the quantum numbers of the spectator electron.

We diagonalize the matrix representation of $H - H_{\text{L,X}}$, assuming $N_L \gg 1$. This matrix is block-diagonal with respect to m_l, is complex symmetric, and is of the Floquet type [126–131]. The eigenvector coefficients obtained by matrix diagonalization are denoted as $c_{n,l,\mu,F}^{(m_l)}$, so that the laser-dressed core-excited states are given by

$$|F^{(m_l)}\rangle = \sum_{n,l,\mu} c_{n,l,\mu,F}^{(m_l)} |n, l, m_l\rangle |N_L - \mu\rangle |N_X - 1\rangle. \tag{27}$$

The corresponding eigenvalue is $E_F^{(m_l)}$.

3.2.3 X-Ray Absorption Cross-Section

As shown in Ref. [19], the non-Hermitian Rayleigh–Schrödinger perturbation theory of Ref. [132] may be applied to calculate the x-ray absorption cross-section in the presence of the laser, i.e., the cross-section for an $H_{\text{L,X}}$-induced one-photon transition from $|I\rangle$ [Equation (25)] to the manifold spanned by the $|F^{(m_l)}\rangle$ [Equation (27)]. The result is

$$\sigma(\omega_X, \vartheta_{LX}) = \frac{4\pi}{3} n_K \alpha \omega_X \sum_{m_l=-1}^{1} \varkappa_{m_l}(\vartheta_{LX})$$

$$\times \operatorname{Im}\left[\sum_F \frac{(d_F^{(m_l)})^2}{E_F^{(m_l)} - E_K - \omega_X}\right], \tag{28}$$

where $n_K = 2$ accounts for the number of electrons in the K-shell,

$$\varkappa_{m_l}(\vartheta_{LX}) = \begin{cases} (1/2)\sin^2(\vartheta_{LX}), & m_l = \pm 1, \\ \cos^2(\vartheta_{LX}), & m_l = 0, \end{cases} \tag{29}$$

expresses the dependence of the x-ray absorption cross-section on the angle between the laser and x-ray polarization vectors,

$$d_F^{(m_l)} = \sum_n c_{n,1,0,F}^{(m_l)} \int_0^\infty u_{n,1}(r) r u_{1,0}(r) dr \tag{30}$$

is a radial transition dipole matrix element, and $u_{n,l}(r)$ is the radial eigenfunction associated with $|n, l, m_l\rangle$.

3.2.4 Three-Level Model

As discussed in Section 5.2, laser dressing leads to a controllable suppression (electromagnetically-induced transparency or EIT) of x-ray absorption on a 1s → np transition when the laser resonantly couples the core-excited $1s^{-1}n$p state to a $1s^{-1}n$s (or a $1s^{-1}n'$d) state. In order to understand the reason for this effect, we reduce the *ab initio* theory described above to a three-level model. To this end, we identify $1s^{-1}n\mathrm{p}_z$, $1s^{-1}n$s, and the atomic ground state with levels a, b, and c, respectively. (Dipole selection rules dictate that the laser can couple only $1s^{-1}n\mathrm{p}_z$ – not $1s^{-1}n\mathrm{p}_x$ or $1s^{-1}n\mathrm{p}_y$ – to $1s^{-1}n$s.) Levels a and b are decaying states, due to both Auger decay (and x-ray fluorescence) and laser ionization, with widths Γ_a and Γ_b. The energy spacing between a and b is symbolized by ω_{ab}; similarly, ω_{ac} denotes the spacing between a and c. The Hamiltonian H [Equation (20)] is represented in the quasi-degenerate direct-product basis $\{|c\rangle|N_L\rangle|N_X\rangle, |b\rangle|N_L+1\rangle|N_X-1\rangle, |a\rangle|N_L\rangle|N_X-1\rangle\}$. The strong laser coupling between a and b is treated by analytically diagonalizing H in the corresponding two-dimensional subspace.

The two dressed states can be probed by the x-rays at $\vartheta_{LX} = 0°$. The x-ray absorption cross-section obtained from Equation (28) is

$$\frac{\sigma(\omega_X, 0°)}{\sigma_0} = \frac{4\Gamma_a^2 \Delta_{LX}^2 + \Gamma_a \Gamma_b (\Omega_{ab}^2 + \Gamma_a \Gamma_b)}{[\Omega_{ab}^2 + \Gamma_a \Gamma_b - 4\Delta_{LX}(\omega_{ac} - \omega_X)]^2 + 4[\Gamma_a \Delta_{LX} + \Gamma_b(\omega_{ac} - \omega_X)]^2}. \tag{31}$$

Here, Ω_{ab} is the Rabi frequency [133] associated with levels a and b, σ_0 is the cross-section obtained on resonance ($\omega_X = \omega_{ac}$) without laser ($\Omega_{ab} \to 0$), and $\Delta_{LX} \equiv (\omega_{ac} - \omega_X) - (\omega_{ab} - \omega_L)$. Equation (31) is identical to the absorption cross-section given in Ref. [134] for EIT in a three-level model. Because the lifetime of the laser-dressed core-excited states considered here is only of the order of 1 fs or less, we neglect other decoherence mechanisms such as collisional broadening. If the width Γ_b in Equation (31) were zero ($\Gamma_a \neq 0$, $\Omega_{ab} \neq 0$), then the absorption cross-section would vanish completely provided the resonance condition $\Delta_{LX} = 0$ is satisfied. For finite Γ_b, there is in general a suppression of absorption on resonance. This suppression of absorption characteristic of EIT is due to the destructive interference of the excitation pathways from the atomic ground state to the two laser-dressed core-excited states.

3.3 Laser-Induced Molecular Alignment

In this subsection, the focus is on x-ray absorption by laser-aligned molecules. The laser intensity is assumed to be low enough that the molecular electronic structure remains essentially unperturbed. The alignment mechanism is based on the dependence of the AC (or dynamic) Stark effect on the spatial orientation of the molecular polarizability tensor with respect to the laser polarization axis. The potential energy of the molecule in the laser field is minimized by aligning the most polarizable axis along the laser polarization axis. Since x-ray absorption resonances depend on the orientation of the x-ray polarization vector in the molecular frame, laser-induced alignment may be employed to control resonant x-ray absorption [135,136].

To arrive at computationally tractable expressions, we exploit the fact that, for x-ray absorption near the K edge of a heavy element such as Br (cf. Section 5.3), three different time scales may be distinguished. First, the fastest time scale is defined by the x-ray oscillations. At the K edge of a bromine atom, the x-ray period is ~0.3 as. Such a short period means, for instance, that the rotational degrees of freedom of the molecule are unable to respond to the presence of the x-rays. The second time scale is related to the relaxation of the core hole formed by x-ray absorption. A K vacancy in a bromine atom relaxes with a decay time of ~0.3 fs. This relaxation defines the coherence time scale, i.e., the time over which x-ray absorption amplitudes instead of intensities must be integrated and thus interference effects in x-ray absorption may occur. Also relevant are the characteristic times on which the rotational response occurs and the laser and x-ray pulse envelopes change. The shortest of these times defines the third time scale. We assume this third time scale to be in the range from a few tens of femtoseconds to several picoseconds. This allows us to treat the processes on all three time scales separately [136].

3.3.1 Laser Alignment of Molecules

In view of the experimental study presented in Section 5.3, we limit the discussion to prolate symmetric-top molecules [137]. The Hamiltonian describing free rotational motion in the rigid-rotor approximation reads

$$H_r = BJ^2 + (A - B)J_c^2, \tag{32}$$

where A and B are rotational constants ($A > B$). The Hamiltonian H_r commutes with the square of the angular momentum operator J^2, the c component of the angular momentum in the body-fixed frame J_c [137] (c refers to the symmetry axis of the molecule), and the z component of the angular momentum in the space-fixed frame J_z. The eigenfunctions $|JKM\rangle$ ($J = 0, 1, \ldots$, $K, M = -J, -J + 1, \ldots, J - 1, J$) of J^2, J_c, and J_z are also eigenfunctions of H_r [138]. The associated energy eigenvalues are [137,138]

$$E_{JK} = BJ(J + 1) + (A - B)K^2. \tag{33}$$

Let T_{rot} be the rotational temperature (in units of energy) of the molecules before their interaction with the laser and x-ray pulses. Then, the initial density operator of the freely rotating symmetric-top molecules is given by

$$\varrho_{\text{rot}} = \sum_{J,K,M} |JKM\rangle g_I(J,K) \frac{e^{-E_{J,K}/T_{\text{rot}}}}{Z} \langle JKM|. \tag{34}$$

The weights $g_I(J, K)$ reflect the nuclear spin statistics [137,139] and

$$Z = \sum_J (2J+1) \sum_{K=-J}^{J} g_I(J,K) e^{-E_{J,K}/T_{\text{rot}}} \tag{35}$$

is the partition function.

The instantaneous laser electric field induces in the molecule an electric dipole moment, which depends on the molecular polarizability components parallel ($\alpha_\parallel$) and perpendicular ($\alpha_\perp$) to the molecular symmetry axis (c axis). As long as the molecule has no low-lying electronic resonances, the static polarizability components differ very little from the dynamic polarizability components at 800 nm. The polarizability components may be calculated utilizing, for instance, coupled-cluster linear response theory [140,141]. The laser-induced electric dipole moment itself interacts with the laser electric field. Thus, after averaging over an optical cycle, an effective laser–molecule interaction potential $H_L(t)$ is obtained [Equation (1)] that is proportional to the cycle-averaged laser intensity $I_L(t) = \mathcal{E}_0^2(t)/(8\pi\alpha)$. Using standard angular momentum techniques, the matrix elements of the interaction Hamiltonian with respect to the $|JKM\rangle$ read

$$\langle JKM|H_L(t)|J'K'M'\rangle = -\frac{4}{3}\pi I_L(t)(\alpha_\parallel - \alpha_\perp)\delta_{KK'}\delta_{MM'}$$
$$\times \sqrt{\frac{2J'+1}{2J+1}} C(J'2J; K0K)C(J'2J; M0M). \tag{36}$$

By numerically solving the time-dependent Schrödinger equation with the Hamiltonian $H_r + H_L(t)$ for each rotational state initially populated in ϱ_{rot} [cf. Equation (34)], one determines the time evolution of the rotational density matrix. In view of Equation (36), $\varrho_{\text{rot}}(t)$ remains diagonal with respect to K and M for all t. To quantify the degree of alignment along the laser polarization axis, one may use the expectation value

$$\langle \cos^2 \vartheta \rangle = \text{Tr}[\varrho_{\text{rot}}(t) \cos^2 \vartheta]. \tag{37}$$

Here, ϑ is the angle between the molecular symmetry axis and the laser polarization axis.

3.3.2 X-Ray Absorption

Making the Born–Oppenheimer approximation, the complete Hamiltonian in the presence of both laser and x-ray fields reads

$$H(t) = [H_r + H_L(t)] \otimes \mathbb{I}_{\text{el}} + \mathbb{I}_{\text{rot}} \otimes H_M + H_X(t), \tag{38}$$

where $\mathbb{I}_{\text{el}}$ is the unit operator in the space of electronic states and $\mathbb{I}_{\text{rot}}$ is the unit operator in the manifold of symmetric-rotor states $|JKM\rangle$. Further, H_M denotes the fixed-nuclei molecular Hamiltonian [142], and

$$H_X(t) = -(\boldsymbol{d} \cdot \boldsymbol{e}_X)\sqrt{8\pi\alpha\omega_X J_X(t)} \cos(\omega_X t) \tag{39}$$

describes the interaction with the x-rays. Here, $J_X(t)$ is the x-ray flux density. We evaluate the scalar product between the dipole operator $\boldsymbol{d}$ and the x-ray polarization vector $\boldsymbol{e}_X$ in the molecule-fixed coordinate system, so that $\boldsymbol{e}_X$ is an operator acting on the $|JKM\rangle$. The molecule-fixed dipole operator $\boldsymbol{d}$, on the other hand, is independent of the Euler angles connecting the molecule- and space-fixed frames.

A wavepacket evolving under the action of $H(t)$ may be expanded in the direct-product basis

$$|JKM, i\rangle = |JKM\rangle \otimes |i\rangle, \tag{40}$$

where the electronic states $|i\rangle$ are eigenstates of H_M with energy E_i. Initially, the molecule is in the electronic ground state $|0\rangle$. The initial density operator of the system is therefore given by

$$\varrho = \varrho_{\text{rot}} \otimes |0\rangle\langle 0|. \tag{41}$$

One can determine the time evolution of the density matrix from the time-dependent Schrödinger equation with $H(t)$ by applying first-order perturbation theory with respect to $H_X(t)$ and exploiting the separation between the three time scales mentioned earlier. In this way, the instantaneous x-ray absorption cross-section of a molecule in a laser field is obtained [136]:

$$\begin{aligned}\sigma(t) &= \Gamma_K \sum_{JKM, i \neq 0} \frac{\langle JKM, i|\varrho(t)|JKM, i\rangle}{J_X(t)} \\ &= 4\pi\alpha\omega_X \sum_{JJ'KM} \langle JKM|\varrho_{\text{rot}}(t)|J'KM\rangle\end{aligned}$$

$$\times \sum_{J''K''M'',i''} (\langle 0|d|i''\rangle \cdot \langle J'KM|e_X|J''K''M''\rangle)$$

$$\times (\langle i''|d|0\rangle \cdot \langle J''K''M''|e_X|JKM\rangle)$$

$$\times \operatorname{Im} \frac{1}{E_{J''K''} + E_{i''} - i\Gamma_K/2 - E_{J'K} - E_0 - \omega_X}. \quad (42)$$

This expression depends on the angle between the laser and x-ray polarization vectors via $\varrho_{\mathrm{rot}}(t)$ and e_X. For comparison with experiment (see Section 5.3), Equation (42) may be used to calculate the x-ray absorption probability as

$$P_i(\tau) = \int_{-\infty}^{\infty} \sigma_i(t) J_X(t-\tau) dt, \quad i = \|, \perp, \text{th}. \quad (43)$$

Here, τ is the time delay between the laser and x-ray pulses, '$\|$' and '$\perp$' denote, respectively, parallel and perpendicular laser and x-ray polarization vectors, and 'th' refers to a thermal (laser-free) ensemble.

4. GENERAL EXPERIMENTAL CONSIDERATIONS

In addition to the theoretical challenges outlined above, experiments combining strong optical fields with x-ray beams [13,144] provide several practical challenges as well. To understand the problems, let us consider the laser parameters necessary to produce strong optical fields. Assume a typical pulse energy of 1 mJ. When focused to 1 cm^2, with a pulse length of 100 ps, that produces a rather modest intensity of only 10^7 W/cm^2. That same pulse energy however, when focused to (100 μm)^2 and compressed to 100 fs, produces an intensity of 10^{14} W/cm^2, sufficiently strong to ionize even rare gas atoms. This means, in order to utilize optical fields strong enough to affect atomic x-ray processes, one must work with very small interaction volumes, and very short time scales.

To probe atoms, ions, and molecules in such small interaction volumes with x-rays requires spatial and temporal overlap of x-ray, laser, and the atoms or molecules being investigated. This necessitates microfocusing x-ray beams to sizes smaller than the laser focus, and a means of arranging for the temporal overlap of the x-ray and laser pulses in the interaction region. Just recently we reported such an x-ray microprobe [13]. That apparatus incorporates the use of microfocused x-rays into a laser-pump/x-ray-probe configuration. This represents a gain of a factor of 10^4 in x-ray flux density over standard laser/x-ray pump–probe experiments [145], thus enabling gas-phase experiments with only 10^7–10^9 atoms in the focal volume. We have carried out such experiments with an effusive gas jet [13], a flowing gas cell [115], and a supersonic gas jet [135]. With these

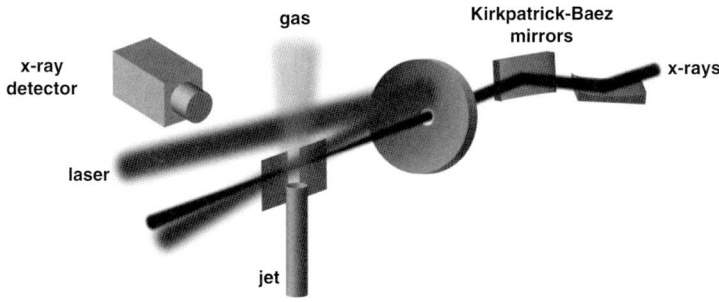

FIGURE 1 Schematic view of a typical apparatus in which microfocused x-rays probe atoms in the focus of a high-intensity laser [13]. Reprinted with permission from Ref. [13] © 2006, American Physical Society

gas sources, we can produce target gas densities spanning the range of 10^{12}–10^{16}/cm^3 in the interaction region.

The interaction region is defined by the overlap in the gas volume with the focus of a Ti:sapphire ultrafast laser system that is phase-locked to the radio-frequency (RF) clock of the synchrotron's electron storage ring. Typical laser parameters are 800 nm, 1–2 mJ/pulse, pulse length 65 fs, and focal width <100 μm FWHM. More detailed descriptions of the laser system and its synchronization with the x-ray pulses can be found in Refs. [146,147].

In our microprobe (see Figure 1), we use monochromatized x-rays (∼10–20 keV) from an undulator at Sector 7 of the Argonne Advanced Photon Source. These are focused efficiently by a large-aperture, dynamically bent Kirkpatrick–Baez (KB) mirror pair [148] to a FWHM of 10 μm. The x-ray flux after focusing is ∼10^6 photons/pulse, with a repetition rate of 271 kHz and energy resolution $\Delta E/E \sim 10^{-4}$.

To detect x-ray absorption in the gas, the interaction region is viewed by detectors to collect photons produced therein (see Figure 1). Hard x-rays, produced predominantly by fluorescence following K-vacancy formation (elastic scattering is minimized by detection at 90° in the scattering plane), are recorded by either energy-dispersive detectors (SiLi or Si-drift) or fast timing detectors (scintillators or avalanche photodiodes) depending on the needs of the particular experiment being run.

Spatial overlap of the laser and x-ray beams is initially achieved by scanning opaque metallic knife edges through the x-ray and laser beams and observing the transmitted x-rays with an ion chamber and the laser with a photodiode. Translating the knife edges horizontally, vertically, and longitudinally along the beams, we determine the separate trajectories of laser and x-ray beams and adjust the former for maximal overlap with the x-rays focused in the interaction region. The knife edges can be scanned with 1 μm resolution and for a typical crossing angle of the two beams of ∼25 mrad, the overlap region will extend ∼4 mm.

Temporal overlap is achieved in two steps. Initial overlap is established by observing the laser and x-ray pulses on an avalanche photodiode downstream of the interaction region and noting the time, relative to the 271 kHz rf timing signal of the storage ring. This is adequate to achieve a sufficient degree of overlap between the fs laser pulses and the ∼100-ps x-ray pulses so that a physical effect can be observed that is only present upon overlap. As an example, consider the 1s → 4p excitation in Kr^+. In neutral Kr, the 4p shell is filled, but in the Br-like Kr^+ ions created by a strong-field laser ($>10^{14}$ W/cm^2), there is a 4p vacancy created. Thus, in the region of overlap between the laser and x-rays, there is a subthreshold 1s → 4p resonance which can be excited by the x-ray beam (see Figure 2). The height of this resonance, in comparison to the same excitation region when the laser is off, is a measure of the degree of overlap in both space and time (Figure 3). That signal can then be used to fine tune the overlap of the beams with the detection viewing region.

These are the essential features common to all experiments in which a laser is overlapped with a beam of x-rays. A detailed description of experimental methods is given in Ref. [14]. We describe below examples where such techniques were exploited and x-rays were used to probe atoms, ions, and molecules exposed to strong optical fields.

5. CASE STUDIES

5.1 Orbital Alignment in Strong-Field Ionization

When atoms are exposed to strong optical fields ($\sim 10^{14}$–10^{15} W/cm^2), they rapidly ionize losing their least bound electrons through tunnel ionization [12]. For example, using the adiabatic strong-field ionization theory from Section 3.1, and defining the saturation intensity in accordance with Ref. [149], we calculate that for Kr, single ionization saturates at an intensity of 1.6×10^{14} W/cm^2, in agreement with experiment [118,150]. When a gaseous assembly of atoms is exposed to such fields, it creates a transient laser-produced plasma. We recently showed how the microprobe technique described in Section 4 can be used to investigate the ions and electrons created in such laser-produced plasmas [13,14,115].

In experiments with linearly polarized laser light, we discovered a strong degree of atomic alignment persisting for times far beyond the spin–orbit time scale (6.2 fs) in Kr [13,15]. The non-relativistic tunnel ionization model predicts a strong propensity for removing a 4p, $m_l = 0$ electron [151], where the quantization axis is taken along the linear polarization direction of the laser field. A dependence of the 1s → 4p cross-section on the angle between the linear polarization vectors of the laser and x-rays is expected and observed (see Figure 4), but the measured

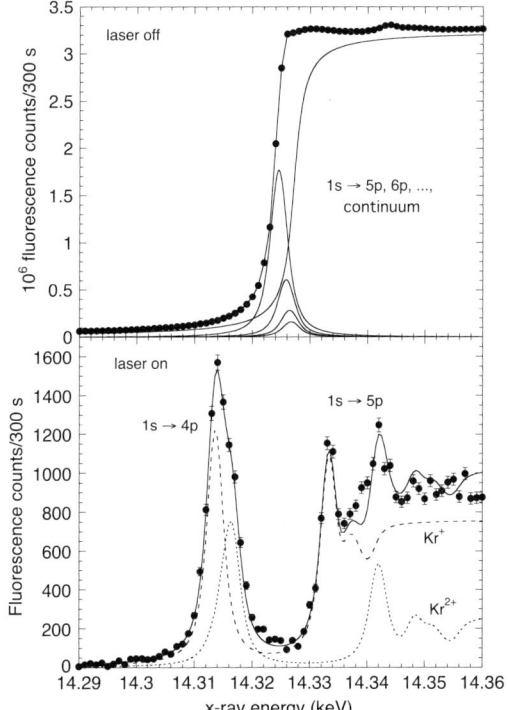

FIGURE 2 K-edge photo-absorption spectra determined from x-ray fluorescence yields for neutral Kr (Top) and laser-ionized Kr ions (bottom) [14]. The ions were generated by circularly polarized laser light, and the residual neutral Kr component has been subtracted. Reprinted with permission from Ref. [14] © 2007, American Physical Society

parallel-to-perpendicular cross-section ratio (≈ 2) is much smaller than that predicted by the non-relativistic models by nearly a factor of 20 [13]! Inclusion of spin–orbit coupling, as described in Section 3.1, leads to calculated cross-section ratios close to the measured ratio [15]. These results highlight the roles of orbital alignment and spin–orbit coupling in the laser ionization process.

A more detailed experimental test of our laser ionization theory (Section 3.1) is given by determining the Kr$^+$ quantum state populations $\rho_{j,|m|} = \rho_{j,j}^{(m)} + \rho_{j,j}^{(-m)}$ [Equation (8)]. Experimental $\rho_{j,|m|}$ were obtained by fitting the measured parallel and perpendicular cross-sections to 1s → 4p$_{3/2}$ and 4p$_{1/2}$ components (see Figure 5). The fit was based on Equations (11) and (12), and the normalization condition $\rho_{3/2,1/2} + \rho_{3/2,3/2} + \rho_{1/2,1/2} = 1$. Resonant x-ray absorption cross-sections for unaligned Kr$^+$ were required for Equations (11) and (12). These were

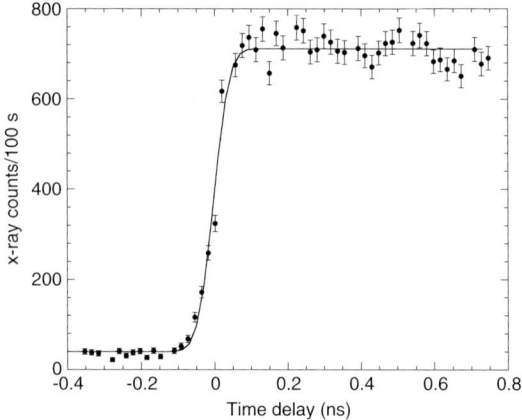

FIGURE 3 Cross correlation scan obtained by measuring x-ray fluorescence counts excited at the $Kr^+ 1s \to 4p$ resonance vs time delay of the x-ray pulse relative to the laser pulse [14]. Only x-rays impinging *after* the laser ionization can excite the resonance. The fitted curve is a Gaussian error function with 98 ± 5-ps FWHM, which is a measure of the x-ray pulse duration in this experiment. Reprinted with permission from Ref. [14] © 2007, American Physical Society

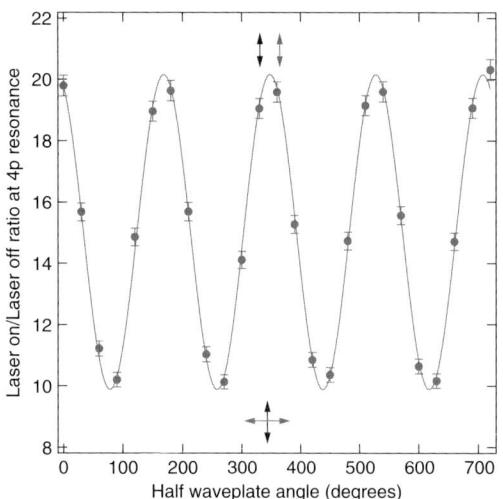

FIGURE 4 The laser-on/laser-off ratio of x-ray fluorescence yield at the $1s \to 4p$ resonance as a function of the angle between laser and x-ray polarization directions [13]. That ratio for parallel polarizations is twice as large as the transverse case showing evidence of residual orbital alignment in strong-field-ionized Kr. Reprinted with permission from Ref. [13] © 2006, American Physical Society

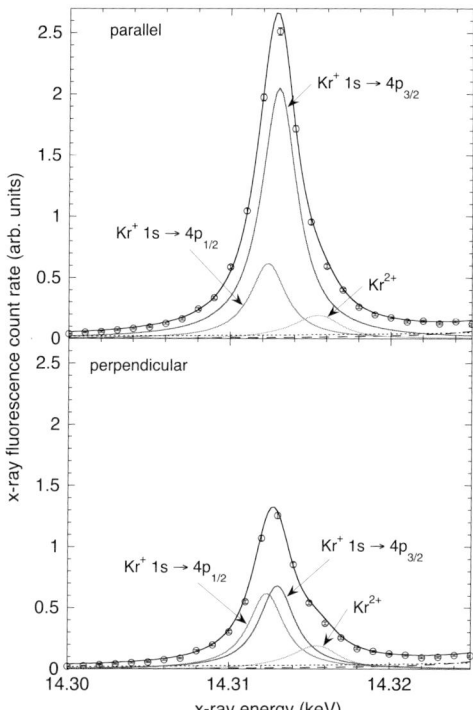

FIGURE 5 The 1s → 4p transitions determined from x-ray fluorescence counting rates (circles) for ions generated with linearly polarized laser light with polarization parallel (top) and perpendicular (bottom) to the x-ray polarization [14]. The major fitted components are $Kr^+ 1s \rightarrow 4p_{1/2}$, $Kr^+ 1s \rightarrow 4p_{3/2}$ and Kr^{2+}. The tails of the higher Rydberg excitations (dashed curves) and continuum (dotted curves) of Kr^+ are also plotted but make minimal contributions in this region of the spectrum. The solid black curves are the total fits. Reprinted with permission from Ref. [14] © 2007, American Physical Society

taken from the calculated transition rates of Ref. [114]. We determined theoretical $\rho_{j,|m|}$ by solving Equations (6) and (7) for a Gaussian temporal laser pulse profile with a peak intensity of 1.9×10^{14} W/cm², which roughly matches the experimental conditions for the data in Figure 5.

Table 1 is a comparison of the experimentally and theoretically determined quantum state populations of strong-field-generated Kr^+. It is evident that the adiabatic strong-field ionization model we use reproduces the general trend displayed by the experimental state populations. The results of theory and experiment are in agreement that the majority of Kr^+ ions produced are in the $j = 3/2$ state. Theory predicts a noticeable degree of alignment in this state: $\rho_{3/2,1/2}/\rho_{3/2,3/2} = 17.8$. The experimental data also are consistent with Kr^+ alignment. However, the experimental ratio

TABLE 1 Ion quantum state populations $\rho_{j,|m|}$, following strong-field ionization of Kr [14] and Xe [16], respectively.

	$\rho_{3/2,1/2}$ (%)	$\rho_{1/2,1/2}$ (%)	$\rho_{3/2,3/2}$ (%)
Kr$^+$, experimental	59 ± 6	35 ± 4	6 ± 6
Kr$^+$, theoretical	71	25	4
Xe$^+$, experimental	75 ± 6	12 ± 3	13 ± 6
Xe$^+$, theoretical	83	14	3

Reprinted with permission from Refs. [14] and [16]
© 2007 American Physical Society

$\rho_{3/2,1/2}/\rho_{3/2,3/2} \approx 10$ is on the low side of the theoretical ratio, which may be an indication of a partial failure of the adiabatic picture. The theoretical population of the $j = 1/2$ state is also smaller than the measured value. We note that it requires the same number of photons (at 800 nm) to ionize either 4p$_{3/2}$ or 4p$_{1/2}$, and the experimental populations are consistent with the 2:1 statistical ratio. The adiabatic strong-field ionization model, which assumes an infinite wavelength, artificially suppresses the ionization of the more strongly bound 4p$_{1/2}$ orbital.

It is interesting to compare the results for Kr with a recent study of strong-field laser ionization of Xe in which extreme ultraviolet high-harmonic radiation was used as a 4d → 5p probe of Xe$^+$ [16]. Similar theoretical calculations were employed in both cases to determine quantum state populations. In Xe, it takes one more 800-nm photon to ionize 5p$_{1/2}$ than for 5p$_{3/2}$. In that case, the adiabatic strong-field ionization model predicts the $j = 1/2$ population quite accurately (see Table 1). However, the alignment of the $j = 3/2$ state, as characterized by $\rho_{3/2,1/2}/\rho_{3/2,3/2}$, was overestimated by more than a factor of four.

These data provide an experimental test of an adiabatic strong-field ionization theory that explicitly treats spin–orbit coupling [15]. The theoretical results are in general accord with the measurements, but limitations of the adiabatic model are indicated. This experiment and that of Ref. [16] demonstrated how spectroscopic probes of ions can yield new insight into strong-field ionization.

In the absence of interactions with the plasma environment, and neglecting spontaneous magnetic-dipole decay from the $j = 1/2$ state, the ion quantum state populations are stationary. Experiments have been carried out to investigate the persistence of the ion alignment as the laser-produced plasma evolves [115]. As seen in Figure 6, we found, in addition to an overall decay in x-ray absorption because of the expansion of the localized plasma cloud, the parallel-to-perpendicular cross-section ratio decays with time from about two to one. At a gas density of 1.3×10^{14} cm^{-3}, the orbital alignment disappears after about 20 ns. With increasing density, the quenching of alignment is enhanced. We were able to demonstrate that the cause of this relaxation of the magnetic sublevel population was

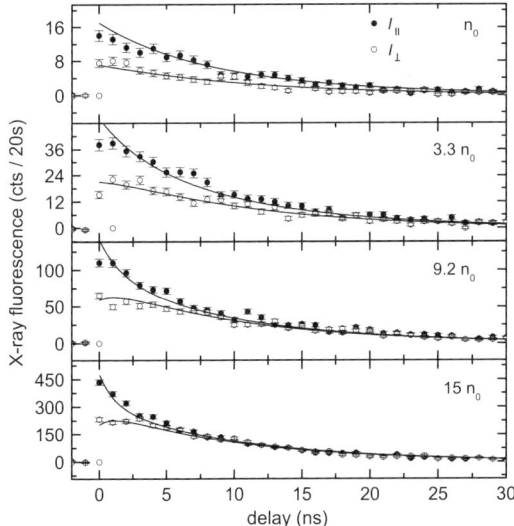

FIGURE 6 Decay of orbital alignment in Kr$^+$ ions for different gas densities ($n_0 = 1.3 \times 10^{14}$ cm^{-3}) [115]. Plotted are the Kr$^+$ 1s → 4p resonance fluorescence vs the time delay of the x-rays from the laser pulse for the linear laser polarization parallel and perpendicular to the x-ray polarization. Laser intensity was 10^{15} W/cm^2. Solid curves show the theoretical results. Reprinted with permission from Ref. [115] © 2007, American Physical Society

e$^-$–Kr$^+$ collisions. A model was developed as described in Section 3.1 and shows good agreement (see Figure 6) with the experimental results. By applying an external magnetic field of ~400 G, coherent spin precession of the Kr ions may be induced [115].

5.2 Electromagnetically-Induced Transparency

In this subsection, we apply the theory described in Section 3.2 to investigate applications to x-ray absorption by laser-dressed noble-gas atoms. To illustrate the differences between heavy and light noble-gas species, we discuss results for both krypton [19] and neon [21]. Unless noted otherwise, the laser is assumed to operate at an optical wavelength of 800 nm and an intensity of 10^{13} W/cm^2, and both laser and x-ray fields are assumed to be linearly polarized. Because of their high binding energy, the core and valence electrons of Kr and Ne remain essentially unperturbed at 10^{13} W/cm^2. However, laser dressing of the core-excited states introduces strong-field physics: For the laser parameters employed here, the ponderomotive potential is 0.60 eV, and the energy needed to ionize the Rydberg electron in the $1s^{-1}nl$ state is at most a few eV. Hence, the Keldysh parameter is of the order of one. In addition, the laser photon energy can resonantly couple at least some of the Rydberg

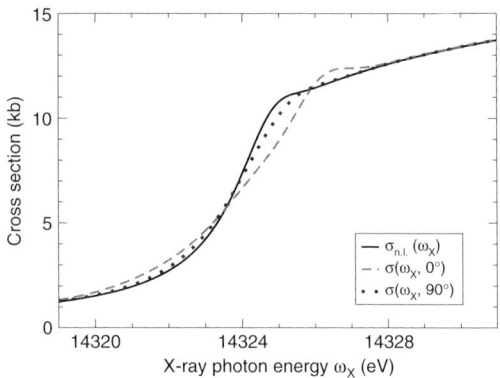

FIGURE 7 x-ray photo-absorption cross-section of the krypton atom near the K edge with laser dressing, $\sigma(\omega_X, \vartheta_{LX})$ and without it, $\sigma_{n.l.}(\omega_X)$ [19]. Here, ϑ_{LX} is the angle between the laser and x-ray polarization vectors. The laser operates at a wavelength of 800 nm with an intensity of 10^{13} W/cm^2. Reprinted with permission from Ref. [19] © 2007, American Physical Society

states. Therefore, neither perturbation theory (which requires a Keldysh parameter much greater than one) nor an adiabatic description (which implies an infinite wavelength) is adequate.

5.2.1 Krypton

The calculated x-ray photo-absorption cross-section of krypton is plotted in Figure 7 for three cases: (a) The cross-section without laser dressing, $\sigma_{n.l.}(\omega_X)$, (b) the cross-section for parallel polarization vectors, denoted by $\sigma^{\|}(\omega_X) = \sigma(\omega_X, 0°)$, and (c) the cross-section for perpendicular polarization vectors, denoted by $\sigma^{\perp}(\omega_X) = \sigma(\omega_X, 90°)$. In view of Equations (28) and (29), the x-ray absorption cross-section for arbitrary angle between the laser and x-ray polarization vectors can be constructed from $\sigma^{\|}(\omega_X)$ and $\sigma^{\perp}(\omega_X)$.

As may be seen in Figure 7, $\sigma^{\|}(\omega_X)$ is suppressed with respect to the laser-free curve $\sigma_{n.l.}(\omega_X)$ between ~14 323 eV and ~14 326 eV. This interval is centered at the 1s → 5p transition [19]. (The cross-section $\sigma^{\perp}(\omega_X)$ behaves in a similar way, though the deformation relative to $\sigma_{n.l.}(\omega_X)$ is much less pronounced.) Within the 2.7-eV decay width of the Kr core hole, 1s^{-1}5p is in resonance with 1s^{-1}5s, 1s^{-1}4d, and 1s^{-1}5d with respect to the emission or absorption of one laser photon. At $\vartheta_{LX} = 0°$, the laser couples 1s^{-1}5p mainly to 1s^{-1}5s and 1s^{-1}4d. This coupling leads to the suppression of $\sigma^{\|}(\omega_X)$ at the 1s → 5p transition.

Qualitatively, the three-level model introduced in Section 3.2 predicts this manifestation of electromagnetically-induced transparency (EIT) in resonant x-ray absorption. The three-level model also allows us to

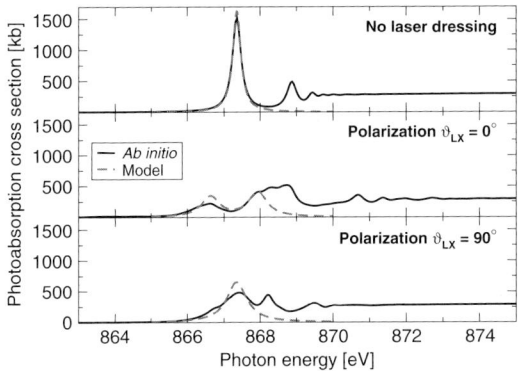

FIGURE 8 x-ray photo-absorption cross-section of neon near the K edge [21]. Results from *ab initio* calculations and a three-level model are shown. Reprinted with permission from Ref. [21] © 2007, American Physical Society

estimate the laser intensity required to observe a significant EIT effect. EIT appears when the two dressed states (see Section 3.2) are energetically separated by more than the core-hole decay width Γ_K (laser broadening of the lines is neglected here). The energy separation is given by $|\Omega_{ab}|$, i.e., the magnitude of the Rabi frequency associated with the two upper, core-excited levels in the three-level model. Hence, $|\Omega_{ab}|$ must be greater than Γ_K. In the case of Kr, this criterion implies a minimum intensity of about 10^{14} W/cm². As we saw in Section 5.1, Kr atoms get strong-field-ionized at such a high intensity, so that the assumptions underlying the theory developed in Section 3.2 become invalid. In order to create the right conditions for EIT in the x-ray regime, we have to choose an atom that either has a smaller K-shell decay width or can stand higher intensities. Neon satisfies both criteria.

5.2.2 Neon

The calculated x-ray photo-absorption cross-section of a neon atom is displayed in Figure 8 for the same three cases already discussed in connection with Figure 7. The prominent absorption feature at 867.4 eV, in the laser-free case, arises due to excitation into the $1s^{-1}3p$ state. The 1s ionization threshold lies at 870.2 eV, and the decay width of core-excited Ne is 0.27 eV [143], i.e., Γ_K is smaller than in Kr by an order of magnitude. In the presence of the laser, the absorption peaks are broader by about a factor of two in comparison to the laser-free case. This is because the laser ionizes the Rydberg electron at a rate that is comparable to the Auger decay rate of the K-shell hole. The most eye-catching impact of the laser dressing can be seen in the vicinity of the 1s → 3p resonance, which is highly suppressed for $\vartheta_{LX} = 0°$. This is the EIT phenomenon we will

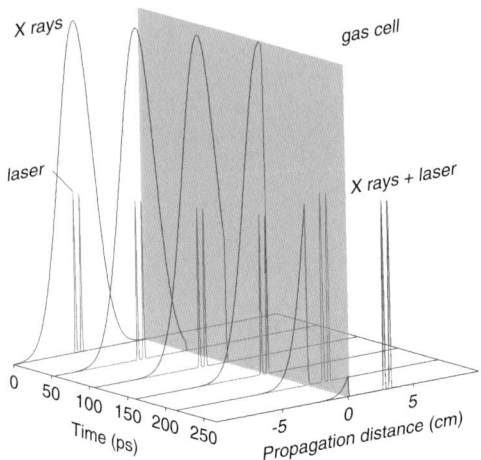

FIGURE 9 Generation of ultrafast x-ray pulses using laser dressing of Ne. A 100-ps x-ray pulse and two 3-ps laser pulses are incident on a gas cell. The transmitted x-ray beam contains two ultrashort pulses overlapping the laser pulses

focus on in the following. The suppression for $\vartheta_{LX} = 90°$ is merely a consequence of laser-induced line broadening.

The transition between $1s^{-1}3p$ and $1s^{-1}3s$ is, within the decay width of the core-excited states, in one-photon resonance with the laser. These form the levels a and b within the three-level description of EIT. The x-ray photo-absorption cross-section of the three-level model is displayed in Figure 8 together with the *ab initio* data. The three-level model and *ab initio* calculations use the same level energies and transition dipole matrix elements. Within the three-level model, we account for the ionization broadening in the presence of the laser by assuming effective linewidths $\Gamma_a = 0.68$ eV and $\Gamma_b = 0.54$ eV. These parameters give reasonable agreement with the *ab initio* results for $\sigma^{\|}(\omega_X)$ and $\sigma^{\perp}(\omega_X)$ in the vicinity of the $1s \rightarrow 3p$ resonance. The model reproduces the double-hump structure in $\sigma^{\|}(\omega_X)$ and, in particular, the suppression of the $1s \rightarrow 3p$ resonance. This agreement leads us to conclude that the dominant physics here is EIT.

The ability to control x-ray absorption in Ne at the $1s \rightarrow 3p$ resonance using a strong laser field allows one to imprint pulse shapes of the optical-dressing laser onto the x-rays. The idea is illustrated in Figure 9. Using a 2-mm long gas cell filled with one atmosphere of neon, an x-ray pulse at the $1s \rightarrow 3p$ resonance energy is practically completely absorbed in the gas cell [21]. The typical duration of an x-ray pulse from a third-generation synchrotron is ~100 ps. Such an x-ray pulse may be overlapped in time and space with one or several ultrashort, intense laser pulses. Those portions of the x-ray pulse that overlap with the laser are transmitted

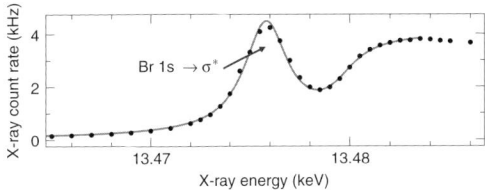

FIGURE 10 Experimental x-ray absorption spectrum of CF_3Br near the Br K edge. The Br 1s → σ^* resonance of CF_3Br at 13.476 keV is indicated. The solid line is the result of a fit. Reused with permission from Ref. [135] © 2008, American Institute of Physics

through the gas cell. In the case shown in Figure 9, where the two dressing laser pulses have a peak intensity of 10^{13} W/cm^2, the intensity of the two transmitted x-ray pulses is about 60 per cent of the incoming x-ray intensity. The time delay between the two x-ray pulses can be controlled by changing the time delay between the two laser pulses, opening a route to ultrafast all-x-ray pump–probe experiments.

5.3 Laser-Induced Molecular Alignment

The molecular alignment process discussed in Sections 2.2 and 3.3 is of intrinsic interest but also of interest in applications to spectroscopy and photophysics, quantum control of molecular dynamics, high-harmonic generation, chemical reactivity, liquids and solvation, and structural determinations by x-ray or electron diffraction. In the adiabatic regime, the degree of alignment that is achievable experimentally is a strong function of the rotational temperature of the target molecules, because only rotationally cold molecules can be localized near the bottom of the angular potential energy well created by the laser field [cf. Equation (1)]. A higher rotational temperature cannot simply be compensated by higher laser intensity, because intensities greater than $\sim 10^{12}$ W/cm^2 can modify the electronic structure (see the previous two subsections) and in this way distort the molecular geometry. In the experiment described in this section, a jet of rotationally cooled CF_3Br molecules was produced by expanding a mixture of 10% CF_3Br and 90% He at 5–9 bar backing pressure through 25 μm or 50 μm pinhole nozzles.

The laser-free x-ray absorption cross-section of CF_3Br across the Br K edge is shown in Figure 10. A two-component fit to the spectrum indicates a 10 per cent background under the 1s → σ^* resonance at 13.476 keV. Relative absorption cross-sections were measured by counting Br Kα (2p → 1s) fluorescent x-rays with two silicon drift detectors that viewed the laser/x-ray/gas-jet overlap region in a setup similar to the one displayed in Figure 1. The degree of laser-induced molecular alignment is usually determined by using additional laser pulses to dissociatively ionize the molecules in an ion spectrometer that projects the fragments onto a position-sensitive detector to record asymmetric fragmentation patterns [4,101]. Instead of a laser ionization probe, we have observed laser alignment of CF_3Br using x-ray absorption at the Br 1s $\rightarrow \sigma^*$ resonance [135]. A theory of x-ray absorption by laser-aligned molecules [136] was presented in Section 3.3. This theory was used to quantitatively model the experimental measurements [135].

Closed-shell Hartree–Fock calculations [141] show that the highest occupied molecular orbitals of CF_3Br are a set of doubly-degenerate π orbitals with large atomic Br $4p_x$ and $4p_y$ components, where the z axis is taken along the C–Br axis of the molecule-fixed frame. The lowest unoccupied molecular orbital has σ^* symmetry (A_1 symmetry in the C_{3v} point group) and largely consists of the atomic Br $4p_z$ orbital. The Br 1s $\rightarrow \sigma^*$ x-ray absorption resonance therefore has its transition dipole vector directed along the C–Br axis (the x and y components of this vector vanish), so the absorption cross-section is sensitive to the angle between the x-ray polarization vector and the C–Br axis. Using coupled-cluster linear response theory [141], the polarizability anisotropy $\alpha_\parallel - \alpha_\perp$ of CF_3Br was calculated to be 12 a.u. The C–Br axis therefore tends to align parallel to the polarization axis of the aligning laser pulses. Hence, the resonant x-ray absorption cross section is sensitive to the angle ϑ_{LX}, defined as the angle between the polarization vectors of the laser and x-ray pulses. ϑ_{LX} was controlled by using a half-wave plate to rotate the laser polarization with respect to the fixed polarization of the x-rays.

In contrast to the laser pulses used for strong-field ionization (Section 5.1), the laser pulses here were left uncompressed (95-ps FWHM) in order to facilitate maximal temporal overlap of the aligned molecular ensemble with the x-ray pulses. The laser pulse duration was measured by cross correlation with 50 fs laser pulses in a BBO crystal. Within the laser/x-ray/gas-jet overlap region viewed by the x-ray fluorescence detectors, the laser peak intensity was spatially quite uniform with an average value of 0.85×10^{12} W/cm^2.

A laser/x-ray cross correlation signal was recorded by tuning the x-ray energy to the σ^* resonance and measuring $P_\parallel(\tau)/P_\perp(\tau)$ [cf. Equations (42) and (43)], the ratio of parallel ($\vartheta_{LX} = 0°$) to perpendicular ($\vartheta_{LX} = 90°$) x-ray absorption probabilities as a function of the laser/x-ray time delay τ. The data were corrected for the 10 per cent non-resonant background under the σ^* resonance, and the resulting ratio is plotted in Figure 11(a). The ratio is seen to increase from 1 to $\sim$1.22 when the laser and x-ray pulses are temporally overlapped. The alignment signature is distinct but relatively small due to the high rotational temperature of the molecules achieved in the jet expansion. The result of theoretical modeling of the data is shown as the curve in Figure 11(a). Two adjustable parameters in the calculated time-delay scan were determined. These are the x-ray pulse length of 122 ± 18 ps and the initial rotational temperature of 20 ± 2 K.

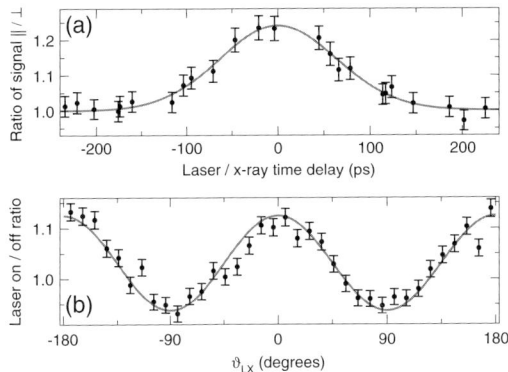

FIGURE 11 (a) Laser/x-ray cross correlation for laser-aligned CF_3Br, $\parallel / \perp$ absorption as a function of laser/x-ray delay. Theory for a 95-ps laser pulse, 0.85×10^{12} W/cm^2, $T_{rot} = 20$ K is overlaid. (b) Laser-on/laser-off ratio of x-ray absorption by CF_3Br, as a function of ϑ_{LX}. $\vartheta_{LX} = 0°$ and $90°$ correspond to the parallel and perpendicular configurations for laser and x-ray polarizations. Theory for $T_{rot} = 24$ K is the overlaid curve. Reused with permission from Ref. [135] © 2008, American Institute of Physics

Calculation of the time evolution of $\langle \cos^2 \vartheta \rangle$ shows that the alignment under these conditions was adiabatic, i.e., $\langle \cos^2 \vartheta \rangle$ followed the laser pulse envelope.

In a second experiment the temporal overlap was maximized ($\tau = 0$), and the laser-on/laser-off ratio of resonant x-ray absorption probabilities was measured as a function of ϑ_{LX} by rotating the half-wave plate. A distinct modulation was observed as shown in Figure 11(b). The data were modeled by calculating the laser-on/laser-off ratio as [136]

$$R(\vartheta_{LX}) = \frac{P_{\parallel}(0)}{P_{th}(0)} \cos^2 \vartheta_{LX} + \frac{P_{\perp}(0)}{P_{th}(0)} \sin^2 \vartheta_{LX}, \qquad (44)$$

where $P_i(\tau)$, $i = \parallel, \perp$, th is the x-ray absorption probability at time delay τ for parallel ($\parallel$) and perpendicular ($\perp$) laser and x-ray polarizations, and for a laser-off thermal (th) ensemble of molecules (see also Section 3.3). The theoretical curve in Figure 11(b) is for an x-ray pulse duration of 122 ps, determined from the time-delay scan, and a rotational temperature of 24 K.

6. SUMMARY AND OUTLOOK

In this article we have described new methods to exert control of x-ray absorption processes in atoms and molecules using strong optical fields. This emerging area combining x-ray and strong-field physics has great potential for the future. At intensities of $\sim 10^{14}$ W/cm^2 we have observed

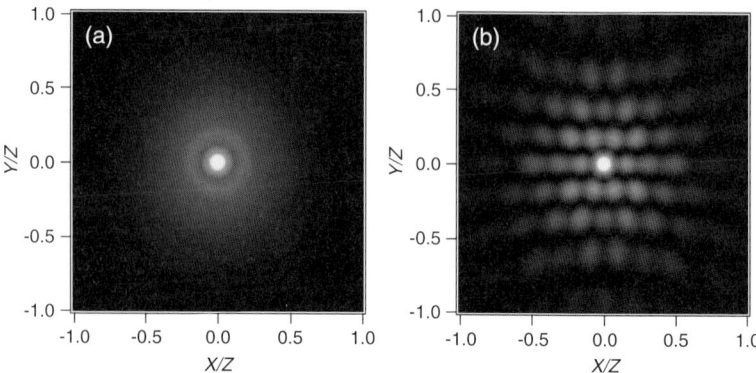

FIGURE 12 Diffraction of 20-keV x-rays from (a) isotropic and (b) aligned 3, 4 dibromothiophene. Incoming x-rays are assumed to propagate in the z direction and are assumed to be linearly polarized in the x direction. In (b), the molecule is assumed to have its symmetry axis aligned along the x axis, with the molecular plane parallel to the x–y plane

orbital alignment in laser-produced ions. In the future, we plan to study the ultrafast orbital dynamics in laser-produced ions [15]. Of particular interest is neon, where the spin–orbit coupling is small and thus the orbital dynamics is observable in real time using femtosecond lasers. A potential application of an x-ray spectroscopic probe of ion orbital alignment is the ability to measure plasma magnetic fields *in situ* [115]. In the regime of $\sim 10^{13}$ W/cm^2 we have calculated the x-ray spectra of laser-dressed atoms which have revealed the phenomenon of electromagnetically-induced transparency for neon. Experiments to demonstrate this effect on the neon 1s → 3p resonance are underway using soft x-ray pulses at 870 eV with pulse durations of $\sim$ 100 fs [152,153]. The short pulse duration of the soft x-rays facilitates a well-defined probe of neon atoms exposed to the requisite laser-dressing intensity, 10^{13} W/cm^2 at 800 nm. Theoretical studies to extend EIT to the hard x-ray regime are in progress. Through these studies it is possible that, with an analogous strategy, controlled amplitude shaping of short-wavelength pulses might become a reality.

At intensities of $\sim 10^{12}$ W/cm^2 we have demonstrated control of resonant x-ray absorption via molecular alignment. In the future we plan to increase the degree of alignment significantly by improved rotational cooling as shown in Ref. [101]. Our calculations for 1 K predict $\langle \cos^2 \vartheta \rangle = 0.80$, $P_\parallel(0)/P_\perp(0) = 4.7$, and an alignment duration of 140-ps FWHM for a 95-ps laser pulse at a peak intensity of 10^{12} W/cm^2. The alignment will then be better matched to the time duration of the x-ray pulses. A highly aligned molecular ensemble would be useful for coherent diffraction imaging of molecular structure [28]. To illustrate this, we have calculated x-ray diffraction patterns for both randomly oriented (Figure 12(a)) and fixed-in-space (Figure 12(b)) 3, 4 dibromothiophene molecules [76].

Coherent control of the laser pulse or a second laser pulse conceivably could align and induce a dynamical process while recording the molecular structure by x-ray diffraction. Reduction of the x-ray pulse duration to ~1 ps is expected to be achieved using RF deflecting cavities at the Advanced Photon Source [29]. This will enable studies of coherently controlled rotational dynamics and photo-ionization of fixed-in-space molecules under field-free conditions. Soon the dream of imaging of laser-controlled molecular processes on atomic time and length scales may become a reality.

ACKNOWLEDGMENTS

We would like to thank D.A. Arms, C. Buth, E.M. Dufresne, C. Höhr, D.L. Ederer, E.C. Landahl, E.R. Peterson, S.T. Pratt, N. Rohringer, J. Rudati, and D.A. Walko for fruitful collaboration. This work and use of the Advanced Photon Source were supported by the U.S. Department of Energy, Office of Science, Office of Basic Energy Sciences, under Contract No. DE-AC02-06CH11357.

REFERENCES

[1] S.A. Rice and M. Zhao (2000), *Optical Control of Molecular Dynamics*, Wiley, New York.
[2] M. Shapiro and P. Brumer (2003), *Principles of the Quantum Control of Molecular Processes*, Wiley-Interscience, New York.
[3] H. Rabitz, R. de Vivie-Riedle, M. Motzkus and K. Kompa (2000), *Science*, **288**, 824.
[4] H. Stapelfeldt and T. Seideman (2003), *Rev. Mod. Phys.*, **75**, 543.
[5] T. Brabec and F. Krausz (2000), *Rev. Mod. Phys.*, **72**, 545.
[6] K. Yamanouchi (2002), *Science*, **295**, 1659.
[7] J.M. Schins, P. Breger, P. Agostini, R.C. Constantinescu, H.G. Muller, G. Grillon, A. Antonetti and A. Mysyrowicz (1994), *Phys. Rev. Lett.*, **73**, 2180.
[8] J.M. Schins, P. Breger, P. Agostini, R.C. Constantinescu, H.G. Muller, G. Grillon, A. Antonetti and A. Mysyrowicz (1995), *Phys. Rev. A*, **52**, 1272.
[9] T.E. Glover, R.W. Schoenlein, A.H. Chin and C.V. Shank (1996), *Phys. Rev. Lett.*, **76**, 2468.
[10] M. Hentschel, R. Kienberger, Ch. Spielmann, G.A. Reider, N. Milosevic, T. Brabec, P. Corkum, U. Heinzmann, M. Drescher and F. Krausz (2001), *Nature*, **414**, 509.
[11] F.J. Wuilleumier and M. Meyer (2006), *J. Phys. B*, **39**, R425.
[12] M.V. Ammosov, N.B. Delone and V.P. Krainov (1986), *Sov. Phys. JETP*, **64**, 1191.
[13] L. Young, D.A. Arms, E.M. Dufresne, R.W. Dunford, D.L. Ederer, C. Höhr, E.P. Kanter, B. Krässig, E.C. Landahl, E.R. Peterson, J. Rudati, R. Santra and S.H. Southworth (2006), *Phys. Rev. Lett.*, **97**, 083601.
[14] S.H. Southworth, D.A. Arms, E.M. Dufresne, R.W. Dunford, D.L. Ederer, C. Höhr, E.P. Kanter, B. Krässig, E.C. Landahl, E.R. Peterson, J. Rudati, R. Santra, D.A. Walko and L. Young (2007), *Phys. Rev. A*, **76**, 043421.
[15] R. Santra, R.W. Dunford and L. Young (2006), *Phys. Rev. A*, **74**, 043403.
[16] Z.-H. Loh, M. Khalil, R.E. Correa, R. Santra, C. Buth and S.R. Leone (2007), *Phys. Rev. Lett.*, **98**, 143601.

[17] R.R. Freeman and P.H. Bucksbaum (1991), *J. Phys. B*, **24**, 325.
[18] J. Ullrich, R. Moshammer, A. Dorn, R Dörner, L.Ph.H. Schmidt and H. Schmidt-Böcking (2003), *Rep. Prog. Phys.*, **66**, 1463.
[19] C. Buth and R. Santra (2007), *Phys. Rev. A*, **75**, 033412.
[20] L. Pan, L. Armstrong Jr. and J.H. Eberly (1986), *J. Opt. Soc. Am B*, **3**, 1319.
[21] C. Buth, R. Santra and L. Young (2007), *Phys. Rev. Lett.*, **98**, 253001.
[22] E. Arimondo (1996), in: E. Wolf (Ed.), Progress in Optics, Vol. XXXV, Elsevier, Amsterdam, p. 257.
[23] S.E. Harris (1997), *Phys. Today*, **50** (7), 36.
[24] M. Fleischhauer, A. Imamoğlu and J.P. Marangos (2005), *Rev. Mod. Phys.*, **77**, 633.
[25] B. Friedrich and D. Herschbach (1995), *Phys. Rev. Lett.*, **74**, 4623.
[26] J. Stöhr (1996), *NEXAFS Spectroscopy*, Springer, New York.
[27] R. Neutze, R. Wouts, D. van der Spoel, E. Weckert and J. Hajdu (2000), *Nature*, **406**, 752.
[28] J.C.H. Spence and R.B. Doak (2004), *Phys. Rev. Lett.*, **92**, 198102.
[29] M. Borland (2005), *Phys. Rev. ST Accel. Beams*, **8**, 074001.
[30] J. Arthur, et al. (2002). Linac Coherent Light Source (LCLS) Conceptual Design Report, SLAC-R-593.
[31] T. Tanaka and T. Shintake, (2005). SCSS X-FEL Conceptual Design Report, RIKEN Harima Institute/SPring-8.
[32] M. Altarelli, et al. The Technical Design Report of the European XFEL, DESY 2006-097, 2006.
[33] B. Crasemann (2006), in. G.W.F. Drake (Ed.), *Handbook of Atomic, Molecular, and Optical Physics*, Springer, New York, p. 915.
[34] C.T. Chantler (1995), *J. Phys. Chem. Ref. Data*, **24**, 71.
[35] C.T. Chantler (2000), *J. Phys. Chem. Ref. Data*, **29**, 597.
[36] B.E. Warren (1990), *X-Ray Diffraction*, Dover, New York.
[37] M.J. Cooper, P.E. Mijnarends, N. Shiotani, N. Sakai and A. Bansil (2004), *X-ray Compton Scattering*, Oxford University Press, Oxford.
[38] M. Jung, R.W. Dunford, D.S. Gemmell, E.P. Kanter, B. Krässig, T.W. LeBrun, S.H. Southworth, L. Young, J.P.J. Carney, L. LaJohn, R.H. Pratt and P.M. Bergstrom Jr. (1998), *Phys. Rev. Lett.*, **81**, 1596.
[39] B. Krässig, R.W. Dunford, D.S. Gemmell, S. Hasegawa, E.P. Kanter, H. Schmidt-Böcking, W. Schmitt, S.H. Southworth, T. Weber and L. Young (1999), *Phys. Rev. Lett.*, **83**, 53.
[40] L. Young, R.W. Dunford, E.P. Kanter, B. Krässig, S.H. Southworth, R.A. Bonham, P. Lykos, C. Morong, A. Timm, J.P.J. Carney and R.H. Pratt (2001), *Phys. Rev. A*, **63**, 052718.
[41] A. Plech, M. Wulff, S. Bratos, F. Mirloup, R. Vuilleumier, F. Schotte and P.A. Anfinrud (2004), *Phys. Rev. Lett.*, **92**, 125505.
[42] K.J. Gaffney and H.N. Chapman (2007), *Science*, **316**, 1444.
[43] H. Winick (Ed.), (1994), *Synchrotron Radiation Sources: A Primer*, World Scientific, New Jersey.
[44] T. Åberg and B. Crasemann (1994), in: G. Materlik, C.J. Sparks and K. Fischer (Eds.), *Resonant Anomalous X-Ray Scattering: Theory and Applications*, North Holland, Amsterdam, p. 431.
[45] P.L. Cowan (1994), in: G. Materlik, C.J. Sparks and K. Fischer (Eds.), *Resonant Anomalous X-Ray Scattering: Theory and Applications*, North Holland, Amsterdam, p. 449.
[46] S.H. Southworth, L. Young, E.P. Kanter and T. LeBrun (2000), in: C.-Y. Ng (Ed.), Advanced Series in Physical Chemistry: Photoionization and Photodetachment, Vol. 10B, World Scientific, River Edge, NJ, p. 1289.
[47] G.B. Armen, H. Aksela, T. Åberg and S. Aksela (2000), *J. Phys. B*, **33**, R49.
[48] J.C. Levin and G.B. Armen (2004), *Radiat. Phys. Chem.*, **70**, 105.
[49] R.W. Dunford, E.P. Kanter, B. Krässig, S.H. Southworth and L. Young (2004), *Radiat. Phys. Chem.*, **70**, 149.

[50] N. Rohringer and R. Santra (2007), *Phys. Rev. A*, **76**, 033416.
[51] W. Bambynek, B. Crasemann, R.W. Fink, H.-U. Freund, H. Mark, C.D. Swift, R.E. Price and P.V. Rao (1972), *Rev. Mod. Phys.*, **44**, 716.
[52] M.O. Krause (1979), *J. Phys. Chem. Ref. Data*, **8**, 307.
[53] M.O. Krause and J.H. Oliver (1979), *J. Phys. Chem. Ref. Data*, **8**, 329.
[54] M.H. Chen, B. Crasemann and H. Mark (1979), *Atom. Data and Nucl. Data Tables*, **24**, 13.
[55] J.H. Hubbell, P.N. Trehan, N. Singh, B. Chand, D. Mehta, M.L. Garg, R.R. Garg, S. Singh and S. Puri (1994), *J. Phys. Chem. Ref. Data*, **23**, 339.
[56] M. Breinig, M.H. Chen, G.E. Ice, F. Parente, B. Crasemann and G.S. Brown (1980), *Phys. Rev. A*, **22**, 520.
[57] B.K. Agarwal (1991), *X-Ray Spectroscopy*, Springer-Verlag, New York.
[58] R.D. Deslattes, E.G. Kessler Jr., P. Indelicato, L. de Billy and E. Lindroth (2003), *J. Anton, Rev. Mod. Phys.*, **75**, 35.
[59] J.H. Scofield (1969), *Phys. Rev.*, **179**, 9.
[60] J.D. Mills and P.W. Langhoff (1996), in: D.L. Ederer and J.H. McGuire (Eds.), *Raman Emission by X-Ray Scattering*, World Scientific, River Edge, NJ, p. 169.
[61] F. Gel'mukhanov and H. Ågren (1999), *Phys. Rep.*, **312**, 87.
[62] D. Strickland and G. Mourou (1985), *Opt. Commun.*, **56**, 219.
[63] G.A. Mourou, T. Tajima and S.V. Bulanov (2006), *Rev. Mod. Phys.*, **78**, 309.
[64] P.B. Corkum, N.H. Burnett and F. Brunel (1989), *Phys. Rev. Lett.*, **62**, 1259.
[65] P.B. Corkum (1993), *Phys. Rev. Lett.*, **71**, 1994.
[66] See for a popular review H.C. Kapteyn, M.M. Murnane and I.P. Christov (2005), *Phys. Today*, **58** (3), 39.
[67] P.B. Corkum and F. Krausz (2007), *Nature Phys.*, **3**, 381.
[68] M.H. Mittleman (1993), *Introduction to the Theory of Laser–Atom Interactions*, Plenum Press, New York.
[69] A. Ludlow, M.M. Boyd, T. Zelevinsky, S.M. Foreman, S. Blatt, M. Notcutt, T. Ido and J. Ye (2006), *Phys. Rev. Lett.*, **96**, 033003.
[70] P. Agostini, F. Fabre, G. Mainfray, G. Petite and N. Rahman (1979), *Phys. Rev. Lett.*, **42**, 1127.
[71] B. Friedrich and D. Herschbach (1995), *J. Phys. Chem.*, **99**, 15686.
[72] B. Friedrich and D. Herschbach (1996), *Chem. Phys. Lett.*, **262**, 41.
[73] J.J. Larsen, I.W. Larsen and H. Stapelfeldt (1999), *Phys. Rev. Lett.*, **83**, 1123.
[74] H. Sakai, C.P. Safvan, J.J. Larsen, K.M. Hilligsoe, K. Held and H. Stapelfeldt (1999), *J. Chem. Phys.*, **110**, 10235.
[75] J.J. Larsen, H. Sakai, C.P. Sarvan, I.W. Larsen and H. Stapelfeldt (1999), *J. Chem. Phys.*, **111**, 7774.
[76] J.J. Larsen, K. Held, N. Bjerre, H. Stapelfeldt and T. Seideman (2000), *Phys. Rev. Lett.*, **85**, 2470.
[77] F. Rosca-Pruna and M.J.J. Vrakking (2001), *Phys. Rev. Lett.*, **87**, 153902.
[78] K. Hoki and Y. Fujimura (2001), *Chem. Phys. Lett.*, **267**, 187.
[79] E. Skovsen, M. Machholm, T. Ejdrup, J. Thogerson and H. Stapelfeldt (2002), *Phys. Rev. Lett.*, **89**, 133004.
[80] S. Guérin, L.P. Yatsenko, H.R. Jauslin, O. Faucher and B. Lavorel (2002), *Phys. Rev. Lett.*, **88**, 233601.
[81] E. Peronne, M.D. Poulsen, C.Z. Bisgaard, H. Stapelfeldt and T. Seideman (2003), *Phys. Rev. Lett.*, **91**, 043003.
[82] M. Leibscher, I.Sh. Averbukh and H. Rabitz (2003), *Phys. Rev. Lett.*, **90**, 213001.
[83] V. Renard, M. Renard, S. Guérin, Y.T. Pashayan, O. Faucher and H.R. Jauslin (2003), *Phys. Rev. Lett.*, **90**, 153601.
[84] E. Peronne, M.D. Poulsen, H. Stapelfeldt, C.Z. Bisgaard, E. Hamilton and T. Seideman (2004), *Phys. Rev. A*, **70**, 063410.
[85] V. Renard, M. Renard, A. Rouzée, S. Guérin, H.R. Jauslin, B. Lavorel and O. Faucher (2004), *Phys. Rev. A*, **70**, 033420.

[86] M.D. Poulsen, E. Peronne, H. Stapelfeldt, C.Z. Bisgaard, S.S. Viftrup, E. Hamilton and T. Seideman (2004), *J. Chem. Phys.*, **121**, 783.
[87] K.F. Lee, I.V. Litvinyuk, P.W. Dooley, M. Spanner, D.M. Villeneuve and P.B. Corkum (2004), *J. Phys. B*, **37**, L43.
[88] C.Z. Bisgaard, M.D. Poulsen, E. Peronne, S.S. Viftrup and H. Stapelfeldt (2004), *Phys. Rev. Lett.*, **92**, 173004.
[89] J.G. Underwood, M. Spanner, M.Yu. Ivanov, J. Mottershead, B.J. Sussman and A. Stolow (2003), *Phys. Rev. Lett.*, **90**, 223001.
[90] J.G. Underwood, B.J. Sussman and A. Stolow (2005), *Phys. Rev. Lett.*, **94**, 143002.
[91] S. Kaziannis, P. Siozos and C. Kosmidis (2005), *Chem. Phys. Lett.*, **401**, 115.
[92] E. Hamilton, T. Seideman, T. Ejdrup, M.D. Poulsen, C.Z. Bisgaard, S.S. Viftrup and H. Stapelfeldt (2005), *Phys. Rev. A*, **72**, 043402.
[93] K. Miyazaki, M. Kaku, G. Miyaji, A. Abdurrouf and F.H.M. Faisal (2005), *Phys. Rev. Lett.*, **95**, 243903.
[94] R. Torres, R. De Nalda and J.P. Marangos (2005), *Phys. Rev. A*, **72**, 023420.
[95] V. Renard, O. Faucher and B. Lavorel (2005), *Opt. Lett.*, **30**, 70.
[96] S. Ramakrishna and T. Seideman (2005), *Phys. Rev. Lett.*, **95**, 113001.
[97] S. Ramakrishna and T. Seideman (2006), *J. Chem. Phys.*, **24**, 034101.
[98] D.H. Parker and R.B. Bernstein (1989), *Ann. Rev. Phys. Chem.*, **40**, 561.
[99] H.J. Loesch and A. Remsheid (1991), *J. Phys. Chem.*, **95**, 8194.
[100] B. Friedrich and D. Herschbach (1991), *Nature*, **353**, 412.
[101] V. Kumarappan, C.Z. Bisgaard, S.S. Viftrup, L. Holmegaard and H. Stapelfeldt (2006), *J. Chem. Phys.*, **125**, 194309.
[102] M.W. Walser, D.J. Urbach, K.Z. Hatsagortsyan, S.X. Hu and C.H. Keitel (2002), *Phys. Rev. A*, **65**, 043410.
[103] J.C. Slater (1951), *Phys. Rev.*, **81**, 385.
[104] J.C. Slater and K.H. Johnson (1972), *Phys. Rev. B*, **5**, 844.
[105] F. Herman and S. Skillman (1963), *Atomic Structure Calculations*, Prentice-Hall, Englewood Cliffs, N.J.
[106] U.V. Riss and H.-D. Meyer (1993), *J. Phys. B*, **26**, 4503.
[107] R. Santra and L.S. Cederbaum (2002), *Phys. Rep.*, **368**, 1.
[108] J.G. Muga, J.P. Palao, B. Navarro and I.L. Egusquiza (2004), *Phys. Rep.*, **395**, 357.
[109] M.E. Rose (1995), *Elementary Theory of Angular Momentum*, Dover, New York.
[110] A.R. Edmonds (1996), *Angular Momentum in Quantum Mechanics*, Princeton University Press, Princeton, New Jersey.
[111] A.J.F. Siegert (1939), *Phys. Rev.*, **56**, 750.
[112] R.M. More and E. Gerjuoy (1973), *Phys. Rev. A*, **7**, 1288.
[113] O.I. Tolstikhin, V.N. Ostrovsky and H. Nakamura (1998), *Phys. Rev. A*, **58**, 2077.
[114] L. Pan, D.R. Beck and S.M. O'Malley (2005), *J. Phys. B.*, **38**, 3721.
[115] C. Höhr, E.R. Peterson, N. Rohringer, J. Rudati, D.A. Arms, E.M. Dufresne, R.W. Dunford, D.L. Ederer, E.P. Kanter, B. Krässig, E.C. Landahl, R. Santra, S.H. Southworth and L. Young (2007), *Phys. Rev. A*, **75**, 011403(R).
[116] R. Santra and C.H. Greene (2003), *Phys. Rev. A*, **67**, 062713.
[117] H. Friedrich (1998), *Theoretical Atomic Physics*, Springer, Berlin.
[118] J.L. Chaloupka, R. Lafon, L.F. DiMauro, P. Agostini and K.C. Kulander (2001), *Opt. Express*, **8**, 352.
[119] P. Kálmán (1989), *Phys. Rev. A*, **39**, 3200.
[120] A. Cionga, V. Florescu, A. Maquet and R. Taïeb (1993), *Phys. Rev. A*, **47**, 1830.
[121] D.P. Craig and T. Thirunamachandran (1998), *Molecular Quantum Electrodynamics*, Dover, Mineola, N.Y.
[122] N. Moiseyev (1998), *J. Phys. B*, **31**, 1431.
[123] U.V. Riss and H.-D. Meyer (1998), *J. Phys. B*, **31**, 2279.
[124] H.O. Karlsson (1998), *J. Chem. Phys.*, **109**, 9366.
[125] N. Moiseyev (1998), *Phys. Rep.*, **302**, 212.
[126] J.H. Shirley (1965), *Phys. Rev.*, **138**, B979.

[127] S.-I Chu and W.P. Reinhardt (1977), *Phys. Rev. Lett.*, **39**, 1195.
[128] S.-I Chu and J. Cooper (1985), *Phys. Rev. A*, **32**, 2769.
[129] P.G. Burke, P. Francken and C.J. Joachain (1991), *J. Phys. B*, **24**, 751.
[130] M. Dörr, M. Terao-Dunseath, J. Purvis, C.J. Noble, P.G. Burke and C.J. Joachain (1992), *J. Phys. B*, **25**, 2809.
[131] S.-I. Chu and D.A. Telnov (2004), *Phys. Rep.*, **390**, 1.
[132] C. Buth, R. Santra and L.S. Cederbaum (2004), *Phys. Rev. A*, **69**, 032505.
[133] P. Meystre and M. Sargent III (1991), *Elements of Quantum Optics*, Springer, Berlin.
[134] K.-J. Boller, A. Imamoğlu and S.E. Harris (1991), *Phys. Rev. Lett.*, **66**, 2593.
[135] E.R. Peterson, C. Buth, D.A. Arms, R.W. Dunford, E.P. Kanter, B. Krässig, E.C. Landahl, S.T. Pratt, R. Santra, S.H. Southworth and L. Young (2008), *Appl. Phys. Lett.*, 094106.
[136] C. Buth and R. Santra (2008), *Phys. Rev. A*, **77**, 013413.
[137] H.W. Kroto (1975), *Molecular Rotation Spectra*, John Wiley & Sons, London.
[138] R.N. Zare (1988), *Angular Momentum*, John Wiley & Sons, New York.
[139] C.H. Townes and A.L. Schawlow (1955), *Microwave Spectroscopy*, McGraw-Hill, New York.
[140] O. Christiansen, A. Halkier, H. Koch, P. Jørgensen and T. Helgaker (1998), *J. Chem. Phys.*, **108**, 2801.
[141] DALTON, A Molecular Electronic Structure Program, Release 2.0 (2005). See http://www.kjemi.uio.no/software/dalton/dalton.html.
[142] A. Szabo and N.S. Ostlund (1989), *Modern Quantum Chemistry: Introduction to Advanced Electronic Structure Theory*, McGraw-Hill, New York.
[143] V. Schmidt (1997), *Electron Spectrometry of Atoms Using Synchrotron Radiation*, Cambridge University Press, Cambridge.
[144] M.P. Hertlein, H. Adaniya, J. Amini, C. Bressler, B. Feinberg, M. Kaiser, N. Neumann, M.H. Prior and A. Belkacem (2006), *Phys. Rev. A*, **73**, 062715.
[145] L.X. Chen, W.J.H. Jäger, G. Jennings, D.J. Gosztola, A. Munkholm and J.P. Hessler (2001), *Science*, **292**, 262.
[146] L. Young, R.W. Dunford, C. Höhr, E.P. Kanter, B. Krässig, E.R. Peterson, S.H. Southworth, D.L. Ederer, J. Rudati, D.A. Arms, E.M. Dufresne and E.C. Landahl (2006), *Radiat. Phys. Chem.*, **75**, 1799.
[147] M.F. DeCamp, D.A. Reis, D.M. Fritz, P.H. Bucksbaum, E.M. Dufresne and R. Clarke (2005), *J. Synch. Rad.*, **12**, 177.
[148] P.J. Eng, M. Newville, M.L. Rivers and S.R. Sutton (1998), *Proc. SPIE Int. Soc. Opt. Eng.*, **3449**, 145.
[149] S.M. Hankin, D.M. Villeneuve, P.B. Corkum and D.M. Rayner (2001), *Phys. Rev. A*, **64**, 013405.
[150] H. Maeda, M. Dammasch, U. Eichmann, W. Sandner, A. Becker and F.H.M. Faisal (2000), *Phys. Rev. A*, **62**, 035402.
[151] R. Taïeb, V. Véniard and A. Maquet (2001), *Phys. Rev. Lett.*, **87**, 053002.
[152] R.W. Schoenlein, S. Chattopadhyay, H.H.W. Chong, T.E. Glover, P.A. Heimann, C.V. Shank, A.A. Zholents and M.S. Zolotorev (2000), *Science*, **287**, 2237.
[153] S. Khan, K. Holldack, T. Kachel, R. Mitzner and T. Quast (2006), *Phys. Rev. Lett.*, **97**, 074801.

CHAPTER 6

Optical Trapping Takes Shape: The Use of Structured Light Fields

K. Dholakia and **W.M. Lee**

SUPA, School of Physics and Astronomy, University of St Andrews, North Haugh, Fife, Scotland KY16 9SS

Contents			
	1.	Introduction	262
	2.	Single beam optical tweezers	264
	2.1	Optical Tweezers: Theoretical Treatment of Trapping Forces	264
	2.2	Design Considerations of an Optical Tweezers System	267
	2.3	Other Incarnations of Optical Trapping	272
	3.	Applications within biophysics and the colloidal sciences	274
	3.1	Molecular and Cell Biology	275
	3.2	Examples of Studies of Colloidal Systems with Single Beam Optical Tweezers	281
	4.	Optical trapping with structured light fields and its applications	282
	4.1	Structured Light Fields	282
	4.2	Large Arrays of Optical Traps	284
	4.3	Non-zero Order Laser Modes	303
	5.	Optical binding	323
	6.	Conclusions	324
	Acknowledgments		324
	References		325

Abstract Optical micromanipulation is a powerful and versatile technique based upon the light–matter interaction. Whilst the forces exerted by optical traps are naturally very small, they are sufficient to realize non-invasive mechanical control over mesoscopic particles within atomic, biological and colloidal systems. The inherent compatibility with modern microscopy enhances the reconfigurability of the trap while the accuracy achieved in a calibrated optical trap presents itself as a quantitative force probe. Thus forces can be applied in a controlled manner to biological systems including cells and molecular motors and processes measured with high precision. The impact is not limited to biology. Optical traps have provided seminal studies in colloidal and optical physics including the phase dynamics of thermodynamic systems, Brownian diffusion, aspects of microfluidics, and fundamental issues related to optical angular momentum. This article aims to focus upon the emergent theme of optical trapping with structured light fields. By structured light fields we refer to the generation of multiple arrays of traps and the use of specialist light fields such as Laguerre-Gaussian beams and Bessel beams. Structured light fields are making a major impact on optical trapping and on subsequent applications including those in biomedicine.

1. INTRODUCTION

Laser microscopy has unraveled a wealth of complex processes in cell and molecular biology. The prominent imaging methods of laser scanning confocal fluorescence imaging and two-photon fluorescence excitation are established techniques for visualizing cells and molecules. The light–matter interaction has had a profound impact upon fundamental science and has enabled a deeper understanding not only in biology but in the domain of atomic and molecular theory. Notable advances in the past few decades have included the concept of light exerting mechanical forces upon atomic ensembles paving the way for the very powerful methods of laser cooling (Chu, 1998; Cohen-Tannoudji, 1998; Phillips, 1998) and the achievement of ultracold quantum gases and onset of Bose–Einstein condensation (Anderson et al., 1995; Davis et al., 1995; O'Hara et al., 2000). The forces of light have made an impact at size scales that are much larger than that of an atom (Angstrom) or molecule: importantly light forces can readily manifest themselves upon objects up to the size of a single biological cell (micrometer).

How do such forces arise? A light field may be considered as a large number of individual photons. Each photon may be considered as quanta of energy that possesses a well-defined momentum. Overall, the exchange of momentum with matter may result in physical motion: in this manner, light may move, hold and exert a very controllable force upon material objects and importantly for our purposes to objects the size of a single cell or smaller. This form of optical momentum transfer and subsequent force is at the heart of the area of optical trapping or optical micromanipulation.

The use of laser sources is crucial to the field though one must acknowledge that we are limited to exerting forces in the order of piconewtons. The interaction between light and the particle (microscopic or smaller) produces a change of momentum at a rate that leads to small forces that are sufficient enough to move or hold a microparticle. At such a size level, this concept can be utilized particularly by biological sciences in the pursuit of studying several macromolecular and cellular processes in a quantifiable and sterile manner. The field of fundamental physics too has benefited in numerous ways using optical traps: seminal studies in the past fifteen years include a deeper understanding of the optical angular momentum of light, exploration of colloidal hydrodynamics, microfluidics and the study of thermodynamics systems at small scales, this latter topic being related to the exciting topic of non-equilibrium physics. Such experimental studies of light–matter interaction have, in turn, advanced physicists toward a more complete appreciation of the theoretical basis for optical forces.

It is now over thirty five years since the first optical trapping experiments and over twenty years since the inception of the popular 'optical tweezers' (Ashkin et al., 1986). Beginning in 1970, Ashkin (Ashkin, 1970) performed experiments with a visible Argon-ion laser. In a single horizontally propagating beam, microspheres were observed to align along the propagation axis and propelled along the propagation direction of the beam: this is now commonly known as optical guiding. He then added a second beam (of equivalent optical power) in counterpropagating geometry that halted the motion of a sphere along the beam axis whilst retaining its position within the bright region of the two beams: the first optical trap was formed. This counterpropagating (or dual) beam optical trapping geometry (Ashkin, 1970) has been realized with optical fibers (Constable et al., 1993) and is making a comeback as we shall discuss later in the form of the optical stretcher (Guck et al., 2001) and for longitudinal optical binding (Tatarkova et al., 2002; Singer et al., 2003; Metzger et al., 2006b, 2007b, 2006c). In subsequent studies Ashkin and Dziedzic (Ashkin and Dziedzic, 1974) also investigated the stability of trapping of hollow and glass spheres with optical beams by balancing optical forces against gravity. Indeed the levitation of hollow spheres constituted the first use of 'structured light fields' in optical traps. An output higher order mode TEM^*_{01} from a laser was needed as such a hollow particle is repelled from regions of high light intensity (its refractive index is lower than its surroundings) and thus needs to be 'caged' by light (Ashkin and Dziedzic, 1974). Sixteen years after the dual beam trap, Ashkin et al. (1986) realized the single beam gradient trap (popularly known as *optical tweezers*) that is the simplest incarnation of optical forces for moving microscopic particles and indeed forms the basis for most of the discussion within this article. This trap has now been well recognized as having the largest impact within the field of optical micromanipulation.

This article aims to give the reader an insight into some of the most exciting areas of this research field with the spotlight firmly upon the topics of multiple trapping or use of specific light modes within optical micromanipulation. We broadly group these topics under the heading of ***structured light fields or beams***: that is moving away from a standard single Gaussian light field. The types of structured light beams include multiple, reconfigurable Gaussian arrays of traps, as well as non-zero-order light modes such as Laguerre-Gaussian modes and Bessel modes. We refer the reader elsewhere for more concise reviews (Molloy and Padgett, 2002; Grier, 2003; Neuman and Block, 2004; Dholakia and Reece, 2006; McGloin et al., 2008; Dholakia et al., 2008) of this topical area. As a precursor to the discussion of these structured light fields in this article, we shall begin with an introduction to optical tweezers, its operation and a brief consideration of some of the experimental issues for its implementation. This is followed by a review particularly how this powerful optical concept is applied to biology namely the study of molecular motors and other biological macromolecules as well as studies of cells and some aspects of colloidal science. Optical tweezers have produced some seminal studies within single molecule biophysics and have allowed an insight into this field in a manner hitherto unforeseen (Block et al., 2007). We then move specifically to the topic of structured light fields and give examples of how such fields may be generated and implemented as well as their applications. The field is dynamic and though in some senses established, it is still very contemporary and if anything expanding its remit across the natural sciences. Many of the applications within biology and chemistry are very active current areas of research which are continually evolving. So the aim will be to give the reader a proficient grounding in the various techniques to facilitate the reasoning behind the future use of such structured light beams.

2. SINGLE BEAM OPTICAL TWEEZERS

2.1 Optical Tweezers: Theoretical Treatment of Trapping Forces

The light–matter interaction is at the heart of optical trapping and the subsequent forces may be understood in a number of ways. Perhaps the most straightforward case is to consider an object in the Mie regime: where the particle radius, r, is much larger than the trapping wavelength, λ. When trapping in the Mie regime, one can invoke the use of geometrical ray optics to picture the forces involved. Ashkin shows a clear calculation of how to derive the forces present in this instance by ray tracing and use of the Fresnel equations at the sphere–medium boundary (Ashkin, 1992),

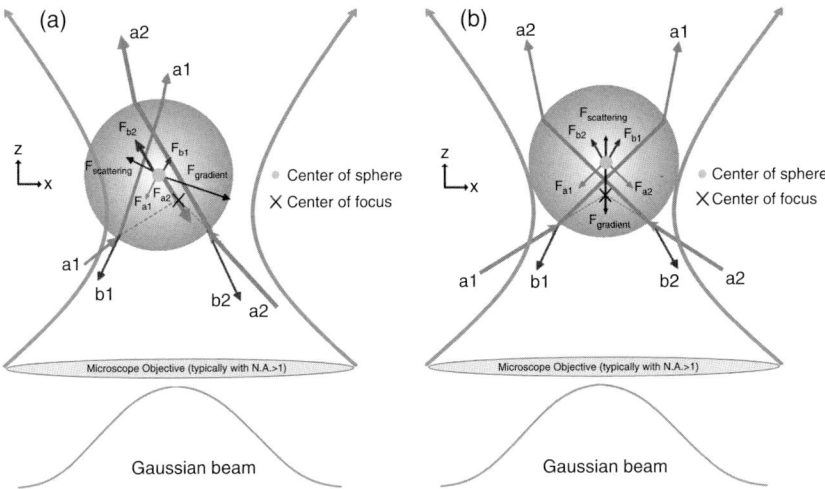

FIGURE 1 Force diagrams for Mie transparent sphere. The lateral and axial trapping forces involved in the three-dimensional tweezing of a single transparent dielectric microsphere (Mie regime) as illustrated with ray optics. (a) Shows the microsphere positioned away from the center of the focus beam, (b) shows the microsphere trapped slightly above the center of the focus beam where the gradient and scattering forces are balanced

by considering the scattering and gradient forces as arising from the reflection and refraction of light. A trapped microsphere (if of higher refractive index than its surroundings and transparent to the incident light) acts as a microscopic lens that focuses the light field. This focusing effect in turn causes the particle to move to the position of highest light intensity as may be seen in Figure 1. Figure 1 shows how we may visualize optical trapping forces in the Mie regime with a high numerical aperture (N.A.) microscope objective (Ashkin, 1992). Both lateral trapping (Figure 1(a)) and axial trapping (Figure 1(b)) in the transverse plane (**x**) and axial direction (**z**) are considered. In this diagram, we can see that gradients of light intensity in both the lateral and axial directions are key to the operation of this technique. Here, the microsphere has a higher refractive index than its surrounding medium. The single Gaussian beam tweezers, is being focused through a high numerical aperture (typical N.A. > 1) microscope objective and in front of the microsphere. In Figure 1(a), the refracted rays (solid line arrows) from two input rays (**a1, a2**), enter a transparent sphere that is positioned off the center of the focused Gaussian beam, where **a1** is the one with lower intensity. The dotted line indicates their unrefracted paths. Each ray produces a force ($F_{a1} F_{a2}$) upon the microsphere that draws it toward the direction of each ray. The net force ($F_{gradient}$) from F_{a1} and F_{a2} draws the microsphere toward the center of the focus. On the other hand, each of the reflected light rays (**b1**, **b2**) creates a force (**F**$_{b1}$, **F**$_{b2}$) that pushes the microsphere away from each of the rays. The net force (**F**$_{scattering}$) from **F**$_{b1}$ and **F**$_{b2}$ acts to push the particle away from the center of the focus. With a high enough gradient force from the highly focused beam (higher number of refracted rays), the microsphere would be drawn toward the center of the focus. In Figure 1(b), the microsphere is trapped stably at a position that is slightly above the upward propagating and focused Gaussian beam when the net forces from the refracted rays (**F**$_{gradient}$) and reflected rays (**F**$_{scattering}$) are balanced in both the transverse and axial planes (**x**, **z**).

Any light scattering should result in a force opposing the direction of light propagation and any refraction of the light results in a force attracting the particle to the region of highest light intensity. This physical picture cannot be used to explain optical forces exerted upon objects that are far smaller than the wavelength: in this Rayleigh regime ($r \leq \frac{\lambda}{20}$), it is more appropriate to consider the particle as a dipole that minimizes its energy in the field gradient created by the light (Malagnino et al., 2002; Harada and Asakura, 1996). A full theoretical treatment of how an optical tweezers performs for a given object in a light field is a non-trivial question. In the Rayleigh regime, the object may be considered as a point dipole and the contributions of the time-averaged gradient force $F_{gradient}$ and scattering forces $F_{scattering}$ may be readily separated as follows:

$$F_{gradient} = \frac{2\pi\alpha}{cn_m^2}\nabla I \qquad (1)$$

$$F_{scattering} = \frac{I\sigma n_m}{c} \qquad (2)$$

where α denotes the polarizability of the sphere which for a standard non-absorbing dielectric object is proportional to the volume of the particle equal to $n_m^2 r^3 \left(\frac{m^2-1}{m^2+2}\right)$, σ is the scattering cross-section of the sphere equal to $\frac{128\pi^5 r^6}{3\lambda^4}\left(\frac{m^2-1}{m^2+2}\right)^2$, r is the particle radius, I is the intensity, n_m is the refractive index of the surrounding medium, c denotes the speed of light, m refers to the ratio of the refractive indices of particle (n) to that of the surrounding medium (n_m) and λ is the wavelength of the trapping laser used. We can see from this that the scattering force is directly proportional to the trapping laser intensity and the gradient (or dipole) force upon the object is due to the inhomogeneous field gradient created by tightly focusing the light beam (Svoboda and Block, 1994b; Neuman and Block, 2004). Chaumet and Nieto-Vesperinas (2000) derived the expressions for the time-averaged forces for the Rayleigh regime. In general, we see the gradient force as being proportional to the polarizability which, when considering a dielectric particle, scales with the volume of the particle.

This means it is quite difficult in practice to trap very small dielectric objects (e.g. diameter of 50 nanometers) but one can readily hold dielectric objects from diameter of ~0.5 to 5 micrometer in optical tweezers. These considerations also explain why gold nanoparticles (with their very large polarizability) may be readily trapped at sizes of 100 nm and below (Hansen et al., 2005) though absorption is a key consideration to be noted (Seol et al., 2006).

Other studies have looked at different aspects of the problem. Barton and co-workers (Barton and Alexander, 1989; Barton et al., 1989) derived fifth-order corrections to the focused Gaussian beam so as to compute the forces using a Maxwell stress tensor approach. Rohrbach and Stelzer (Rohrbach and Stelzer, 2002) extended the Rayleigh theory to make it valid for large particles by inclusion of second-order scattering terms. The incident field is expanded in terms of constituent plane waves allowing apodization and aberration transformations (due to the high numerical aperture microscope objective) to be incorporated in the theoretical model to yield the resultant optical forces upon the dipole, in this instance without resorting to the use of the Maxwell stress tensor method. The vast majority of optical tweezers and trapping experiments are performed where the particle size is comparable to the wavelength of the trapping laser beam. In this region key studies by Rohrbach (2005) reported good quantitative agreement between the theory calculations and experimental measurements pertaining to the strength of the optical tweezers. His theoretical approach for trapping forces computed the Lorentz force density. He found that the optimal trapping performance is reached when the diameter of the particle d is comparable to the wavelength of light, λ, (within the viscous medium of refractive index n), $d \approx \frac{\lambda}{n}$ (Rohrbach, 2005). Overall, it is important to note that the numerical and theoretical modeling of optical forces at these scales still present many challenges.

2.2 Design Considerations of an Optical Tweezers System

Constructing an optical tweezers that is calibrated to measure forces involves some important physical optics principles and due consideration of certain mechanical components. A more comprehensive technical discussion of this may be found elsewhere (Lee et al., 2007). The basic optical tweezers requires only a single laser beam for its operation and may be based around a standard microscope which already comes equipped with a high N.A. objective lens (Lee et al., 2007). Alternatively one may build an optical tweezers from standard cage plate or other opto-mechanical components (Rohrbach et al., 2004; Appleyard et al., 2007). Many of the deliberations are upon the types of lasers, choice of microscope objective and the beam delivery by the optical setup which are all crucial for three-dimensional (x, y and z) trapping.

The vast majority of optical trapping experiments use monochromatic continuous wave (CW) sources that have the key properties of low absorption of light by the subject being trapped, especially when using biological materials. The laser wavelength is chosen in the near-infrared region to avoid damage to biological specimens. Wavelengths of 830 nm and 970 nm have been shown to be particularly favorable to these studies (Neuman et al., 1999). In the near-infrared there are ample laser sources with good beam quality (denoted by $M^2 \sim 1$ (Kogelnik and Li, 1966; Siegman, 1998) and beam pointing stability notably the new generation of near-infrared fiber lasers. In the domain of colloidal science, there has been a widespread use of visible lasers notably at 532 nm due to the low absorption of water that reduces heating effects (Hale and Querry, 1973). Ultrashort pulsed lasers have also been used when one wishes to combine optical trapping with other multi-photon processes. For the case of pulses of ~100 femtosecond duration if the repetition rate of the laser is high enough (e.g. 80 MHz), any anomalies due to particle diffusion in between pulses are not significant and the forces exerted equal their continuous wave counterparts (Agate et al., 2004). Temporal coherence of the light field is not such an issue (except of course where the laser bandwidth encroaches upon an absorption band within the trapped sample). Some recent experiments have even used 'white light' laser sources based on a supercontinuum generation (Li et al., 2005) to combine optical trapping with spectroscopy.

To achieve a diffraction-limited beam spot (radius = $0.61\lambda/N.A.$) at the focal plane, a high numerical aperture (min. N.A. ≈ 1) microscope objective lens is a natural choice. Today, microscope objectives are designed to produce an image at infinity. With such 'infinity corrected' objectives, researchers may add an additional telescope or other optical components (i.e. spatial light modulator) to relay or modulate the laser source to the microscope objective (within the 'infinity space') without affecting the bright-field or fluorescence imaging aspect of the system. The optical telescope is desirable for achieving a diffractive-limited spot for true three-dimensional tweezing. One needs to slightly overfill the back aperture of the microscope objective (Lee et al., 2007). The use of optical conjugates for the trapping beam enables the optimum steering of beam before entering the microscope objective. The input beam needs to be translated across the sample stage whilst ensuring the beam does not 'walk off' or is deviated in a way that might cause it to miss or clip any apertures in the beam path, notably the back aperture of the microscope objective. Ensuring that the input beam is centered upon the back aperture during steering would mean a good quality trapping beam at the focal plane at all times. In Figure 2 we illustrate the idea of optical conjugates for the steering of the trapping beam. The back aperture of the microscope objective is imaged onto a mirror: thus positioning the light

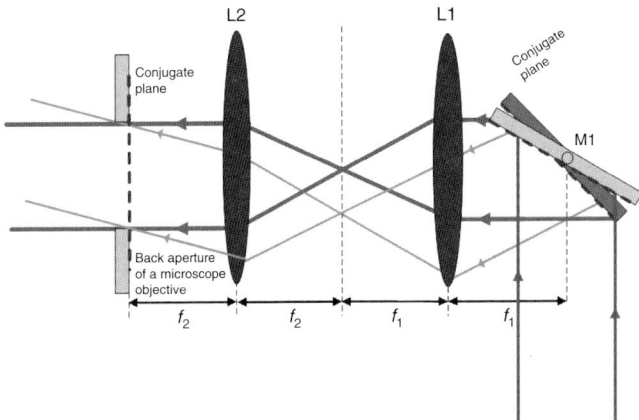

FIGURE 2 Beam steering lens system. The diagram illustrates a beam steering lens system that is used in an optical trapping system and the concept of conjugate planes

onto that mirror is equivalent to positioning the beam onto the objective: tilting the mirror in the lateral plane now moves the beam at the sample plane but does not move it across the back aperture of the microscope objective. We now look at Figure 2 in some detail to expand upon these points. The beam steering system is made of up two lenses as seen in the figure, denoted as L1 and L2, and a rotating mirror M1. Each lens is placed at one focal length away from the other and the rotating mirror M1 is positioned at one focal length away from one of the lenses L1. By placing the back aperture of a microscope objective at one focal length away from lens L2, a beam steering system is formed. Here, the rotating mirror M1 and the back aperture of the microscope objective form conjugate planes. In Figure 2, we illustrate just two ways in which the rays are being redirected by the mirror M1 at an angle with respect to the principle axis of the beam steering lens systems: green thin line and red thick line. With the beam steering lens system, both rays maintain their spatial position onto the back aperture of the microscope objective. The central purpose of the conjugate planes is to ensure that a collimated beam (light (blue line)) before the mirror M1 maintains overfilling of the back aperture of microscope objective without any 'walk off'.

Preparing a sample for tweezing also requires some careful thought. Colloidal particles are usually dispersed in de-ionized water and approximately 10 microliters is placed on the sample, while biological cells are often suspended in a suitable buffer solution. In both cases, when employing an oil-immersion microscope objective for tweezing, the beam would propagate through a layer of oil, a thin piece of glass (coverslip) and then into the sample solution which is of a lower refractive index than the glass and oil. Thus the beam will, unsurprisingly, experience a large change in refractive index that generates spherical aberrations, broadening the beam spot along the axial plane. Thus, well-aligned optical tweezers may still suffer inherent aberrations due to refractive index mismatches. As given by the results, the trap performance degrades when operating in a deeper region of the sample (Rohrbach and Stelzer, 2002; Lee et al., 2007). With the use of immersion oils with varying refractive indices, Reihani and Oddershede (Reihani and Oddershede, 2007) succeeded in minimizing spherical aberrations to allow efficient axial trapping over a longer axial distance using oil-immersion objectives. On the other hand, water-immersion objectives are corrected for the refractive index of water (removing the requirement of using immersion oil and a coverslip) which in turn reduces the spherical aberration (Rohrbach and Stelzer, 2002; Vermeulen et al., 2005). Alternative adaptive optical techniques to obviate these aberrations are available in the form of dynamic holographic elements (Wulff et al., 2006) or by the use of deformable mirror (Theofanidou et al., 2004) technology by imposing correction terms of the appropriate Zernike polynomials upon the input wavefront of the tweezing beam.

Trapping a particle is of course a key step but how do we then use this trapped object to make force measurements? Converting an optical tweezers into a highly accurate force measurement system requires an understanding of the underlying physical principles of its operation with relationship to particle dynamics and Brownian motion. The crucial tenet is that an optical tweezers may be modeled as a near ideal highly overdamped simple harmonic oscillator. In particular, the particle resides in a parabolic potential well created by the optical trapping beam. When the particle is at equilibrium, it tends to rest at the center of the potential well (trap) where it has the minimum energy. Once the particle is perturbed, it will experience a restoring force that is proportional to the distance away from the center of the well (the resting position) akin to a minuscule spring that obeys Hooke's law: this perhaps innocuous statement embodies some of the most powerful principles of optical trapping as illustrated in Figure 3. This idealized case may be represented by a one-dimensional equation of motion in direction x for a Brownian particle in a harmonic potential:

$$m\frac{\partial^2 x}{\partial t^2} + \gamma \frac{\partial x}{\partial t} + \kappa x = F_f(t). \tag{3}$$

If we examine the left-hand side of Equation (3), the first term represents the inertial force component for a particle mass m; the second term is the velocity dependent viscous damping force, where γ is the drag coefficient; and the final term is the optical restoring force for

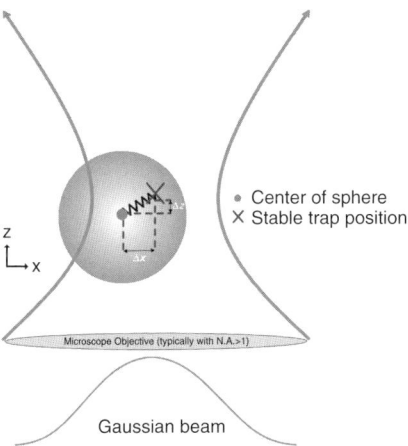

FIGURE 3 Stiffness of an optical tweezers. A basic diagram that illustrates a microsphere being perturbed (i.e. brownian motion) where the restoring forces are proportional to the displacements (Δx & Δz) of the particle from the stable trapping position

the trap stiffness, κ. The right-hand side of the equation represents the fluctuating force, $F_f(t)$, due to the Brownian motion. By monitoring either the distribution of the particle position within the trap or the frequency response of the particle motion we may obtain two independent methods for determining the trap stiffness, κ. Under typical trapping conditions (which is the low Reynolds number regime (Squires and Quake, 2005)), the viscous damping forces dominate and inertial effects (i.e. $m\frac{\partial^2 x}{\partial t^2}$ in Equation (3)) may be neglected. As illustrated in Figure 3, the particle in this potential well essentially behaves like a spring obeying Hooke's law where the force on the object is proportional to the displacement (Δx and Δz as seen in Figure 3) of its position from the center of the trap with the constant of proportionality denoted as the trap stiffness. The particle position may be recorded as a histogram that may be described by a Boltzmann distribution that in turn will be directly related to the shape of the trapping potential, which is harmonic over the size range of a fraction of the trapping wavelength (Berg-Sorensen and Flyvbjerg, 2004; Svoboda and Block, 1994a).

To use optical tweezers for measuring small forces, we require the addition of an opto-electronic sensor capable of measuring the mean position of the trapped object in all the three dimensions to extract the corresponding trap stiffness, κ. The sensor and detection circuitry are calibrated by moving the trapped objects over a known distance. With the relationship of signal (voltage) and the distance (nanometer), the sensor can be calibrated for specific distances. Once in this position calibration factor and the stiffness κ of the optical tweezers, which is calculated from analysis of the thermal motion of the trapped object or by application of known viscous drag forces, are quantitively collected, the optical tweezers can then be implemented as a force transducer. A popular method of detection is the imaging of the back focal plane interference pattern of the trapped object (Allersma et al., 1998) onto an opto-electronic detector (quadrant photodiode or position-sensitive detector) (Berg-Sorensen and Flyvbjerg, 2004). Such a detector has a high bandwidth (>Megahertz) and fast response (nanoseconds), and the signal can be collected from either back-scattered light or forward-scattered light with the aid of a probe laser beam (Volpe et al., 2007a; Huisstede et al., 2005). The detector measures the geometrical center of the trapped bead, in three dimensions, over a given time scale (e.g. typical acquisition time of 4 seconds). In the absence of any biological forces, the motion of a trapped microsphere is determined by the balance between the thermal Brownian motion and the trapping force. We refer the reader to the paper by Lee et al. (Lee et al., 2007) that explains in detail the assembly and calibration of an optical tweezers and the choices regarding the detection system. Neuman and Block (2004) also made pertinent comments regarding the choice of a quadrant photodiode and position-sensitive detector detection system. A recent direct comparison of video and quadrant photodiode detector techniques presented by Keen et al. (Keen et al., 2007) shows the viability of using a high-speed video camera to measure trap stiffness with relevance to multi-particle trapping. Recent techniques in optical tweezers have started to make use of evanescent light scattering to determine weaker long-range forces between a single colloidal particle and a flat surface as a function of separation (Clapp et al., 1999). This type of geometry has been successfully applied for the direct determination of weak critical Casimir forces (Hertlein et al., 2008). We remark in concluding this section that most excitingly the particle position may be determined to dimensions that are at least one to two orders of magnitude smaller than that of the trapping wavelength: this is the crux of the reason why such small displacements (and thus forces) may be discerned with this method.

2.3 Other Incarnations of Optical Trapping

Significant attention in the field has been invested into the designing of alternative trapping geometries to that of single beam optical tweezers. Many of the technical developments aim to accommodate specific trapping geometry and add flexibility for force analysis, simultaneous spectroscopy or the trapping of larger objects. Here we describe a few of these developments.

An elegant and inexpensive approach to create two adjacent but independent optical tweezers has been desired by the biophysics community for stretching and holding macromolecules: the dual beam optical

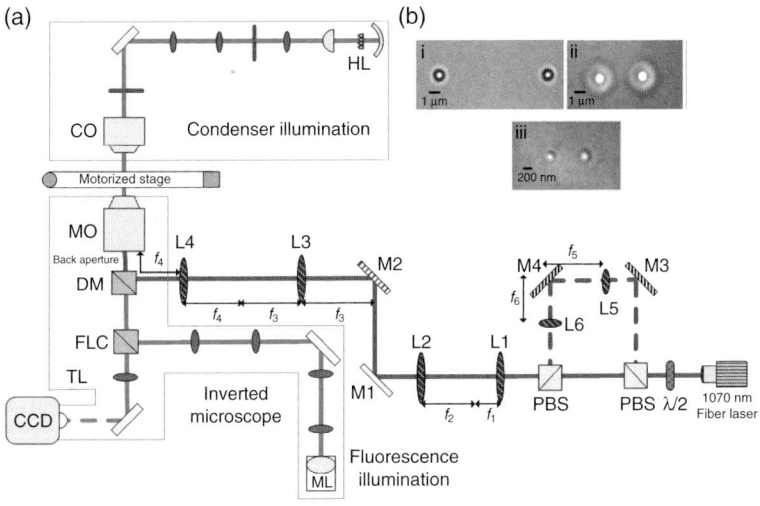

FIGURE 4 Dual beam optical tweezers setup. (a) Here, a dual beam optical trapping setup is built with a single fiber laser (wavelength 1070 nm) and inverted fluorescence biological microscope system. The linearly polarized collimated beam from the 1070 nm fiber laser is directed onto a half waveplate ($\lambda/2$) before being split up into two independent beams by the first polarization beam splitter cube (PBS). The optical power in each beam can be controlled by rotating the waveplate. Both beams are recombined at the second polarization beam splitter cube (PBS) and expanded by a beam telescope formed by lenses L1 and L2. The beams expand by a ratio of f_2/f_1. The expanded beams are redirected by a dielectric mirror (M1) onto the steering mirror (M2). A beam steering lens system, lenses L3 and L4, ensures that the expanded beam is well aligned onto the back aperture of the microscope objective (MO) when the steering mirror (M2) is rotated (see Figure 2). NIR dichroic mirror (DM) provides a reflectivity efficiency >90% at 1070 nm. A fluorescence illumination module with a mercury lamp (ML) is directed onto a fluorescence filter cube (FLC) that illuminates the sample with the same microscope objective (MO). A bright-field illumination from a halogen lamp (HL) and a condenser (CO) illuminates the samples. The diffracted light from the sample is collected by the microscope objective and relayed onto a CCD camera (CCD) via the tube lens (TL). The second optical tweezers is independently controlled with mirrors (M3) and (M4). (b) Shows two particles trapped in each of the optical tweezers in a dual beam optical trapping system. (i) and (ii) Two polymers of 1 μm in diameter trapped and viewed with bright-field and phase contrast microscope objectives respectively. (iii) Two gold nanoparticles, each of 200-nm diameter, trapped and viewed with a differential interference contrast microscope objective

tweezers (Fallman and Axner, 1997) achieves this goal as shown in Figure 4. Figure 4(a) shows a simple illustration of the dual beam optical trapping setup using a single near-infrared fiber laser and an inverted fluorescence microscope system. Using a set of polarization optics, a single beam can be split into two beams of equal power but each orthogonally polarized (to minimize optical interference effects) with the aid of a Mach–Zehnder interferometer setup, with each beam creating an independently steerable optical tweezers. Here the two tweezers are formed in the same transverse plane, as seen with the tweezing of polystyrene microspheres and gold nanoparticles in Figure 4(b). In some instances, this form of dual beam tweezers may used together with the high-speed steering provided by an acousto-optical modulator (Greenleaf et al., 2005; Shaevitz et al., 2003; Vossen et al., 2004). We shall describe the acousto-optic modulator and its use in optical micromanipulation later in this article.

Other forms of trapping arrangements exist amongst which the dual beam optical fiber trap (Constable et al., 1993; Guck et al., 2000; Singer et al., 2003; Metzger et al., 2006b; Jess et al., 2006) has taken prominence. As mentioned earlier Ashkin's first trap (Ashkin, 1970) was made with two counterpropagating Gaussian beams. A powerful and now common variant of that experiment is the dual beam fiber trap where the (divergent beam) output of the two single mode optical fibers are meticulously aligned to face each other coaxially at a separation of around 100 micrometers. Here the two emergent beams from the fiber create a counterpropagating geometry for a single trap. In this trapping geometry, the scattering force dominates due to the divergence of the output beam. The total gradient force from the two beams acts to draw a high refractive index particle into the overlap region between the two beams, whilst the scattering force from each of the beams balances to hold the object in between the two beams. Here, the risk of optically-induced thermal effects i.e. photodamage from two-photon absorption (Konig et al., 1996), especially when manipulating biological cells, is much lower due to the absence of tightly focused light fields. One is able to immobilize a single particle and collect scattered signal i.e. fluorescence or Raman scattering through an auxiliary microscope objective (Jess et al., 2006). The system has led to the development of the optical stretcher that controllably deforms cells held in the trap. The deformation has a dependency upon the cytoskeleton of the trapped cell and ultimately gives a signature of cancerous vs non-cancerous cells (Guck et al., 2001; Lincoln et al., 2004; Guck et al., 2005). The dual fiber optical trapping system may not be readily integrated into a commercial microscope system when compared to the single beam/dual beam optical tweezers system, but is notably compatible with microfluidic flow chambers (Lincoln et al., 2007).

3. APPLICATIONS WITHIN BIOPHYSICS AND THE COLLOIDAL SCIENCES

Optical tweezers are not usually used to move macromolecules directly. However they offer a powerful method that can indirectly manoeuvre such macromolecules with the trapped microparticles acting as 'handles'. By labeling DNA (deoxyribonucleic acid) strands with biotin and mixing them with polystyrene microspheres (coated with streptavidin which this binds to biotin with high affinity (Appleyard et al., 2007)), one can obtain for example DNA strands tethered to microspheres, which form the 'handles' required to indirectly manipulate the DNA strands. From the preceding sections we know that for a single beam tweezers to work well, we must use optically transparent dielectric particles (usually of size 0.5–5 microns in diameter). The determination of the position of the microsphere when trapped is the crucial step to ultimately make quantitative measurements. The behavior of an optical tweezers transducer is very different from a macroscopic force transducer; this is because it works in the very low force regime, where thermal forces are significant. It is also rather different from the atomic force microscopes (AFM) because the motion of an AFM probe is dominated by its relatively high mass and high stiffness so that the probe tip shows resonant behavior.

3.1 Molecular and Cell Biology

Prior to the use of optical tweezers much of the knowledge of mechanical processes in biological systems was gleaned from bulk samples. In many senses this was unsatisfying and gave insufficient insight into the exact mechanisms of processes at the single molecule level. The emergence of optical traps made available non-invasive biological tools with sensitivity and range that are beyond other forms of transducers in this field (Neuman and Nagy, 2008). In terms of force, optical traps exert forces in the piconewton range. So they are ideally suited to unraveling DNA, exploring the world of protein–protein interactions and importantly the work cycles of motor proteins. However such forces fall well short of those required to break covalent bonds (1 nN) (Grandbois et al., 1999).

The initial studies with optical tweezers were on biological samples that were relatively large in size. Ashkin and colleagues first employed optical tweezers to trap and move bacteria and small numbers of tobacco mosaic virus (Ashkin and Dziedzic, 1987) and then to manipulate single cells (Ashkin and Dziedzic, 1989b), cell organelles (Ashkin and Dziedzic, 1989a) and finally to measure the force of cell organelle movement inside living cells (Ashkin et al., 1990). However such studies did not involve truly quantitative measurements which later followed, opening up the realm of quantifiable single molecule dynamic. We now proceed to describing some of these studies.

3.1.1 Single Molecule Studies

Understanding the living cell is a fascinating aspect of modern biology: the cell has a range of *molecular motors* that operate by converting chemical energy to mechanical work. Exploring the work cycles of such motors and their behavior has become a major application of optical

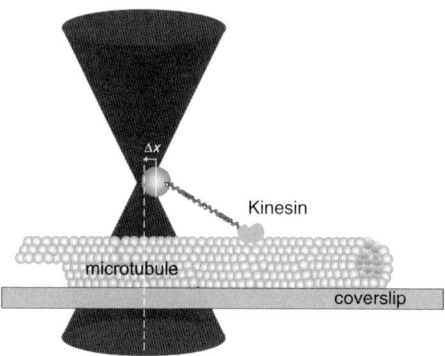

FIGURE 5 Operation of a surface based assay optical force measurement system. The diagram illustrates an optical force clamp recording displacements of single kinesin molecules while maintaining an average load. Adapted figure with permission from (Visscher et al., 1999) © 1999, Macmillan Publisher Ltd: Nature

tweezers in the past two decades. We may broadly group such motors into the two categories of rotary motors and linear motors. Membranes possess what are termed 'rotary motors' which derive energy from ionic flow across transmembrane electrochemical gradients. Linear motors use chemical reactions as their power source, for example the hydrolysis of the chemical, adenosine triphosphate (ATP) to adenosine diphosphate (ADP) and phosphate. Optical tweezers are not just able give insights into the forces exerted by both of these types of motors but also into the manner of their motion (Block, 1996).

A major step forward in the use of optical tweezers for single molecule studies occurred when researchers (Block et al., 1990) measured the individual steps taken by kinesin, a molecular *porter* (a subclass of a linear motor described above) during its motion along a fixed microtubule track created upon a microscope slide. The single kinesin molecule was attached with appropriate surface chemistry as described earlier to a microsphere and this was then held close to its microtubule track (Visscher et al., 1999). A buffered salt solution containing the chemical fuel ATP constituted the sample medium. The use of the optical trap and a quadrant photodiode (for imaging) permitted detailed observations of the kinesin motor and microtubule track interaction. The trapped sphere was pulled along by the kinesin and approximately eight nanometer-sized steps taken by the kinesin molecule as seen in Figure 5 were observed during the experiment. The behavior of the kinesin molecule is attributed to it pausing for a random interval after taking each step as it waited for a fresh ATP molecule to arrive.

The optical tweezing geometry needed to be adapted to enable them to be applied to the study of the actin–myosin system, which is the molecular motor associated with muscle action. This interaction occurred in an intermittent fashion and to understand this system a dual trapping arrangement was required. Finer et al. (1995) adapted the dual trap system (Fallman and Axner, 1997) described earlier to hold the two proteins, actin and myosin, in close proximity so that several interactions could be recorded. The dual trap system enabled them to hold a single actin filament suspended between two trapped microspheres. This allowed them to maneuver the actin filament into close proximity of a third (fixed) myosin-coated microsphere. When the proteins make contact with a single molecule of fuel (the sample solution contained the chemical fuel, ATP), the protein breaks down resulting in a single kick, or displacement of actin. This movement was measured by monitoring the position of one of the trapped beads. Subsequent studies by Molloy and colleagues (Molloy et al., 1995) shed more light upon this process and indeed this system is one of the most widely studied to date using optical tweezers. The studies on motor proteins are likely to be paradigms for future single molecule mechanical studies. Readers who would like further information on optical tweezers based studies of rotary and linear motor proteins are directed to the excellent recent reviews (Greenleaf et al., 2007; Mehta et al., 1999) and the references therein.

The determination of the double-helix structure of Deoxyribonucleic acid DNA by Watson and Crick is an undoubted landmark for structural biology (Watson and Crick, 1953). DNA has been studied extensively with optical traps. For example, a length of DNA was attached to a microsphere and then extended by dragging through a viscous medium in order to measure its mechanical properties (Perkins et al., 1994). This gave useful information about its polymer mechanics. Knowledge of its mechanical bend-persistence length (about 50 nanometers) informs the biologist about the probability of adjacent regions of DNA forming some contact with one another. For instance, we know that if two pieces of the sequence are 20 nanometers apart then the intervening length of DNA is relatively stiff, if they are 500 nanometers apart then the linking region will be flexible enough to allow the two regions to come together. Later, researchers (Smith et al., 2003; Bustamante et al., 2003) applied very large forces (>60 pN) to DNA and found that it suddenly overstretches and breaks the double-helical B-form to give a parallel ladder. Figure 6(a) illustrates force vs extension for dsDNA and ssDNA molecules, obtained with a micropipette and an optical trap. Figure 6(b) shows the changes in extension observed at a constant extension during polymerization (Poly) or force-induced exonuclease activity (Exo) (Wuite et al., 2000).

A major advance in displacement and force studies with optical tweezers was recently achieved by Abbondanzieri (Abbondanzieri et al., 2005) where they achieved resolution of Angstrom level motion, as seen in Figure 7. In this figure we see a single, transcriptionally active molecule

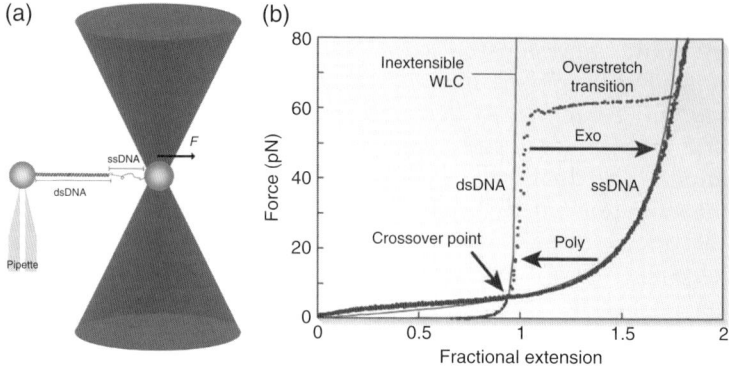

FIGURE 6 DNA force measurements with optical tweezers. (a) Shows a schematic of a DNA molecule that is stretched between beads held in a micropipette and a force-measuring optical trap. The measured extension is the sum of contributions from the single-stranded DNA (ssDNA) and double-stranded DNA (dsDNA) segments. (b) Force vs extension for dsDNA and ssDNA molecules. Reprinted figure with permission from (Abbondanzieri et al., 2005) © 2003, Macmillan Publisher Ltd: Nature

of RNA polymerase (green) attached to a bead (blue) held in trap T_{weak} (the right-hand optical tweezers) and tethered via the upstream DNA (dark blue) to a larger bead held in trap T_{strong} (the left-hand optical tweezers). The right-hand bead is maintained at a position near the peak of the force–extension curve of T_{weak}, where there is no change in the applied force over the displacement (white arrow), creating a force clamp (trap stiffness k). During elongation, the DNA tether lengthens and the beads move apart. Owing to the force clamp arrangement, only the right bead moves: displacement is measured for this bead. With the configuration shown in Figure 7, they measured the force produced by RNA polymerase as it transcribed a DNA gene. They attached a glass bead to one end of a DNA strand and then allowed the other end to bind to an immobilized molecule of RNA polymerase. When a 'transcription buffer' was added (containing all the necessary nucleotides for transcription) RNA polymerase moved along the DNA proceeding with its task of transcribing the gene and producing a new RNA chain. The authors grabbed the bead that was attached to the free end of the DNA and then measured the pulling force produced by the RNA polymerase. The force was about 25 piconewton, perhaps surprisingly this is five times greater than that produced by either of the studied motor proteins myosin or kinesin. The authors measured a mean distance of 3.4 Angstrom in discrete steps taken in the transcription process. The technical advancement reported in this paper included the development of an ultrastable trapping system. The authors encased the optical path in

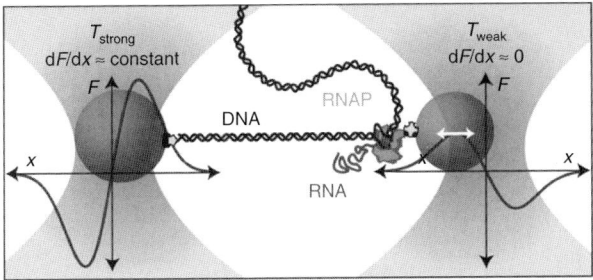

FIGURE 7 Schematic force vs position curves (horizontal axes) shown for both trap beams from the Abbondanzieri study. Reprinted figure with permission from (Abbondanzieri et al., 2005) © 2005, Macmillan Publisher Ltd: Nature

helium gas so as to stabilize the trapping position removing any instability in laser pointing arising from the surrounding environment. The refractive index of helium was closer to unity than air, thus decreasing variations in laser positioning and reducing the level of noise by a factor of ten. Furthermore, the team developed a constant force clamp by placing one of the two trapped beads slightly off-axis on the nonlinear part of the force vs extension (displacement) curve (Abbondanzieri et al., 2005).

One may explore very exciting aspects of statistical mechanics using optical traps and macromolecules. Intriguingly, developments in statistical mechanical theory can be used to solve fundamental problems in experimental thermodynamics. Jarzynski (1997) proved an equality relating the irreversible work to the equilibrium free energy difference. This work put forward the notion that it is possible to obtain equilibrium thermodynamic parameters from processes occurring well away from equilibrium. Liphardt et al. (2002) explored the Jarzynski's equality by using optical tweezers to mechanically stretch a single molecule of RNA reversibly and irreversibly between two conformations. Application of this equality to the irreversible work trajectories recovered the equilibrium free energy difference profile of the stretching process. Excitingly this experiment and interpretation form a link between the statistical mechanics of equilibrium and non-equilibrium systems (Liphardt et al., 2002). A very good account of further experiments and studies of the physics of non-equilibrium systems is given elsewhere (Bustamante et al., 2005).

3.1.2 Cell Biology

As mentioned before, Ashkin's early work dealt with the possibility of using optical tweezers to move and manipulate biological cells and its internal organelles (Ashkin, 1997). The trapped microsphere can also be used to tether onto cell membranes such as blood cells (Dao et al., 2005) which had led to the measurements of its mechanical structures.

Liang et al. (1993) used optical tweezers in combination with pulsed laser cutting (laser scissors), to manipulate and explore the behavior of chromosomes during cell division in view of studying the complexities of mitosis. Seeger et al. (1991) used the same type of technique for isolation of a microdissected gene. An excellent and thorough review of optical microbeams (laser scissors) may be found in the article by Berns (Berns et al., 1998). We have already mentioned the optical stretcher that uses the dual beam counterpropagating geometry trap. Guck et al. (2002, 2001) showed that the forces from two diverging beams (output from optical fiber ends) can steadily measure the mechanical deformation on red blood cells without any tethering. In this way the elastic and viscoelastic properties of red blood cells (RBCs) alter in a manner that is often correlated with the manner in which the cells respond to structural and molecular alternation induced by the onset and progression of a disease within the cell. Hence the cell deformation in the presence of the calibrated force can be used to calculate the mechanical properties (shear modulus). Optical traps have been applied to study infected RBCs with optical tweezers by Mohanty et al. (2005). In hypertonic buffer, a normal RBC rotated by itself when trapped by optical tweezers and the rotational speed increases linearly at lower trap-beam powers and more rapidly at higher powers. In contrast, a malarial-parasite-infected RBC did not rotate under the same experimental conditions. The rotational speeds of other RBCs from the malaria-infected sample were found to be an order of magnitude less than that for normal RBCs. Biological processes i.e. cell division, vesicle trafficking, endocytosis and phagocytosis within the cells are crucial to its proper functionality and may be probed with optical tweezers. Using a calibrated optically trapped latex bead, Kress et al. (2005, 2007) were able to measure direct cell binding of beads to membranes of living cells during the process of phagocytosis. Kress and co-workers were able to measure the F-actin-dependent stepwise retraction of filopodia, which act as cellular tentacles that pull particles into the cells during cell ingestion, with a trapped microsphere. The measured refraction forces can be at high as 19 piconewton and at a mean step size of 36 nanometers.

Optical tweezers operating at 980 nanometers were explored for the movement of mature isolated retinal cells. Such studies may lead to the creation of synaptic circuits *in vitro*. Rod and cone photoreceptors as well as other retinal nerve cell types were manipulated and approximately 60 per cent of the tweezed cells survived for at least 48 hours. The tweezing was observed not to affect the organelle, nuclear and cytoplasmic structure of the manipulated cells. Additionally, in photoreceptors, synaptic vesicles and ribbons were also unaffected. A recent work by Franze et al. (2007) revealed that suspended Müller cells are optically trapped and aligned between a dual beam fiber trap. The Müller cells were able to guide the tweezing light from one fiber to the other with relatively good efficiency, thus substantiating the fact that these cells are the 'optical fibers'of the retina and capable of relaying light from the inner surface to the layer of the photoreceptors with low light scattering loss. Optical micromanipulation has been shown to provide a strong technique to manoeuvre whole neurons (Townes Anderson et al., 1997). Ehrlicher et al. (2002) performed intriguing work showing that neuronal growth could be influenced by laser light. The authors proposed some possibilities to explain this behavior and attributed the guidance of the neurites to the optical trapping of actin monomers to provide nucleation sites for actin polymerization which drives cellular growth. The exact mechanism for such behavior is somewhat elusive and is under investigation (Ehrlicher et al., 2002; Stevenson et al., 2006).

Optical tweezers are also finding their niche in the immobilization of a cell over a probe beam (Xie et al., 2002; Jess et al., 2006; Creely et al., 2005). The probe acts to scatters enough light off the cell (the vibrational energy level) which is then recorded by a detector. This form of optical trapping and detector technique opens up a new range of spectroscopy techniques where the optical traps are used to fix cell in three dimensions. This allows research to choose the specific site for detection i.e. Raman scattering. Combining optical tweezers, Raman spectroscopy and microfluidics allow interesting prospects for cell studies. Ramser et al. (2005) had control over the medium surrounding the cell in their microfluidic chamber that is defined by electron beam lithography that is then molded into rubber silicon. Different buffers flowed through the channels, while the resonance Raman response of an optically trapped red blood cell (RBC) was recorded. The authors were able to monitor the oxygenation cycle of the cell in real time and understand the effects of photo-induced chemistry. This system has prospects for *in vivo* monitoring of cellular drug response (Ramser et al., 2005). Jess et al. (2006) were able to combine the dual beam counterpropagating trap with simultaneous Raman spectroscopy creating a form of Raman flow cytometer. They also were able to take definitive Raman spectra from local parts of the trapped cell and perform multivariate spectra analysis by ascribing the spectra signal from specific regions of the cell.

3.2 Examples of Studies of Colloidal Systems with Single Beam Optical Tweezers

A calibrated optical trapping system has application in areas of colloidal sciences, especially in the measurements of weak interaction forces between small particles suspended in a viscous medium (∼micrometer). Furthermore, in the microfluidic environment (of low Reynolds number), these small interactions become more pronounced. Meiners and Quake (Meiners and Quake, 1999) made use of a dual beam trapping system

(Fallman and Axner, 1997) and measured the hydrodynamic interaction between two trapped spheres. They recorded the position fluctuations of each bead and from this determined both the correlation and anti-correlation functions. They showed that these trapped spheres exhibit a pronounced, time-delayed dip in the cross-correlation. This time delay is linked to the relaxation time of the harmonic well created by the tweezers. Further detailed study by the authors showed that the anti-correlations observed were well understood by using the Langevin equation and invoking a standard Oseen tensor hydrodynamic coupling (Meiners and Quake, 1999).

Microrheology refers to the study of the deformation of viscoelastic materials or fluid flow in response to the applied force. Crocker et al. (2000) introduced a new form of two-point microrheology that is centered on the measurement of the cross-correlation of the thermal motion of pairs of tracer particles. Thus it is not surprising to infer the possibility of using optical traps for such local viscosity measurements (Nemet et al., 2002; Bishop et al., 2004; Pesce et al., 2005). A more recent paper by Brau et al. (2007) summarizes the range of microrheology applications for which optical tweezers can be employed. The non-invasive nature of optical tweezers lends itself very well to exploring the local viscoelastic properties of solutions. The optical trap may act both in an active or passive mode for microrheology applications. In the passive mode one simply monitors the position of the particle and relates these observations with the predictions of the Stokes–Einstein equations. A more active role is also possible: the trapped particle may be dragged through the sample, thus mechanically deforming the medium under study.

4. OPTICAL TRAPPING WITH STRUCTURED LIGHT FIELDS AND ITS APPLICATIONS

4.1 Structured Light Fields

The previous sections have shown how a single focused light beam in the form of optical tweezers can be a remarkable tool for a diverse range of biological and colloidal sciences. We firstly consider and indeed question why we might wish to deviate from the use of a single Gaussian beam for a single optical trap and use structured light fields (or beams), either as traps on their own or replicated in arrays. Though the optical tweezers have proven themselves as a general interdisciplinary tool, serious drawbacks and limitations remain. The basic tweezing geometry uses a microscope objective lens and a standard Gaussian laser beam. This arrangement can only provide a single trap with an approximately ellipsoidal focal volume, thus constraining the ratio of trap stiffness in lateral and axial directions. These conventional techniques offer little flexibility for tailoring the trap (and its stiffness) in a three-dimensional space. A Gaussian beam expands and diffracts as one knows upon high focusing and potentially limits the types of applications, e.g. optical guiding or the even transportation of particles separated over large distances (millimeter). Rotation would be beneficial to the study of molecular motors if one can gain precise rotational control within optical tweezers, a calibrated rotational torque (La Porta and Wang, 2004; Volpe et al., 2007b) can be applied upon bio-molecules. In the field of microrheology, a rotating particle may be used to measure a range of viscoelastic behavior in different media. In a microfluidic environment, rotating single particles or groups of particles can induce pumping action within a laminar flow (Friese et al., 2001; Galajda and Ormos, 2001; Terray et al., 2002) which has potential for controlling flow rates and pumping and mixing small volumes of analyte. By increasing the number of the optical tweezers – creating multiple trap arrays – researchers may start to explore different trapping experiments in parallel or work with larger array of cells or colloids. One can also view a large array of optical tweezers as an *optical potential energy landscape* (a topographic array of potential energy wells) which can thus impose a distribution of optical forces over a large area. These landscapes can in turn optically organize colloidal particles into two- and three-dimensional quasi-crystal structures and can be used to tune the inter-particle interaction over an extended physical space. Such an ensemble of self-assembled colloidal particles provide a strong platform for potential rich scientific studies in areas such as materials science and thermodynamics (Bechinger et al., 2001; Korda et al., 2002a). The full impact of this technology is yet to be realized. In the following section we aim to address these issues and broadly group our discussion under the heading of *structured light fields*. There is a consensus in the community that this area is burgeoning and potentially one of the most powerful for future studies within the field of optical trapping. From the materials science viewpoint for example, a system by a given set of microparticles that can be organized and assembled into any form of geometry would be of great importance to material sciences. There is a wealth of novel science that may be tapped with not just creating large arrays of trap but also controlling each trap site independently. Additionally one opens up the prospect of collective or co-operative effects between objects held at individual trap sites within a potential landscape. Self-assembling of colloidal particles (Dillen et al., 2004) plays a vital role in the understanding of phase (liquid–solid) transitions and depending on the optical properties of the particles, one can form composite materials with these colloids. We start by detailing the different ways of multiplexing a single beam optical tweezers into many traps. We then proceed to mentioning some emerging applications, in colloidal and biological sciences, that have employed such arrays of optical traps and where appropriate compare and contrast different approaches. After this we discuss two popular forms of non-zero-order light beams: firstly Laguerre-Gaussian beams which are light fields with inclined wavefronts and accompanying orbital angular momentum. Then we look at Bessel light fields that show a 'non-diffracting' or propagation invariant nature; Bessel beams that offer interesting dynamics for guiding and sorting in particular. Both Laguerre-Gaussian and Bessel modes may provide an added ability within different geometries of optical micromanipulation.

4.2 Large Arrays of Optical Traps

In this section we give a brief introduction to four major techniques that allow crafting arrays of optical traps and look at each of these methods involved and comment upon some of the experiments that have been performed using each technique. A direct and straightforward way of obtaining multiple trapping sites is with the use of interference. A second method capitalizes on the fact that optically trapped particles are typically suspended within a viscous medium (i.e. water, glycerol and buffer solution) and operate in a highly overdamped regime. The outcome of such damping is that it brings about relatively slow rate of diffusion of particles over a given space. Thus if a single optical trap is scanned onto discrete locations at a rate and returns to its original position quickly enough before the particles diffuses out of the given area. It would be possible to simultaneously tweeze multiple particles at each of the discrete sites by time sharing the beam between each trapping location. Another technique uses the principles of holography and, for example, imprints the phase of several beam splitters onto a single holographic (photo-sensitive) element: one beam can then be split into multiple beams at the same time but at different spatial points. The beams can then be relayed directly onto the trapping plane forming individual traps. Finally, it is possible to convert phase structure of any form into its corresponding intensity with high optical efficiency by using the generalized phase contrast technique, which increases the optical efficiency of the output power. We remark before continuing that trap multiplexing may be achieved by means other than the four we describe, for example the use of vertical cavity emitting lasers (Birkbeck et al., 2003; Flynn et al., 2002; Shao et al., 2006). Both near-field optical micromanipulation (Garces-Chavez et al., 2005, 2006; Reece et al., 2006; Righini et al., 2007) and light-induced dielectrophoresis (Chiou et al., 2005) present themselves as powerful methods toward the manipulation of arrays of particles over a large area (e.g. ~ 1 mm^2). However, here we concentrate on the four areas mentioned as they are the most prominent. Let us now look at these in some more detail.

4.2.1 Interferometric Trapping

Optical interference is inherently linked to the temporal and spatial properties of the generating light field. Laser sources typically offer very high spatial and temporal coherence that may be used in a number of ways. In the area of optical trapping, interferometric patterns can instantaneously provide an optical potential energy landscape over a large area. If we overlap two light fields taken from a single laser source with an appropriate coherence length, two-dimensional interference patterns can be formed which is possibly the most straightforward way to generate multiple beams (traps). To achieve this, one can directly modify the dual beam optical trap, described earlier in Figure 4 (Fallman and Axner, 1997) with non-polarizing optics. This creates a Mach–Zehnder interferometer and it is possible to create periodic linear intensity fringes from the two trapping beams from this system to form an extended optical potential landscape (Chiou et al., 1997; MacDonald et al., 2001). In this manner, Chiou et al. (1997) made use of the translating intensity line traps formed with the help of a microscope objective and the interference fringes from two coherent laser beams, where high refractive index microspheres are trapped and transported. MacDonald et al. (2001) manipulated low index particles of appropriate size relative to the fringe spacing in between the bright fringes of such an interference pattern. Such two beam interference was also employed by Chowdhury et al. (1985) in studies of laser-induced freezing.

With three or more beams, it would be possible to form three-dimensional optical lattices for tweezing large arrays of particles (Burns et al., 1989; MacDonald et al., 2001). Indeed past studies in optical binding, which we will discuss later (Burns et al., 1990, 1989), have shown how multi-beam interference may be used for studies in optical binding, where particle–particle interaction arising from multiple scattering is taken into consideration. As shown in Figure 8, a basic three-dimensional optical interference pattern or lattice can be made up from three interfering beams each having a complex amplitude denoted by ψ_1, ψ_2 and ψ_3. The complex amplitude of the final interference light field, $\psi_{int}(r)$ can be described by

$$\psi_{int}(r) = \psi_1 + \psi_2 + \psi_3$$
$$= A_0 e^{ik_0 \cdot r} + A_1 e^{ik_1 \cdot r} + A_2 e^{ik_2 \cdot r} \tag{4}$$

where r is a position vector, A_0, A_1, A_2 are arbitrary complex amplitude constants, and k_0, k_1, k_2 are the wavevectors in the plane and $lk_0 + mk_1 + nk_2 = 0$ where l, m, n are the optical lattice indices. In Figure 8, we show that three beam interference that can generate an array of discrete high intensity spots (inset) is distributed in an overall Gaussian envelope. In fact one finds that with only four non-co-planar light beams fourteen different types of three-dimensional optical lattices (Bravais lattices) can be actually formed, as indicated by Cai et al. (2002).

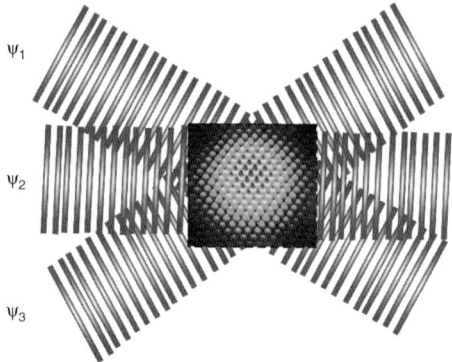

FIGURE 8 Illustration of an interference intensity pattern (central inset) from three interfering waves. Three light fields, ψ_1, ψ_2 and ψ_3, overlap to produce a three-dimensional optical lattice

4.2.2 Time-Shared Optical Trapping

As mentioned in the introduction of Section 4.2, a particle trapped in an optical tweezers in liquid (medium of high viscosity) is equivalent to an overdamped oscillator. The concept of time-shared optical traps is that one may remove the tweezing beam for a very short period of time to tweeze another object and return to this original tweezed particle some time later. Of course, once the optical field is moved away from particle's current position, the particle is left free to diffuse due to Brownian motion. If the tweezing beam returns to its original position within a relatively short time scale (less than a 0.1 millisecond or so) before a micron-sized particle diffuses away due to Brownian motion, the mean position of the particle essentially remains unchanged (it would only have diffused the order of a few nanometers). The characteristic time, τ, that a particle remains within in a given area defined by b, which is the radii that the particle is diffusing, and by a the radius of the particle, is given by

$$\tau^{-1} = v = \frac{2k_B T (\text{erf}^{-1}(\gamma))^2}{3\pi \mu a^3 b^2} \tag{5}$$

where k_B is the Boltzmann constant, T is the temperature, γ is the fraction of particles that must remain within the current diffusion area during a single scan and μ is the viscosity. Over a given area, the smaller the particle the faster the diffusion will be whist the trap is absent (Mio et al., 2000). This characteristic of the time-shared multiple optical traps is also its Achilles' heel and needs careful consideration in the design of all such systems.

Using a pair of fast scanning mirror driven by galvanometers, Sasaki et al. (1991, 1992) demonstrated that by scanning a single trap in a designed path repetitively one may obtain a specific two-dimensional light pattern, where microspheres would find themselves aligned onto the designated light patterns. By rotating the trap in a circular path over a few hundred Hertz, they were also able to cage and transport reflective metallic particles or low refractive index microdroplets. Visscher et al. (1993) integrated a two-axis galvanometer mirror onto a confocal fluorescence microscope system to achieve rotational control of *E. coli bacterium*. Visscher et al. (1996) improved the multiple laser trapping system with acoustic optical deflectors (AODs), where the optical trap can be steered at a rate of a few kHz (>natural frequency response of a micrometer sphere)). This form of automated tweezing technique has greatly improved the precision and repeatability of single molecule experiments (Lang et al., 2004). In essence, these beam steering systems provide a reasonably accurate and reproducible two-dimensional $(x-y)$ intensity pattern by deflecting the input beam.

The modes of scanning can be achieved by the galvanometer controlled mirror (Sasaki et al., 1991, 1992), piezo-driven deflecting mirrors (Mio et al., 2000) or acousto-optic deflector (AOD) (Visscher et al., 1996; Terray et al., 2002). The scanning mirror techniques operate by scanning a pair of oscillating mirrors making use of two highly reflective mirrors aligned such that each provides scanning in each of the axes (x, y). On the other hand, the AOD operates by deflecting an input beam through a diffractive grating formed with an acoustic wave. The relatively slow scan rates (few hundred hertz) offered by the galvanometer mirrors (due to inertia) impede the possibility of being able to form multiple traps. The fastest rate at which a mechanically deflecting mirror can be put into oscillation is around ~12 kHz (GSI-Lumonics Inc) as we need to consider the mechanical inertia of the system. However, an AOD can easily accomplish a repetitive scan rate of the order of a MHz due to the fast traveling speed of the acoustic waves. At such high-speed scanning, a single beam can be 'shared' between different positions. A two-axis acousto-optic deflector is typically made up of two crystals (tellurium dioxide, TEO_2 or Lithium Niobate, $LiNbO_3$) lined up orthogonally along the propagation axis for steering of infrared laser beam. Within one AOD, a standing acoustic wave modulates the refraction index of the crystals which in turn diffracts the light propagating through it much like a phase grating. Tuning the depth of the phase change can also modulate the intensity of the laser beam and by moving the grating you can steer the first diffracted order in one direction. Two AODs scanning in orthogonal directions can be combined in series to provide both x and y deflections of the optical trap as shown in Figure 9. Each deflector consists of a crystal e.g. Tellurium oxide with piezo-electric transducers with an applied RF voltage to create a standing

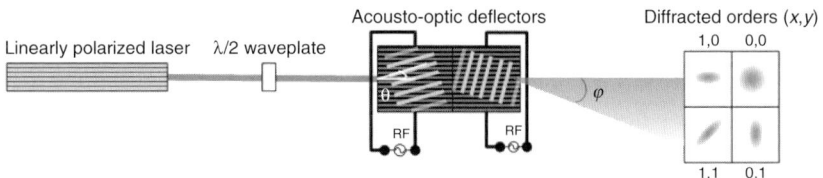

FIGURE 9 Operation of two-axis $(x-y)$ AODs. A single gaussian beam enters a two-axis acousto-optic deflector that subsequently yields the four diffraction orders arranged in a two (x, y) dimensional coordinate system

wave across the crystal. The input laser beam is incident at the Bragg angle (θ), efficiently diffracted by the sound wave grating and deflected by an angle ϕ (see Figure 9).

The AOD system does however suffer from some optical efficiency tradeoff due to its diffractive nature, whereas the galvanometer scanning mirrors have a high optical efficiency owing to the high reflectivity in the mirrors. The first diffracted order which typically contains around 60 per cent of the total input optical power is used for optical tweezing. Furthermore, the AOD optical efficiency gradually decreases with respect to the deflection angle of the grating and thus creates a less than ideal spread of intensity distribution. This can be compensated by imposing an amplitude modulation to even out the intensity which would mean that each AOD needs to be characterized for the change of intensity vs deflection angle. This change of intensity would have a marked effect on the stiffness of the trap generated. This would also mean that the number and quality (trap stiffness) of each of the 'time-shared' optical traps are very much dependent on the response time of all the optoelectronic components: the quality of the electronics and the programming aspects (Wallin et al., 2007). In Figure 9 we illustrate the operation of the two-axis AOD system. A linearly polarized laser beam passes through a half waveplate (controlling the direction of the polarization of the light beam) before entering the pair of AODs. A simple demonstration of the two-dimensional scanning from an AOD can be seen in Figure 9. A slow linear scan is applied on each individual AOD and each sound wave grating from each AOD diffracts the input beam into its individual diffracted orders. In Figure 9 we illustrate the first four diffracted orders (x, y) arranged over a two-dimensional grid where the intensity at each of the diffracted orders (0, 0) is the undiffracted Gaussian spot, (0, 1) and (1, 0) are the vertical and horizontal scanned line intensities (each of the AODs) and (1, 1) is a diagonal scanned line intensity which is the combination of the two scanning axes of AODs, as in Figure 9. A recent work by Valentine (Valentine et al., 2008) shows the uses of an electro-optic deflector, where an external electrostatic field creates a refractive index gradient in an appropriate crystal. This is used to deflect and controllably steer an optical tweezing beam. It potentially offers an improved optical throughput and a reduction in deflection-angle (beam pointing) errors compared to an AOD system leading to increased resolution in force and position measurements.

4.2.3 Holographic Optical Trapping

Dennis Gabor won the Nobel Prize in 1971 for 'his invention and development of the holographic method', where he was able to reconstruct three-dimensional images by way of interferometry. Gabor has also looked into the reconstruction of images for the purpose of microscopic imaging (Gabor, 1951). Holography is a technique where the intensity and phase (complex amplitude) of an object is captured by the interference patterns between the light scattered from the object and a reference light field and recorded upon a light-sensitive medium (hologram). Illuminating the hologram with the reference beam can therefore reveal a virtual image of the object. One of the first implementations of the tweezing of an ensemble of colloids was by Fournier et al. (1995) who illuminated a binary hologram and generated self-images of the grating (Talbot images) in planes that are periodically positioned along the direction of propagation. Holographically generated Laguerre-Gaussian beams were used for trapping and rotation in 1995 by He et al. (1995b,a). This study will be discussed in more detail later.

One can numerically superimpose the complex amplitude of a 'desired' (output) light pattern and a reference (input) beam to form a computer-generated hologram from which the desired light pattern can be reconstructed. As such, one can effectively transform an input Gaussian light field into a more elaborate form at will. This means that the technique of computer-generated holography can be utilized to shape the optical potential of the trap by altering the complex amplitude of the input light field, $|I_{int}(x, y)| e^{i\alpha(x,y)}$, where I_{int} is the amplitude, α is the phase of the light field, (x, y) are the Cartesian coordinates. In modern optics, Fourier methods play a major role in many adaptive optical systems where a lens can be considered as performing a Fourier transform, see Figure 10(a). A complex amplitude of $|I_{int}(x, y)| e^{i\alpha(x,y)}$ at the back focal plane of the lens is Fourier transformed onto the front focal plane as

$$I_{front}(u, v) = \iint |I_{int}(x, y)| e^{i\alpha(x,y)} \cdot e^{-i\frac{2\pi}{\lambda f}(ux+vy)} dx dy \qquad (6)$$

where (u, v) are the corresponding spatial frequency coordinates, f is given as the focal length of the lens and λ is the wavelength of the light. By manipulating the input complex amplitude function $|I_{int}(x, y)| e^{i\alpha(x,y)}$ with a numerical algorithm, the output complex amplitude $I_{front}(u, v)$ can be manipulated. For trapping, one would need to place a microscope

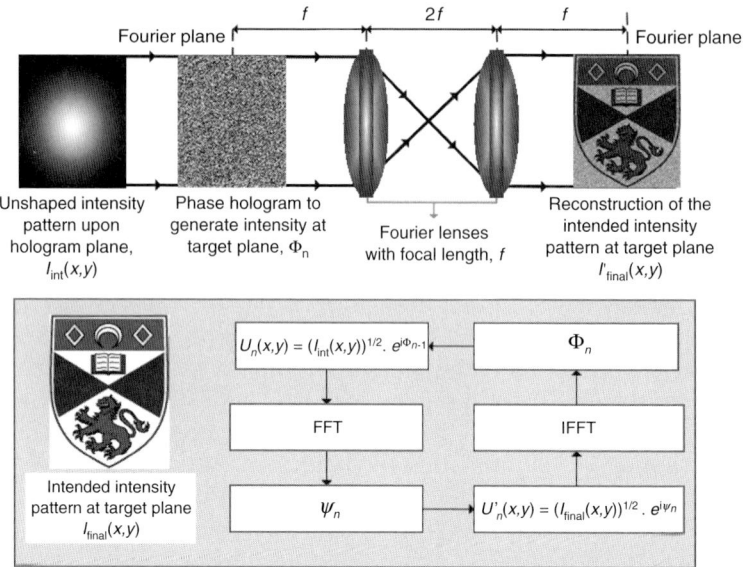

FIGURE 10 Fourier holography and iterative Gerchberg–Saxton (GS) algorithm. (a) Illustrates 4f fourier lens system where each lens can be considered as performing a fourier transform of a phase hologram generated using the GS algorithm. (b) a flowchart that describes the GS algorithm (courtesy of Gabe Spalding and Ryan Smith)

objective with its back focal plane coinciding with the Fourier planes of the lens in Figure 10(a) so as to image the intensity patterns onto the sample plane.

A number of numerical algorithms such as superposition algorithms, Gerchberg–Saxton algorithms, random mask encoding, direct-binary search algorithms and simulated annealing can be used to optimize the holograms (Tricoles, 1987). However, in this section, we choose to give attention to the Gerchberg–Saxton algorithm due to its high conversion efficiency and popularity in the field. In Figure 10(b), a simple flow chart is drawn to illustrate the initial loop on the iterative Gerchberg–Saxton (GS) algorithm with a fast Fourier transform (FFT). In Figure 10(b) we show that the algorithm starts by first obtaining the initial complex field in the hologram plane $U_n(x, y)$ with the input beam $I_{int}(x, y)$ and the estimated phase at the target plane Φ_{n-1}. By using the argument of the FFT of $U_n(x, y)$, one can then obtain the initial calculated phase at the target plane ψ_n. The phase distribution in the target plane, ψ_n is combined with the target intensity, $I_{final}(x, y)$, to give the complex field $U'_n(x, y)$. Lastly, one calculates the corresponding phase distribution in the hologram plane, Φ_n, by taking the argument of the inverse FFT (IFFT) of $U'_n(x, y)$. Over a number of iterations, the actual intensity in the target plane matches the desired intensity there, $I_{final}(x, y)$. Φ_n is the corresponding phase-hologram pattern needed to produce this intended field at the target plane. In Figure 10(a) we see the reconstructed image $I'_{final}(x, y)$ after 50 iterations. The reconstructed pattern does fit the intended intensity up to >90 per cent with the GS algorithm. Such a holographic technique can generate a two-dimensional intensity pattern (image) which will only allow tweezing in a single transverse plane. Whyte and Courtial (2005) showed that this technique can be extended to allow three-dimensional shaping of the beam which forms a three-dimensional holographic optical trap. For the sake of completeness, it is also worth noting that direct-binary search, time sharing and encoding multiple holograms of a Fresnel lens have also been demonstrated to be able to generate three-dimensional array holographic traps (Melville et al., 2003; Jesacher et al., 2004; Leach et al., 2004).

The holographic element can be made with the use of computer-generated holograms and results in a static diffractive optical element, or implemented with a liquid-crystal display, spatial light modulators, to generate arrays of beams at will. The calculated hologram (binary or phase) pattern can be etched into glass or plastic which can only be used to generate one type of complex trap pattern. Due to the recent progress in display technology, programming spatial light modulators can now dynamically change the phase holograms. The resolution and the refresh rates of the programmable spatial light modulators offer the prospect of dynamically changing the hologram (trap patterns) in real time and thus actuating multi-particle motion. Many of the current spatial light modulators (SLM) used for holographic optical tweezers make use of the reflective nematic liquid-crystal (LC) display which usually operates by electrically or optically addressing the individual liquid crystals within the display. In the electrical case, the electrical potential difference causes the liquid-crystal molecules to align themselves along a direction thereby changing its refractive index which in turn modulates the phase of the input light field. Most nematic SLMs typically have 256 phase levels but these have slow refresh rates (typically up to ~60 Hz). Ferroelectric SLMs, in contrast, have been employed in optical trapping and have demonstrated their higher refresh rates of ~10 kHz, but they are limited to only two levels of phase changes (Hossack et al., 2003).

Holographic optical trapping (HOT) remains possibly the most enticing among the upcoming prospects for multiple trapping and generation of structured light fields (Liesener et al., 2000; Dufresne et al., 2001; Curtis et al., 2002; Hossack et al., 2003; Melville et al., 2003; Leach et al., 2004; Sinclair et al., 2004a,b; Polin et al., 2005; Lee and Grier, 2005; Schmitz et al., 2005; Grier and Roichman, 2006; Martin-Badosa et al., 2007). This is due to three overriding issues: firstly, it does not involve time sharing of the light field as observed with acousto-optical devices and all the traps operate simultaneously; the diffractive optic element (DOE) can also be implemented in a manner that allows the beam to overcome any inherent aberration present, i.e. seidel aberrations present in the optical path can be compensated for (Wulff et al., 2006). Finally the use of appropriate algorithms can generate structures in three dimensions. We refer the reader to recent papers that discuss the technicalities of setting up and optimizing a HOT system in detail (Martin-Badosa et al., 2007; Polin et al., 2005). Despite the dynamic potential of the holographic optical trap, a spatial light modulator does typically suffer from relatively poor diffraction efficiency at the first order (~30–40%) and inability to withstand a input beam of high optical power (a few Watts).

The Generalized Phase Contrast (GPC) technique is an alternative that does not use the SLM as a hologram device but rather more directly as a phase element. Since the GPC is not a holographic method, it therefore does not suffer from the potential complexities associated with calculating and updating new holograms for trapping and the motion of many particles (Eriksen et al., 2002; Rodrigo et al., 2002). We describe this briefly in the next section.

4.2.4 Generalized Phase Contrast Technique for Multiple Traps

In 1676, Antonie van Leeuwenhoek became the first man to have observed bacteria with a home-made microscope, which he later reported to the Royal society and ignited the field of microbiology (Pontecorvo, 1963). As we now know, many biological microparticles are often difficult to be seen due to their refractive index being very close to that of their surrounding medium. It took another two hundred years before Frits Zernike invented an imaging technique that circumvented these issues with an elegant physics solution: converting faint phase gradients into intensity variations by the method of phase contrast with phase contrast technique (Zernike, 1955). In a related fashion, Eriksen et al. (2002) reformulated Zernike's phase conversion so as to convert phase structures into intensity patterns. This technique makes use of a phase modulated SLM that is placed at the conjugate (Fourier plane) to the trapping plane with a small phase shift, positioned at the focal plane between the two Fourier lenses. This generalized phase contrast technique (GPC) is a very straightforward and efficient method as no numerical algorithms are used since it is a direct phase to intensity conversion. It is possible to extend the GPC method to three-dimensional trapping with counterpropagating dual beam geometry (Perch-Nielsen et al., 2005) and to develop systems with automatic alignment.

4.2.5 Studies with Structured Light Fields

Now that we have explored the various major methods of multiple trap generation we proceed to describing some of the major experiments using these trap arrays.

4.2.5.1 Brownian diffusion
A microsphere suspended within a viscous medium may be considered a Brownian particle. Optical tweezers generates a parabolic potential well that acts to suppress the Brownian motion of a microsphere. One can imagine that a trapped microsphere is situated at the lowest point of the potential well. By switching the potential off, the microsphere will exhibit diffusive motion due to the Brownian thermal noise. In order for the sphere to escape from the potential well, it would need to generate a potential energy larger than the energy potential of the well. By creating a modulated potential (high and low intensity gradient) over time and a large region (potential landscape), a particle would seek to diffuse toward the region of higher optical gradient. Faucheux et al. (1995b) and colleagues devised an experiment using a pair of galvanometer scanning mirrors to introduce a toroidal trap with an intensity modulation to observe biased forms of Brownian diffusion with optically trapped objects. Such optically-induced thermal ratchets may have some analogy to the biological Brownian motors (Derenyi and Vicsek, 1995). Crocker et al. (1999) and Lutz et al. (2004a,b) studied the diffusive behavior of colloidal particles while they are optically confined within scanning optical tweezers and have observed that single-file colloidal diffusion occurs. Such Brownian ratchet effects can be studied with other structured light fields, namely Laguerre-Gaussian light beams (Roichman et al., 2007).

4.2.5.2 Large scale colloidal dynamics
Colloidal science has undergone a major renaissance in recent times and remains an area where optical trapping continues to make an important impact. At the most basic level, individual colloidal interactions are complex because they are *solvent-mediated*. The Nobel prize winner Jean Perrin pioneered the use of monodisperse colloidal particles as model thermodynamic systems. Perrin's works involved the detailed observation of the different ways in which colloidal particles aggregate and interact. His early experimental studies showed concrete evidence that colloidal systems offer an excellent level of controllability and tunability of inter-particle interaction that can give important insights into the working of atomic systems and that are readily amenable to measurements. Furthermore, the strong and weak interacting forces between chemically treated surfaces or other solvents may illuminate our understanding of surface bonds i.e. van der Waals forces. With the use of multiple optical traps and patterned optical potential energy landscapes, one has unprecedented access to *colloidal interactions*. The pair potential between two particles controls thermodynamic behavior (such as the crystallization vital to photonic applications), is also inherently many-bodied: in principle the potential depends on the positions of *all* the particles. While the two-body part of the potential has been studied using optical tweezers for some time, studies of the many-body

counterpart with trapping is still in its infancy in many respects. We now proceed to look at some of the key experiments in this area.

Chowdhury et al. (1985) imposed a simple periodic two beam interferometric light pattern upon a two-dimensional system of colloidal particles confined between two glass plates. By modulating the period of the interferometric pattern, they were able to examine phase transition (liquid to solid) in such a colloidal structure much like a freezing process which is now termed as 'laser-induced freezing' based on observation with light scattering detection. The effects of optical forces over a large number of colloids are subsequently studied in detail by Golovchenko and Fournier (Burns et al., 1990, 1989) where they arranged particles using a large Gaussian beam and again by interfering multiple beams from a single Argon-ion laser source. Their observations led to the creation of matter (fixed colloidal structures) with just the optical forces. More recently, Brunner and Bechinger (2002) studied an optically-induced 'melting process' with optical lattices formed from three interfering beams to obtain trimers (three colloids) at each trap site as seen in Figure 11. In this figure we see the contour plots of the lateral density distribution ρ and the averaged local particle density $\rho_{\text{loc}}(x, y)$ for different light potentials V_0: (a), (b) $0\,k_BT$, (c), (d) $40\,k_BT$, (e), (f) $60\,k_BT$, and (g), (h) $110\,k_BT$. where k_B is the Boltzmann constant and T is the temperature. The horizontal axes are x and y, respectively. All units are in μm. The optical potential energy landscape was formed by three beams of a linearly polarized Nd : YVO$_4$ laser (wavelength 532 nm with maximum optical power of 5 Watts) which intersected at an angle of 60 degrees in the sample plane, where they formed a triangular interference pattern.

Studying the Brownian dynamics between two colloids (Meiners and Quake, 1999; Crocker, 1997) researchers have detected the weak hydrodynamic coupling between two spheres, each held in close proximity by independent optical tweezers, while suspended in an incompressible fluid. With large potential energy landscapes and high resolution video microscopy, it is possible to extend that study to long-range entropic forces between colloids confined in an extended potential (Crocker and Grier, 1996; Crocker et al., 1999). Lutz et al. (2006) investigated interesting behavior with three colloidal spheres traveling along circular scanning tweezers with a periodic peak potential barrier. Hydrodynamic interactions between the colloids assist the three-body system to surmount the optical potential barrier. With holographic optical tweezers, Korda et al. (2002a) looked at dynamics of hydrodynamic forces from an ensemble of colloids distributed in a single monolayer. With the use of video tracking programs, they were able to determine the evolution of the 'phase transition' of the monolayer of colloidal.

4.2.5.3 Optical sorting
Biomedical scientists are interested in selecting or sorting cells at will. For example when considering a blood sample,

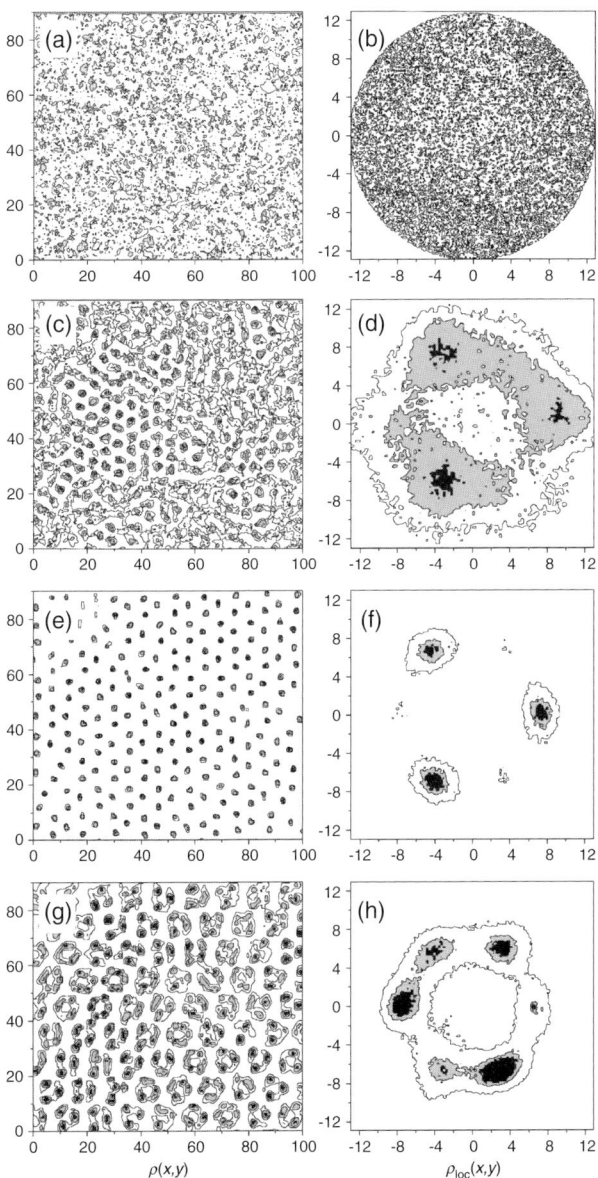

FIGURE 11 Phase behavior of colloidal molecular crystals on optical lattices. Contour plots of the lateral density distribution ρ and the averaged local particle density $\rho_{\mathrm{loc}}(x, y)$ for different light potentials V_0 listed in the text. The horizontal axes are x and y, respectively. All units are in μm. Reprinted figure with permission from Brunner and Bechinger (2002) © 2002, American Physical Society

one may wish to perform a cell count to exactly determine the number of red cells present. A cell biologist or clinician would be very interested in selecting rare cells e.g. stem cells from a large cell population. This has provided the impetus to bone marrow transplantation and a more detailed understanding of diseases of the hematopoietic system. In regenerative medicine sources of stem cells can be exploited to provide new disease-free tissue and in cancer where the key target cells for disease development and thus for successful therapy are the tumor stem cell populations. Recent advances in optically based methods for sorting at the micrometer size scale have shown promising results in the sorting of different types of biological cells. For optical sorting, we can start to look at the microfluidic flow and the optical forces at work. Figure 12 shows a straightforward method to implement optical sorting which is simply to have an optical force 'switch' that drags pre-selected cells into a desired chamber or flow cell (Wang et al., 2005). This is akin to a microscopic fluorescent activated cell sorter (FACS). Cells are aligned to the center of the channel by hydrodynamic focusing. Cells are analyzed and then switched based on their detected fluorescence. Target cells are directed by the laser to the collection output while all other cells flow to the waste output. However, for higher throughput, we need to start looking at higher flow rates or dealing with a more highly dense population. It would also be desirable to sort based upon intrinsic properties (passive sorting) such as size or shape rather than make use of markers (active sorting). If we were to flow the liquid that bears individual particles (colloidal or cells) across such an optical potential energy landscape, we would open up a competition between Stokes forces within the liquid and optical forces imposed on the particle (Korda et al., 2002b). Each particle may have a range of physical responses based on their size, overall refractive index and geometrical shape. With controllability over the landscape, it would be possible to sort particles into sizes and shapes (MacDonald et al., 2003; Lacasta et al., 2005; Ladavac et al., 2004; Cizmar et al., 2006b). Optical potential energy landscapes formed by multiple optical traps could take the form of an array of line traps or two- or three-dimensional array of trap spots (see Figure 8). Such landscapes have intriguing properties when they interact with cells and colloidal particles placed upon them. In the presence of flow over such an optical potential energy landscape particles can follow paths across the landscape that are inclined with respect to the flow direction: indeed different particles experience different potential energy landscapes thus ultimately following different trajectories. This can form the basis for some interesting competitive effects and more generally optical separation in two dimensions in a wholly passive or marker-free manner.

Sorting of colloids is also fascinating and provides monodisperse samples that would be difficult to achieve otherwise. Ladavac et al. (2004)

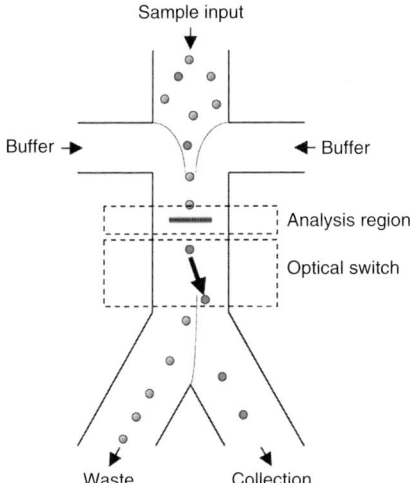

FIGURE 12 Active cell sorter. Microfluidic sorting of mammalian cells by optical force switching. Reprinted figure with permission from Wang et al. (2005) © 2005, Macmillan Publisher Ltd: Nature Biotechnology

made use of a holographic-generated array of optical traps to selectively deflect large silica spheres (diameter 0.79 microns) from small ones (diameter 0.5 microns) within a single flow channel by virtue of their size relying upon a differing response to the landscape. As mentioned before, interferometric traps can form large three-dimensional optical lattices that may be used for this same purpose of passive sorting. MacDonald et al. (2003) created a three-dimensional optical lattice using five light beams independently controlled and focused through an aspheric lens. By flowing particles through this lattice they observed the phenomena of optical sorting (or separation): particles flowing through the lattice were observed to follow trajectories that were dependent upon the particles' intrinsic properties: deflection of particles was observed based both upon size and refractive index. Low index particles in the form of protein microcapsules (ultrasound contrast agent) were also separated by size. Subsequent studies by the authors showed some preliminary evidence for the passive sorting of red and white blood cells in such a system (MacDonald et al., 2004). In a more recent work, a time modulated optical sieve created by interference was also used for sorting of poly-disperse colloids without any flow present (Ricardez-Vargas et al., 2006). Moving interferometric patterns imposed onto the surface can also be applied to the sorting of nanometric particles (Cizmar et al., 2006b). The simultaneous sorting of four different types of colloidal sizes into four parallel laminar flow streams purely using a time-generated optical potential energy landscape with AODs (Milne et al., 2007b), was observed. This optical fractionation

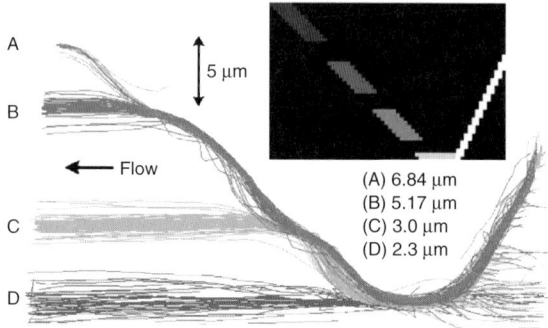

FIGURE 13 AOD sorting. Shows the plotted trajectories of the four different types of silica spheres (A) 6.84 μm, (B) 5.17 μm, (C) 3.0 μm, and (D) 2.3 μm, sorted through the time sharing optical landscape (see insert). Reprinted figure with permission from Milne et al. (2007b) © 2007, Optical Society of America

is shown in Figure 13 where once can see the plotted trajectories of four different types of silica spheres (A) 6.84 μm, (B) 5.17 μm, (C) 3.0 μm and (D) 2.3 μm, and against the direction of the flow. AODs offer certain advantages to the use of SLMs due to their ability to handle higher power and create potentially larger area optical potential energy landscapes. We refer the interested reader to other specific reviews in particle separation using optical fields (Dholakia et al., 2007b,a).

4.2.5.4 Optical organization of particles and cells The assembly function of optical tweezers can readily be achieved through computer controlled spatial light modulators or counter propagating AOD traps. With multiple optical traps, one can also start to look at creating colloidal crystals or cells fixed in a specific geometry. In the materials sciences, the engineering of crystalline structures requires the control of orderly arrangement of the atomic lattices to modify the properties of a material. There are many forms of crystalline structures defined by its crystallographic axes and the angles between the axes i.e. isometric, tetragonal, orthorhombic, monoclinic, triclinic etc. It is also possible to engineer quasi-crystals which have long-ranged orientational order that can be made up of dielectric materials with specific optical properties, thus providing a route to the development of photonic band gap materials. MacDonald et al. (2002) first used interferometric patterns from two co-propagating Laguerre-Gaussian (LG) beams to trap and rotate cubic structures. With holographic optical traps, many researchers (Melville et al., 2003; Leach et al., 2004; Grier and Roichman, 2006; Roichman and Grier, 2005) have looked extensively into this area of generating three-dimensional quasi-crystal structures and unit cells. In Figure 14, we see a large ensemble of particles trapped and arranged in a quasi-crystalline

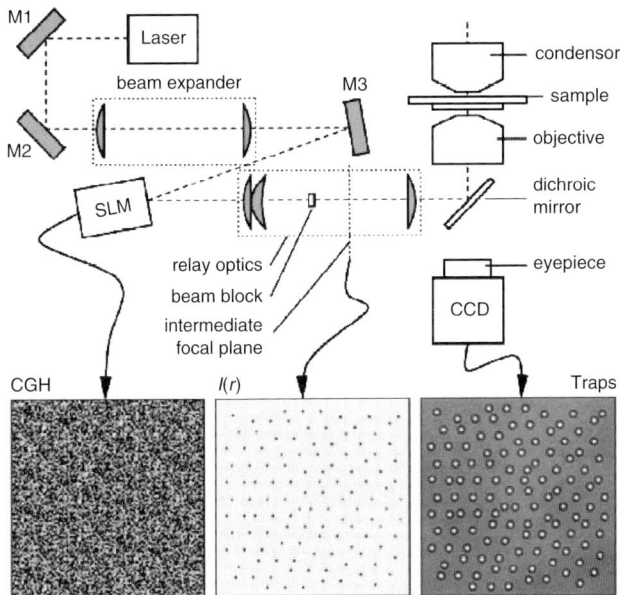

FIGURE 14 Schematic implementation of holographic optical traps. A laser beam is reflected by two mirrors M1, M2 and M3 is expanded and directed onto a liquid crystal spatial light modulation (SLM), which imprints a computer-generated hologram (CGH) onto its wavefronts. The 200×200 pixel region of a CGH shown encodes a pattern, $I(r)$ of 119 optical tweezers that are arranged in a quasi-periodic arrangement. The phase hologram is relayed using relay optics and onto the input pupil of an objective lens that focuses it into holographic optical traps. Each optical tweezers traps a colloidal sphere of 1.5 micrometer in diameter. Reprinted figure with permission from Grier and Roichman (2006) © 2006, Optical Society of America

structure work by Grier and Roichman. In Figure 14, a laser beam is reflected by two mirrors M1, M2 and M3 is expanded and directed onto a liquid crystal spatial light modulation (SLM), which imprints a computer-generated hologram (CGH) onto its wavefronts. The 200×200 pixel region of a CGH shown encodes a pattern, $I(r)$ of 119 optical tweezers that are arranged in a quasi-periodic arrangement. The phase hologram is relayed onto the input pupil of an objective lens that focuses it into holographic optical traps. Each of the optical tweezers traps an individual 1.5 micrometer (in diameter) colloidal sphere in water.

With a time-shared optical tweezing system, Vossen et al. (2004) have shown interesting results where three-dimensional colloidal structures are trapped and organized with two layers of optical traps steered with a single dual axis AOD. To create the two layers of time-shared multiple traps at separate planes, a single laser beam is split into two beam paths that are recombined at the back of microscope objective with different divergences, thus resulting in a different axial trapping position for each tweezing beam. By synchronizing the Pockels cell after the polarizing beam splitter cube with the AOD they were able to create two independent arrays of optical tweezers that trapped fluorescence particles in three dimensions as shown in Figure 15(a). Figure 15(b) shows the confocal images of fluorescent particles trapped in a three-dimensional array of tweezers created by synchronizing the Pockels cell and the AODs. In (i) six particles were trapped in the upper plane and in (ii) nine in the lower plane. In (iii) fluorescence from particles in both planes was detected. The height difference between the two trapping planes was 1.7 micrometer. The upright objective was used for trapping while the inverted objective was used for imaging. The 1.4 micrometer diameter FITC–SiO_2 (fluorescein isothiocyanate silica spheres) particles were dispersed in ethanol, and only their fluorescent cores were imaged and Figure 15(b)(iv) shows a computer generated image based on the confocal data. The three-dimensional fluorescence colloidal structures are optically assembled and imaged with the confocal fluorescence microscopy shown in Figure 15(a).

Recently, studies have organized and permanently fixed both colloid and cells separately using multiple optical traps. For example, Jordan et al. (2004) fixed three-dimensional crystal structures in a polymeric host while held in the traps. Jordan et al. (2005) also applied the technique to move individual *E. coli* cells in the liquid gelatin at predefined positions. When the cells are fixed in place, the lasers were switched off. The three-dimensional configurations of cells within a gelatin sample at predefined positions remained intact for many days. The cells survived within the gelatin matrix for several days when provided with the appropriate nutrients. This technique can be used to help in understanding the role of position, proximity and number of neighboring cells, not only in cell culture, but also in cell differentiation. The arrangement of a variety of different cell types in complex architectures promotes tissue differentiation and growth within the field of cell/tissue engineering.

Akselrod et al. (2006) used multiple optical traps to generate three-dimensional heterotypic networks of living cells in a hydrogel complex (see Figure 16). They showed cell viability after assembling Swiss 3T3 fibroblast cells which are surrounded by a ring of bacteria and also amass hundreds of *Pseudomonas aeruginosa* into two- and three-dimensional arrays. In essence the AOD and its fast dynamical behavior were used to generate the trap array whereas the SLM here was exploited as a device to impose both a diffraction grating (lateral offset of the arrays) and a Fresnel lens (to offset the array along the beam propagation axis) to impart a three-dimensional nature to the arrays. The dynamic optical trapping systems combined the strengths of the spatial light modulator with that of the acousto-optic modulator. Apart from the technical novelty,

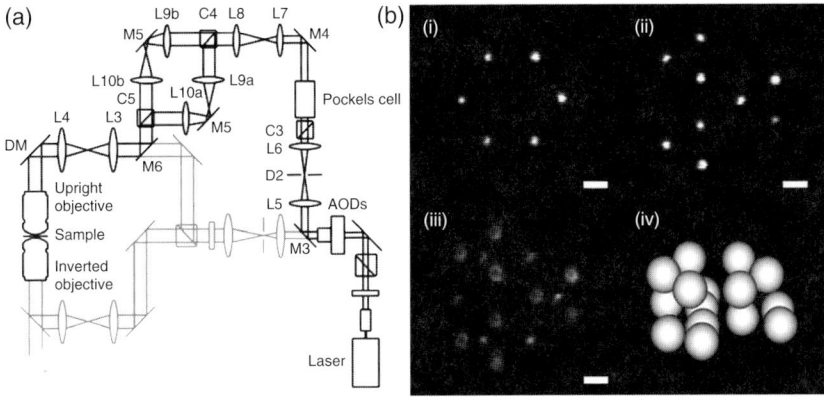

FIGURE 15 Setup of multiple optical traps generated with AODs and the corresponding confocal images of optically assembled three-dimensional colloidal crystals. (a) Shows the setup used for creating arrays of tweezers in two separate planes. A mirror M3 after the AODs reflects the beam into a 1:1 telescope formed by two lenses (L5 and L6, both with focal length of 120 mm). A polarizing beam splitting cube C3 is placed in front of the pockels cell to remove any horizontal component of the polarization introduced by the AODs. A mirror M4, the lenses L7 and L8 (with focal lengths of 65 and 140 mm) expand the beam to overfill the back aperture of the objective. A polarizing beam splitter cube C4 splits the laser beam into two separate paths. In each of the paths the beam passes through a 1:1 telescope formed by a pair of lenses (L9a, b and L10, b all $f = 90$ mm). The lenses L9a, b are positioned in a plane conjugate to the back focal plane of the upright objective. After recombination of the two beams using a polarizing beam splitter cube C5, the combined beam is coupled into the microscope with the mirrors M5, M6 and lenses L3 and L4. A movement of lens L3 results in a collective displacement of the two traps. The upright objective was used for trapping, while the inverted objective was used for imaging. The part of the setup drawn in light gray was not used when two trapping planes were created. (b) Shows the images of the three-dimensional optically assembled fluorescence structures. The scale bars are 1 μm. Reprinted figure with permission from Vossen et al. (2004) © 2004, American Institute of Physics

this particularly shows that living cell microarrays generated with optical traps may pave the way for exploiting the functionality of cells and processes such as cell differentiation and could prove to be powerful tool within systems biology and tissue engineering.

4.2.5.5 Multi-point force transducers
An issue for any multiple trap array is the ability to measure trap stiffness for all the trap sites simultaneously and in real time: such a measurement is of course crucial if one is for example trying to perform complex cell mechanotransduction studies and needs force measurements at various positions upon a given cell or perform multiple single molecule studies in parallel. Whilst the quadrant photodiode has the high bandwidth and rapid response time desired, it is only realistically able to measure when one or two traps

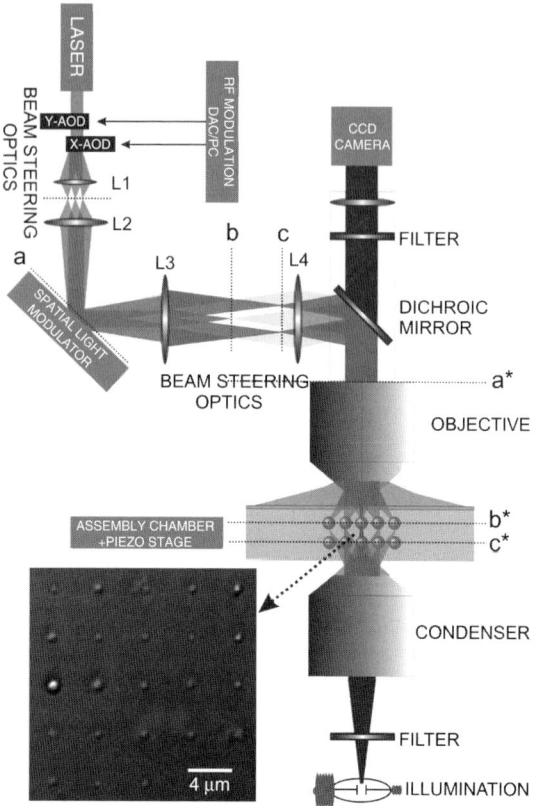

FIGURE 16 Schematic diagram of a time-shared holographic optical trapping apparatus. Trap arrays are formed using a high NA objective in a commercial optical microscope in conjunction with two acousto-optic deflectors and a spatial light modulator (SLM) to produce a time-shared (3D) array of optical traps. The plane of the SLM, a, is imaged into the microscope objective entrance aperture (OEA) a*, and the corresponding planes b and c are imaged into the focal region of the microscope. The same microscope that is used to produce the cell traps is also used for viewing (via the blue beam). The inset in the lower left shows an example of a two-dimensional (5 × 5) array of P. Aeruginosa formed using this apparatus and subsequently embedded in hydrogel. The distances are: AODs–L1 = 165 mm; L1–L2 = 650 mm; L2–SLM = 332 mm; SLM–L3 = 421 mm; L3–L4 = 1400 mm; and L4–OEA = 493 mm, where the focal lengths are L1 = 150 mm, L2 = 500 mm, L3 = 1000 mm, L4 = 400 mm. Reprinted figure with permission from (Akselrod et al., 2006) © 2006, Biophysical Journal

are involved and is not readily suitable for a multiple trap array. Digital video devices (charge couple device (CCD) or complementary metal-oxide–semiconductor (CMOS)) have, in recent years, improved in terms of capture speed and frame rates that have been widely implemented in microscopy imaging systems (Howard, 2007). The image of the trapped particle is relayed onto an imaging plane of the digital camera which captures the particle's fluctuation of its position using an ordinary (~30 frames per second) or even a high-speed camera (~400 frames per second) over a length of time (Polin et al., 2005, 2006). A histogram of positions over a given time period is then plotted so as to measure the variance of the position, $\langle x^2 \rangle$. An object in a harmonic potential would have a potential energy (thermal fluctuation) equal to its trap stiffness, k (position fluctuation). By employing the equipartition theory it would then be possible to determine the trap stiffness,

$$\frac{1}{2}k_B T = \frac{1}{2}k \langle x^2 \rangle \tag{7}$$

where k_B is the Boltzmann's constant, and T is the absolute temperature. This measurement technique does not rely on the value of the viscous drag and can be extended to multiple trapping sites with simply imaging the trapping plane onto the camera's imaging plane. The advantage of using this technique is the possibility of calculating the trap stiffness for each trapped particle. Thus one can envisage that with the equipartition theory and a high-speed camera (>10 thousand frames per second), it would be possible to calibrate multiple force transducers which is a good way to ascertain the several trap stiffnesses in real time. This would likely emerge as a key area in the instrumentation of advanced optical trapping apparatus in the near future. Already we are seeing the first indications of this possibility: Di Leonardo et al. (2007) explored the hydrodynamics of a ring of microparticles held by an array of holographic optical traps. The hydrodynamic coupling in the multi-particle system is measured by analyzing the position fluctuation in the trapped microparticles using a high-speed video camera. Due to the coupling, a set of eigenmodes that described the motion, were observed, each of which exhibited a characteristic decay rate. The authors observed good agreement between their experimentally recorded eigenvalues and the numerical predictions of the Oseen theory.

4.3 Non-zero Order Laser Modes

Whilst multiple trap arrays created with the various methods above are undoubtedly powerful one may also wish sculpt the phasefront of a given light field or alter the wavevectors of the light field to create more elaborate light fields. Light beams emitted from a laser cavity are electromagnetic fields, the wave equation solutions of which satisfy appropriate boundary conditions. Some of the solutions form a complete and orthogonal set of functions (Kogelnik and Li, 1966) and are called the 'higher modes of propagation'. Essentially, any transverse mode optical distribution can be expanded in terms of these modes. These higher order transverse modes are often referred to as non-zero-order light fields. These non-zero-order light fields have come to the fore in recent years in several fields in atomic, molecular and optical physics: optical micromanipulation is no exception. Light fields may impart linear momentum to trapped objects which has been the core of our discussions so far. With higher order modes, in particular the Laguerre-Gaussian (LG) modes, angular as well as linear momentum transfer can take place to trapped objects due to the helical wavefronts of the field. This helical wavefront leads to an unusual trajectory for the Poynting vector: this vector denoting energy flow possesses an azimuthal component, that in turn leads to the concept of orbital angular momentum (Allen et al., 1992). The tunability of the orbital angular momentum $l\hbar$, where l is the azimuthal index of the beam, in such light beams offers itself as a useful tool in the exertion of torques on particles on a toroidal trap by simply increasing the azimuthal index of the beam in contrast to the optical torque arising from spin angular momentum which varies with optical power (Friese et al., 1998) and is limited to $\hbar$ per photon. Another light field that has seen many applications in optical micromanipulation is the Bessel beam. These beams may exhibit propagation invariant or 'non-diffractive' characteristics: thus it maintains its beam waist size without considerable broadening. The beam can even reform or reconstruct beyond obstacles when partly obscured. With such abilities, the Bessel beam definitely brings valuable features to the field of optical trapping and even in the broader area of biophysics (Tsampoula et al., 2007). We commence our discussion of these unusual light modes with the Laguerre-Gaussian beam.

4.3.1 Laguerre-Gaussian Light Beams and their Applications

In the context of the transverse laser mode theory, the circularly symmetric LG laser modes form a complete basis set for paraxial light beams and thus one may express any given light mode as a sum of them. A given mode is usually denoted LG_p^l where l and p are the two integer indices that describe the mode. The azimuthal index l refers to the number of 2π phase cycles around the circumference of the mode and $(p+1)$ indicates the number of radial nodes in the mode profile (see Figure 17). For each increment of p, an additional concentric ring is added to the mode profile while an increment of l results in an increase in the diameter of the dark core, termed optical vortex, located on the beam axis. The azimuthal index l is often referred to as the topological charge of the optical vortex. LG modes with $l \neq 0$ have garnered much interest owing to their azimuthal phase term $\exp(-il\phi)$ and that gives rise to a well-defined orbital angular momentum (OAM), of $l\hbar$ per photon (Allen et al., 1992; Mair et al., 2001). This may exceed any angular momentum associated with the spin angular

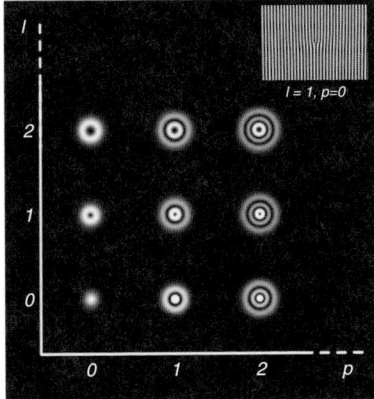

FIGURE 17 Various orders of the laguerre-gaussian beams (inset shows a computer-generated hologram for LG_0^1). In the paraxial approximation, optical vortices can be expressed as laser beam modes, known as LG modes, LG_p^l. The laser modes are characterized by two integer indices namely l, the azimuthal index, and p is the radial index. Inset shows the computer-generated hologram for LG_0^1

momentum state of the field. The physical interpretation of the orbital angular momentum was given earlier, being couched within the idea of an inclined optical wavefront (Allen et al., 1992, 1999) and azimuthal component of the Poynting vector. Under the paraxial approximation, this form of angular momentum in the LG beam can be decoupled from spin angular momentum arising from its polarization state (Barnett, 2002). A general description of the electric field of a LG mode $E(LG_p^l)$ of indices l and p may be written as:

$$E\left(LG_p^l\right) \propto \exp\left[\frac{-ikr^2z}{2\left(z_r^2+z^2\right)}\right] \cdot \exp\left[\frac{-r^2}{\omega^2}\right]$$
$$\cdot \exp\left[-i(2p+l+1)\arctan\left(\frac{z}{z_r}\right)\right] \cdot \exp[-il\phi]$$
$$\cdot (-1)^p \cdot \left(\frac{r\sqrt{2}}{\omega}\right)^l \cdot L_p^l\left(\frac{2r^2}{\omega^2}\right) \qquad (8)$$

where z denotes the distance from the beam waist, z_r is the Rayleigh range, k is the wave number, ω is the radius at which the Gaussian term $e^{\left(\frac{-r^2}{\omega^2}\right)}$ falls to $1/e$ of its on-axis value, r is the radius, ϕ is the azimuthal angle and L_p^l is the generalized Laguerre polynomial. The term $(2p+l+1)\arctan\left(\frac{z}{z_r}\right)$ is the Guoy phase of the LG mode that varies with the mode indices.

LG laser modes may be generated in a number of ways either within or outside a given laser cavity. Generating the LG modes (circular symmetry) from within the laser resonator cavity is possible if the cavity has a certain asymmetry (Okida et al., 2007), however most laser cavities do not possess the symmetry required to output such circularly symmetric transverse laser modes. One of the first techniques of LG mode generation made use of output higher order Hermite-Gaussian (HG) transverse modes (rectangular symmetry) by inserting an intra-cavity cross-wire into a laser cavity. The HG modes may then be converted to a LG laser mode using a mode converter (Beijersbergen et al., 1993). A mode converter consists of two cylindrical lenses of focal length, f, canonically disposed with respect to one another. When placed at a distance of $\sqrt{2}f$ between the two cylindrical lens this system of lenses introduces a Guoy phase shift of $\pi/2$ on an incident HG mode laser beam of indices m and n and transforms it to a LG mode of indices $l = (m - n)$ and $p = \min(m, n)$. Although the purity of the generated LG mode may be high, this requires the careful selection of the HG laser mode desired and the avoidance of any undesired astigmatism in the optical system. A more practical and versatile method would be the generation of LG modes directly from a fundamental TEM_{00} Gaussian beam, external to the laser cavity. Two main methods that satisfy this requirement, each using diffractive optical elements, have been established. These are a spiral phase element or a computer-generated hologram. When considering spiral phase elements, a high refractive index substrate is shaped into the spiral phase ramp (Beijersbergen et al., 1994; Oemrawsingh et al., 2004). With recent microfabrication techniques, the spiral phase element has been miniaturized (Cheong et al., 2004; Lee et al., 2004). The exact output mode here is a superposition of LG modes (Beijersbergen et al., 1994) but the system can yield a high conversion efficiency >80 per cent and uses on-axis optical components. The refractive index of the spiral phase plate is specific to the wavelength of the input beam. The computer-generated holographic generation method requires little in the way of 'fabrication process' as one can mathematically encode the spiral phase with a given input field at an angle onto a computer-generated pattern as seen in the inset of Figure 17. The transmission function of the off-axis hologram T required to generate a single ringed ($p = 0$) LG beam of azimuthal order $\pm l$ may be represented as

$$T = \frac{1}{2}(1 - \cos(k_x x \pm l\phi)) \qquad (9)$$

where k_x define the periodicity of the grating along the x axis and l is the azimuthal order and φ is $\tan^{-1}\left(\frac{y}{x}\right)$. By directing a Gaussian beam, $G(r, \varphi, z)$, upon the hologram with a transmission function of T, a number of diffracted orders, i.e. the zeroth order and the two conjugate orders, would emerge. The Gaussian beam is merely diffracted into different orders of the beam with helical wavefront. It is important to know that the topological orders from the two diffraction orders are of equal but opposite charges (positive and negative).

$$\begin{aligned}
G(r, \varphi, z) \cdot T &= G(r, \varphi, z) \cdot \left[\frac{1}{2}(1 - \cos(k_x x \pm l\varphi))\right] \\
&= G(r, \varphi, z) \cdot \left[\frac{1}{2} - \frac{e^{i(k_x x \pm l\varphi)} - e^{-i(k_x x_1 \pm l\varphi)}}{4}\right] \\
&= \left[\frac{1}{2}G(r, \varphi, z) - \frac{1}{4}G(r, \varphi, z)e^{i(k_x x \pm l\varphi)}\right. \\
&\quad \left. - \frac{1}{4}G(r, \varphi, z)e^{-i(k_x x \pm l\varphi)}\right]
\end{aligned} \tag{10}$$

This holographic technique has gained more popularity due to the ease and versatility of LG beam generation. The orbital angular momentum of these light fields can be observed by careful consideration of the helical wavefronts of an LG beam. The inclined helical wavefront leads one to consider the energy flow in such fields: the Poynting vector moves in a corkscrew-like manner (Allen et al., 1992). This angular momentum is therefore linked with the azimuthal component of the Poynting vector. A trapped particle placed in such a (e.g. LG_0^1) field would rotate continuously around the inclination of wavefront which is illustrated by the curved (red) arrow on top of Figure 18 due to orbital angular momentum transfer by scattering.

However the first experiments to explore this orbital angular momentum looked at transfer by absorption. Two groups (He et al., 1995a; Simpson et al., 1996, 1997; Friese et al., 1996) from Australia and the United Kingdom, have independently observed this mechanical torque exerted by the LG beam by transfer of its orbital angular momentum to an absorptive microparticle. In the first studies in 1995, He and colleagues (He et al., 1995a) set absorptive copper oxide particles into rotation using LG modes: in fact (as already mentioned) this was one of the first ever implementations of holographic optical trapping. The experiment trapped particles in two dimensions and showed that the rotation rate is not due to any asymmetric scattering. By reversing the winding direction of the 2π azimuthal phase term of the LG beam the particle was observed to rotate in the opposite sense. In another experiment, data from which may be seen in Figure 19, Simpson et al. (1997) used the mode converter to generate LG modes which were then used to rotate absorptive objects and experimentally decoupled of the spin angular momentum of light

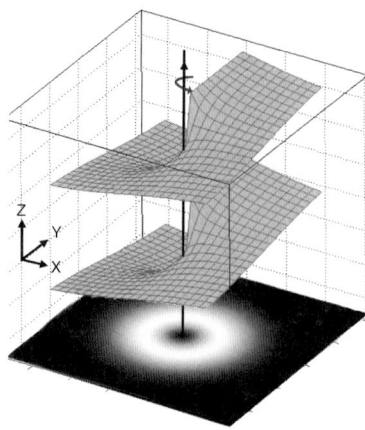

FIGURE 18 Helical wavefront of a LG_0^1 beam. A LG_0^1 beam has a well-defined orbital angular momentum (red arrow) due to the helical wavefront

from the orbital angular momentum of light by observing the rates of rotation due to each form of angular momentum. Friese et al. (1996) achieved analogous results using holographically generated LG modes of azimuthal index $l = 3$. In these first experiments that exploited these LG modes, the particles used were optically absorptive which naturally would not be ideal for biological applications but clearly elucidate the physics of these distinctive light modes. Later studies showed that one could actually transfer orbital angular momentum onto transparent dielectric particles simply by scattering off the inclined wavefronts (O'Neil et al., 2002) as might be inferred from Figure 18. Particles were placed off-axis within the circumference of the LG beams and were observed to show differing forms of motion based upon whether they were responding to the spin or orbital component of the light field. In turn this gives insight into the intrinsic and extrinsic nature of spin and orbital angular momentum.

LG beams have had a significant influence in the advancement of optical trapping. It is important to explore further applications of such modes within the context of biological and colloidal sciences as it is only their phase structure that has gained attention but their annular intensity profile is also of importance for this field. Before we proceed to their applications in optical micromanipulation of colloids and cells, we remark that such light fields are making major in-roads into atomic systems with applications that purely exploit the beam profile for blue-detuned toroidal traps (Kuga et al., 1997), atom guiding (Schiffer et al., 1998; Rhodes et al., 2002) and studies in Bose–Einstein condensation (Wright et al., 2000; Andersen et al., 2006). More details of the use of such light fields for atomic trapping and guiding may be found elsewhere (Grimm et al., 2000).

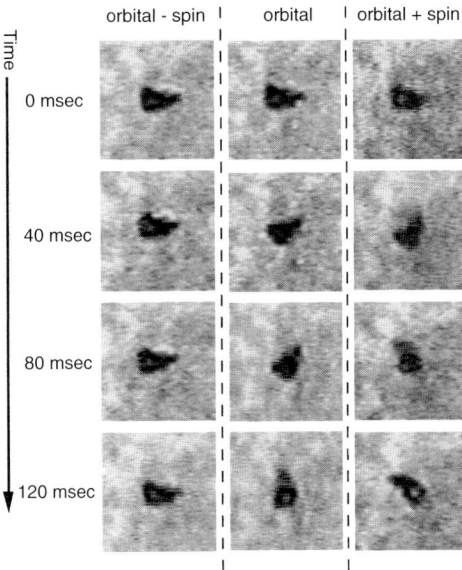

FIGURE 19 Spin and orbital angular momentum transfer with circularly polarized LG_0^1 beam. Successive frames of the video image showing the stop–start behavior of a 2-μm-diameter teflon particle held with an optical spanner. Reprinted figure with permission from (Simpson et al., 1997) © 1997, Optical Society of America

Low index particles are finding ever more applications notably in chemical and biological contexts. Ashkin (1970) first observed that low refractive index particles are repelled from the high intensity region of light while high refractive index particles are drawn into the trap. He later used a high order mode laser beam (TEM_{01}^*) to levitate and hold a low glass sphere against gravity (Ashkin and Dziedzic, 1974; Ashkin, 1997). Using a pair of fast scanning mirrors driven by galvanometers, Sasaki et al. (1991, 1992) demonstrated that by repetitive scanning of a single trap into a specific two-dimensional light pattern microspheres would find themselves aligned onto the designated locations. By rotating the trap over a few hundred Hertz, they were also able to cage and transport reflective metallic particles or low index microdroplets. In the general context of optical tweezing applications, one is typically focused upon the trapping of high refractive index microparticles. When focused to a tight spot through a high numerical aperture microscope objective, these vortex beams form a special type of optical tweezers that are distinguished by their ability in the manipulation of low refractive index microparticles: all higher order LG beam ($p = 0, l > 1$) possesses a smooth annular intensity profile, resembling a donut and as mentioned a low index particle will be repelled from the light region and be confined within the dark core, as witnessed from the studies of Gahagan and Swartzlander (1999). If we now refer back to Figure 1 and revisit the optical tweezing of a higher index particle with a Gaussian beam we see that light coming straight through the center of back aperture of the objective leads to axial scattering forces that act against the gradient forces to destabilize the trap. If the radius of the microsphere matches approximately the size of the beam waist of the LG beam, high refractive index particles can also be tweezed with an LG beam but will experience a much lower on-axis light scattering force compared to the use of a Gaussian beam due to the LG beam profile. This reduces the on-axis scattering force exerted along the axial direction as shown by Simpson et al. (1998).

Optical vortices provide a much needed solution for the manipulation of droplets where the refractive indices of most liquids are essentially lower than their surrounding medium. For selective mixing of droplets without cross-contamination, it requires minor positioning of the droplets. In a recent work, Lee et al. (2005) explored placing a spiral phase element at gradual step away from the center of an incident Gaussian beam and manipulated the position of the dark vortex core: off-axis optical vortex. In this way a low index microparticle can be manipulated around the beam central axis without moving the entire beam. Lorenz et al. (2007) adapted the technique with the use of two such off-axis LG (optical vortex) beams to controllably fuse two aqueous droplets. In Figure 20(a)–(d), a series of off-axis optical vortex beams are shown with the position of the dark core shifted at different distances away from the center of the beam, by displacing a spiral phase element (azimuthal charge $l = 1$) to different transverse locations (y) on the beam waist, ω, and orthogonally across the propagation direction. This action gradually diminishes the overall annular ring intensity pattern into a C-shaped intensity pattern. In Figure 20(e)–(g), a vortex-trap-induced fusion of two aqueous droplets in acetophenone is shown. The images in the inset were obtained by recording the back-scattered laser light from the vortex trap off the interface between the coverslip and water. The scale bar represents 10 µm. In a similar fashion, ultrasound contrast agent may be trapped in the dark core of the vortex and used in combination with exposure to ultrasound for microbubble cavitation and subsequent drug delivery. Such sonoporation was demonstrated by Prentice et al. (2005, 2004). O'Neil and Padgett (2000) also showed the ability to trap and rotate micron-sized metallic particles in three dimensions using an inverted optical vortex trap that demonstrated a decoupling of spin and orbital angular momentum. As such, these vortex tweezing experiments demonstrate that shaping the tweezing beam profile can enhance the ability to trap other types of particles. This technique potentially requires less effort than continuous scanning of a Gaussian beam (Sasaki et al., 1992). By taking advantage of the plasmon

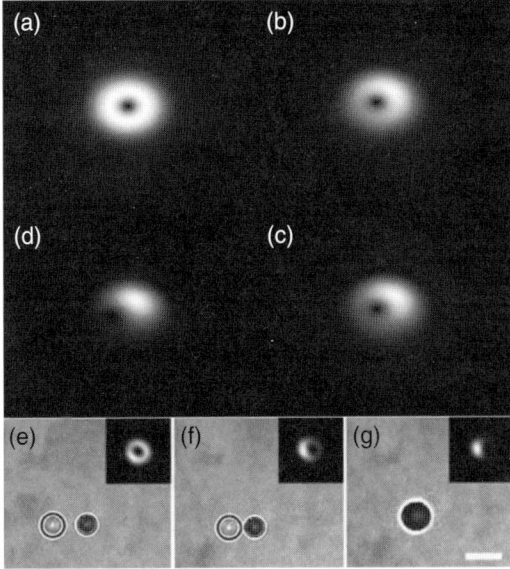

FIGURE 20 Numerical simulation of off-axis optical vortex beam and experimental images of the fusion of droplets with off-axis optical vortex beams. Numerical calculation of the far-field diffraction intensity pattern of an optical vortex beam, LG_{01}, of beam waist, ω, with its azimuthal phase change displaced away from the central axis at different transverse locations (y) on the beam waist and orthogonally across the propagation direction; (a) $y = 0$, (b) $y = \omega/4$, (c) $y = \omega/2$, (d) $y = \omega$. (e)–(g) show sequence of images showing vortex-trap-induced fusion of two aqueous droplets in acetophenone. The scale bar represents 10 μm. Reprinted figure with permission from Lorenz et al. (2007) © 2007, American Chemical Society

resonance of metallic nanoparticles and the dark core of an LG beam, Dienerowitz et al. (2008) were able to optically confine 100-nm diameter gold nanoparticles in a novel form of vortex trap.

At the micrometer size scale, fluid flow can be laminar. Within such a low Reynolds number regime, we need methods to induce rapid direction changes within flow channels to control flow or mixing processes of different species of particles or different fluids. Rotating trapped objects with optical tweezers thus becomes an important consideration for such studies as well as microrheology (Bishop et al., 2004; La Porta and Wang, 2004; Oroszi et al., 2006). As mentioned earlier the rotating optically trapped particles also hold immense promise for microrheology where the rotational Stokes drag reaches equilibrium with the rotating birefringent object permitting a local measurement of viscosity (Bishop et al., 2003b) with only picoliter quantity of liquid (Parkin et al., 2007). Other examples include the spinning of photopolymerized structures (Kelemen et al., 2007) which may be set into rotation by asymmetric scattering. Micropumps may be created by simultaneously trapping and rotating microspheres held by multiple optical tweezers created by acousto-optic modulators (Terray et al., 2002). Naturally the optical angular momentum of light, be it spin or orbital, can induce rotation in trapped objects. If we first consider the case of spin angular momentum one can set a trapped birefringent particle into rotation due to the exchange of spin angular momentum, where the trapped particle acts like a microscopic waveplate. The experiment is a microscopic analog to Beth's famous experiment (Beth, 1936). Friese et al. (1998, 2001) set a calcite particle into rotation with a circularly polarized trapping beam. Two birefringent microspheres may be set into rotation in opposite senses to one another, creating an optical pump though the flow rates and, speed of particle motion is slow (Di Leonardo et al., 2006; Leach et al., 2006b). To obviate the reliance upon intrinsic birefringence Neale et al. (2005) engineered birefringence into SU-8 polymer which could be considered as 'form birefringence' (Bishop et al., 2003a) and allowed rotation of arbitrary structures with circularly polarized light. Notably optical torques can be imparted onto particles with the orbital angular momentum of the LG beams via scattering or absorption and we have mentioned these above. In terms of applications, LG beams too can be extended to the generation of optically driven pumps. Ladavac and Grier (2004) used holographically created rows of alternating single ringed LG beams of very high azimuthal index ($p = 0, l = \pm 21$) to trap and rotate large numbers of microspheres to generate fluid flow as seen in Figure 21. The figure shows a time-lapse composite of 16 images in half-second intervals of colloidal spheres in the 'holographic pump'. The circles identify the trajectory of a single sphere as it moves 25 micrometer to the left in 7 seconds. Its peak speed is 5 micrometers per second. In turn this caused other particles to flow through the system. By pushing particles toward the water–air interfaces (away from a hard surface), Jesacher et al. (2006) observed high rotation rates of particles trapped in holographic optical vortex traps and also demonstrated interactive particle flow steering with arrays of optical vortex pumps.

Applying and controlling a small amount of torque upon biological particles i.e cells or chromosomes, often require the beams to be tailored to the shape of the biological particles. Orientation of particles with optical traps is also a desirable quality in this respect. With higher order laser modes, Sato et al. (1991) have demonstrated that HG modes are capable of rotating elongated biological particles. Can the LG beam offer a possibility to apply optical torques onto different types of microparticles? Interfering LG beams with either plane waves or with other LG beams is an alternative beam shaping technique to create rotating light patterns as demonstrated by Paterson et al. (2001). In this work chromosomes were controllably oriented and spun by controlled adjustment of the relative

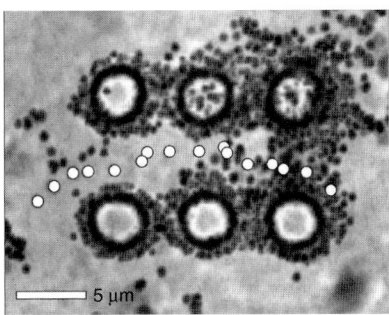

FIGURE 21 Microfluidic flow generated with an array of optical vortices. Time-lapse composite of 16 images in half-second intervals of colloidal spheres in the holographic pump created by an array of LG beams. Reprinted figure with permission from (Ladavac and Grier, 2004) © 2004, Optical Society of America

optical path length in the LG beam interferometer. Interfering two LG beams of equal but opposite azimuthal index can be used to generate an array of spots that again may be rotated with careful adjustment of the relative path length between the two arms of the interferometer. These patterns can create and rotate three-dimensional structures (MacDonald et al., 2002). The interfering spiral pattern of the LG beams can be modified to fit the shape of the object. By applying the angular Doppler technique, the particle of interest can be rotated at high frequency (Arlt et al., 2002).

The inclined wavefront and the annular intensity pattern of a LG beam form in some sense offer an ideal toroidal optical trap with a given driving force. A constant driving force along a toroidal trap can demonstrate interesting behavior as already mentioned (Faucheux et al., 1995a,b; Harada and Yoshikawa, 2004). In a more recent experiment, Lutz et al. (2006) demonstrate that the particles are able to surmount potential barriers due to hydrodynamic interactions amongst the particles circulating in the toroidal trap. Roichman et al. (2007) made use of higher order LG beams ($l = 50$ and 80) and observed similar colloidal interactions, as shown in Figure 22.

4.3.2 Bessel Light Beams and their Applications

Diffraction is inherently linked with the wave nature of light and is one that needs careful consideration in any optical system. Modern laser technology produces highly coherent and monochromatic light fields that have low divergence over a long propagation distance (meters). However, if a Gaussian beam is tightly focused, it would tend to spread rather rapidly over a short propagation distance, beyond what is termed the Rayleigh range Z_R. This range is the typical criteria used to characterize the expansion or spread of a Gaussian light field and denotes the distance

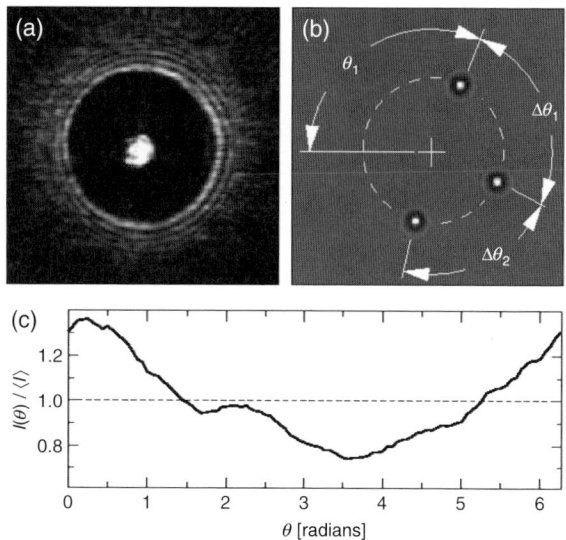

FIGURE 22 Optically driven colloidal ring. (a) Projected intensity pattern for an optical vortex with. (b) Video microscope image of three colloidal silica spheres trapped on the optical vortex. (c) Measured intensity variations around the optical vortex's circumference. Reprinted figure with permission from Roichman et al. (2007) © 2007, American Physical Society

over which a Gaussian beam increases its cross-sectional area by a factor of two.

$$Z_R = \frac{\pi \omega_0^2}{\lambda} \tag{11}$$

where λ is the wavelength and ω_0 is the beam waist size. As one can see in the context of optical tweezers high focusing objectives (large numerical aperture), the Rayleigh range is merely a few microns or so. As the beam spreads and the intensity gradient of the beam diminishes over a certain axial distance, the optical forces will reduce accordingly. If one could overcome such diffractive spreading, then a multitude of applications including several in optical micromanipulation would be possible. For example one would be able to create an elongated optical guide that retained strong transverse confinement over its entire length due to minimal intensity variations. In this manner transport of biological or colloidal material over very long distances with high accuracy would be possible.

Durnin et al. (1987) suggested a potential beam by which diffraction is 'suppressed' during free space propagation in theory and experiment. The beam was the Bessel light beam. Such beams appear to offer a resistance to the nature diffraction over a specific finite distance and thus this seems like an attractive alternative to using Gaussian beams in a number of scenarios. The ideal version of this beam however implied the need for an infinite input aperture (and thus infinite power) and could not be experimentally realized. In subsequent experimental work, Durnin et al. (1987) showed that one could generate an approximation to a Bessel beam (BB) that retained its key propagation invariant features over a limited range. An important point in that when we compare the propagation properties of a BB to a Gaussian beam, the appropriate comparison is between the central core size of the Bessel beam and the spot size of the Gaussian beam (Durnin et al., 1988).

The Bessel beam has proved to be the most popular and widely used 'non-diffracting' light field (McGloin and Dholakia, 2005) but it is not the only light field with these intriguing properties. Higher order Bessel functions can be optically realized with LG beams illuminating an axicon (Arlt and Dholakia, 2000). The higher order Bessel beam, where l ($l >= 0$), possesses an inclined wavefront therefore has a orbital angular momentum rather like the LG beam. The center of the beam then is either a bright spot (index $l = 0$) or possesses an hollow core $l(l >= 1)$ or vortex. Other types of non-diffracting light field modes exist such as Mathieu beams (Chavez-Cerda et al., 2001) which may be considered as elliptical generalizations of Bessel beams. Mathieu beams are described by the 'ellipticity' parameter q, and the integer l, which denotes the order of the mode. It is worth noting that the recent realization of 'non-diffracting' Airy beams by Siviloglou and Christodoulides (2007) may be of interest to the optical micromanipulation community (Dholakia, 2008).

For most beams under consideration, each plane wave component would take up a different phase shift over a given propagation distance with the wave vector constantly changing. The resulting beam is considered as the interference pattern of the plane waves that naturally changes in its complex amplitude. However, the Bessel beam falls into the category of a special beam where in this case the phase shift accrued is the same for each and every plane wave component. Such beams do not change shape on propagation, and are termed propagation invariant or 'diffraction-free'. This also leads us to understanding how we may actually create 'white light Bessel' beams (Fischer et al., 2005; Leach et al., 2006a) where temporal coherence is not an issue. The ideal Bessel beam has an amplitude distribution that may be expressed as:

$$E(r, \phi, z) = A_0 \exp(ik_z z) J_l(k_r r) \exp(\pm il\phi) \qquad (12)$$

where J_l denotes a lth-order Bessel function, k_z and k_r are the longitudinal and radial wavevectors, with $k = \sqrt{k_z^2 + k_r^2} = 2\pi/\lambda$ (λ being the

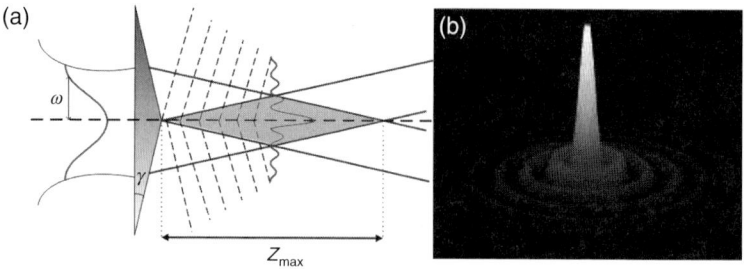

FIGURE 23 Illustration of an axicon-generated bessel beam and its three-dimensional intensity profile. (a) Illustrates the generation of a zero-order bessel beam with a non-diffractive distance of z_{max} from a gaussian beam of beam waist (ω) through the use of an axicon lens element (angle γ). (b) Shows the three-dimensional intensity profile of a bessel beam. (adapted and reprinted with permission from the author (Cizmar, 2006))

wavelength of the electromagnetic radiation used to form the Bessel light beam). The parameters r, ϕ and z are the radial, azimuthal and longitudinal components respectively. The higher order Bessel beam ($l > 0$ is an index equivalent to l, the azimuthal index of the LG beam described earlier) possesses an azimuthal phase variation which results in a dark non-diffracting hollow core located on the beam central axis. A zero-order Bessel is generated from conical wave vectors where the plane waves overlap along the center of the propagation axis as seen in Figure 23(a). The intensity for a zeroth-order Bessel beam in three dimensions is shown in Figure 23(b). One can consider an annulus (infinitely thin for a perfect ideal Bessel beam) in k-space and the optical Fourier transform of a ring will result in a Bessel beam. Using an annulus placed in the back focal plane of a lens was in fact the first way a finite approximation to a Bessel beam was generated in the laboratory (McGloin and Dholakia, 2005; Durnin et al., 1987).

A feature of the Bessel beam that we have mentioned is that it contains a given power that is evenly distributed and equally shared between its constituent rings (Durnin et al., 1988; Lin et al., 1992). Thus though more rings imply a longer 'diffraction-free' propagation distance, we trade this against power in the central maximum. An efficient and popular way to generate a Bessel beam is by use of conical optical element known as an axicon (McLeod, 1954; Indebetouw, 1989). The axicon is a conically shaped transparent optical element that readily imposes a phase shift $\varphi(r, \theta) = k(n - 1)\gamma$ onto an incident Gaussian light beam, as shown in Figure 23 where n is the refractive index of the axicon material, and γ is the internal angle of the element. The diagram also clearly shows the notion that a Bessel beam is a set of waves propagating on a cone. Microfabricated axicons, much like microfabricated spiral phase plates, have also proven to be useful for direct integration into optical trapping systems (Cheong et al., 2005). If the axicon is not illuminated at normal incidence, astigmatism is introduced and non-circularly symmetric patterns can result (Thaning et al., 2003; Bin and Zhu, 1998; Tanaka and Yamamoto, 2000). If the axicon is illuminated with a converging or diverging beam, one can also change the baseline of the beam profile and obtain an offset or tilt across the beam profile, resulting in a type of biased optical potential energy landscape: such a washboard potential can be used for optical micromanipulation (Tatarkova et al., 2003). Now holographic generation of Bessel beams has also been established as a powerful and efficient manner by which to generate such modes. Such holograms can be in the form of static, etched elements (Vasara et al., 1989), or fully reconfigurable when using spatial light modulators (Davis et al., 1996b,a). Interestingly, an obstruction at the center of the Bessel beam would not affect overall, the beam propagation: the beam can actually self-heal! This is due to the conical wavevectors that constitute the beam.

We now proceed to review the experiments that have used Bessel light modes within optical micromanipulation that benefit from the key attributes of this light beam. We note that conical glass axicons and more generally Bessel light modes have been utilized also in the domain of atomic physics for studies involving cold atoms and we refer the interested reader to some relevant papers elsewhere (Arlt et al., 2001a; Schmid et al., 2006).

Arlt and colleagues (Arlt et al., 2001b) implemented a Bessel light mode within an optical trapping system as shown in Figure 24. A telescope was used to expand the output beam of a laser that was subsequently incident upon the axicon. The generated Bessel beam is again telescoped down to the size desired to perform micromanipulation experiments within a sample chamber. It is important to note that the absence of any axial intensity gradient along the beam propagation length meant that this is a two-dimensional trap. The experiments showed trapping of a microsphere and separately a glass rod without the use of a microscope objective.

Subsequently, Garces-Chavez et al. (Garces-Chavez et al., 2002a) used the Bessel light mode's self-healing properties in a trapping system. This experiment used three sample chambers made with multiple cover slips to illustrate the key self-healing aspects of the beam: showing the direct evidence of conical wave propagation. As shown in Figure 25(a), this beam was incident upon different particle samples within each of three chambers where the top and bottom chambers were separated by around 1 mm. The cells (I and II) are 3 mm apart, and 100 mm deep. In parts (i–vi) of the figure, frames from a video taken of objects captured by the Bessel beam (beam spot of radius 5 micrometer) are shown. In (i), a hollow sphere (low refractive index) of 5 μm in diameter is trapped in

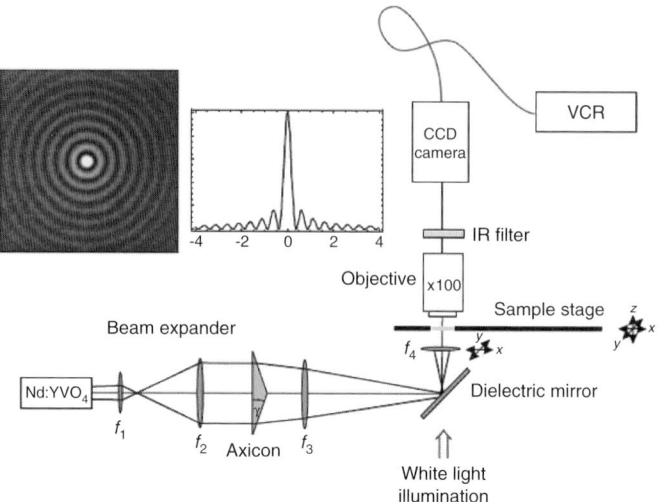

FIGURE 24 Optical trapping setup with a zeroth-order bessel beam. Experimental arrangement for bessel tweezing: lenses $f1 = 50$ mm and $f2 = 250$ mm expand the beam to illuminate the axicon, which generates the BB. Lenses $f3 = 250$ mm and $f4 = 25$ mm reduce this BB to one with a central maximum of ≈ 9.4 μm diameter. Lens f4 is adjusted to manipulate the particles. The inset shows a picture and cross-sectional profile of the bessel light beam which propagates in the vertical direction. Reprinted figure with permission from (Arlt et al., 2001b) © 2001, Elsevier

cell I between the central spot and the first ring of the Bessel beam. (ii), The beam is at a short distance above i, Here the beam has been distorted by the particle. (iii), At some small distance above the first sample cell, the beam has reformed and is no longer distorted. (iv), The beam enters cell II, and is able to stack three 5 μm in diameter solid silica spheres. (v), (vi), The beam profile above the stack of particles, with the beam reformed once more. A trapped particle in the beam center constituted the 'obstacle' and deformed the light field. The reformation of the beam was observed in between each chamber and the beam was able to trap in all three chambers simultaneously, however, one could see the beam exhibiting a two-dimensional trapping effect in each chamber. In one example a low index particle was confined between the central maximum and first bright ring, a calcite particle set into rotation in the beam center. By using a higher density of colloid, particles were trapped in not only the central maximum but some of the outer concentric rings. Even then the beam was observed to reform sufficiently well to trap more colloids in concentric ring patterns in a chamber placed further along the propagation direction as illustrated. As the central core of the Bessel beam facilitates long-range guiding of microparticles, Dholakia et al. (2004) observed two-photon fluorescence signal from the sample medium (with fluorescein

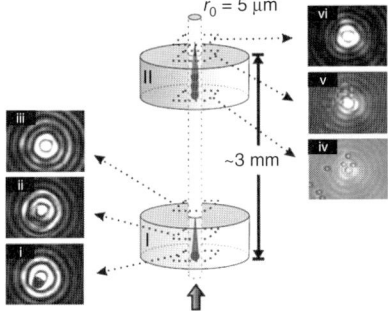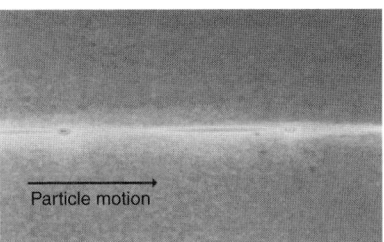

(a) © 2002, Macmillan Publisher Ltd: Nature.   (b) © 2004, Institute of Physics.

FIGURE 25 Observation of the reconstruction of a zeroth-order bessel beam in trapping. Inverted bessel beam optical tweezers experimental setup. Reprinted figure with permission from (Garces-Chavez et al., 2002a). (i–vi) frames from a video taken of captured objects as presented in the text. (b) The central core of the bessel beam facilitates long-range guiding of microparticles. Two-photon fluorescence signal from two fluorescent microspheres of diameter 5 μm that are optically confined and propelled with femtosecond bessel beam (pulse duration = 100 fs, central wavelength = 750 nm, diameter of central core = 4 μm). The beam has reformed approximately 90 μm in front of the trapped sphere. Reprinted figure with permission from (Dholakia et al., 2004)

added) of diameter 5 micrometer optically confined and propelled with a femtosecond Bessel beam. In that experiment, the beam can be observed to reforming approximately 90 μm in front of the trapped fluorescent sphere, as shown in Figure 25(b). This gives an interesting manner by which the optical field may be visualized around trapped or guided microparticles.

A Bessel beam can naturally persist over an extended range. Cizmar et al. (2005) used two counterpropagating zeroth-order Bessel beam modes to create a very long (millimeter) interference pattern along the axis of beam propagation. The interference pattern forms a one-dimensional optical potential energy landscape that resembles a 'conveyor belt', as shown in Figure 26(a). This allowed for three-dimensional trapping at the anti-nodes and subsequent capture and delivery of several submicron particles over a distance of hundreds of micrometers (Figure 26(b)). Precise delivery was achieved by shifting the phase of one of the beams causing the interference pattern to translate in space and thus moving trapped colloidal particles to specific regions. Both theoretical and experimental investigation showed that certain sizes of polystyrene particles jump between neighboring axial traps with higher probability than others. Another geometry that uses extended interferometric landscape and subsequently sorting of colloidal particles with a surface interference pattern was shown by Cizmar et al. (2006b). Experimental

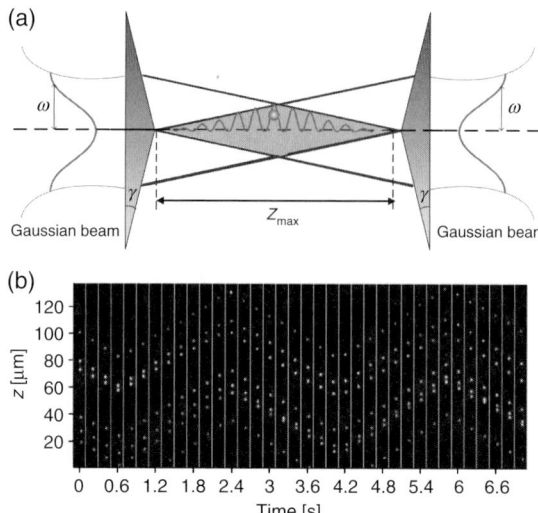

FIGURE 26 Optical conveyer belt. (Top) Counterpropagating geometry with two bessel beams should result in a very long chain of standing wave traps. By varying the phase of one of the beams, the standing wave can be converted into an extended linear array of optical traps for delivery of micro-objects (blue sphere) over large distances creating an 'optical conveyor belt'. (Bottom) Simultaneous optical confinement and delivery of polystyrene particles of radius 100 nm over a distance of 60 μm using the mechanism of optical conveyor belt. (Reprint figure with permission from the author (Cizmar, 2006).)

measurements of the trap stiffness using video tracking (Cizmar and Zemanek, 2007) have revealed that the ratios of longitudinal and lateral optical trap stiffnesses along the standing waves were around one order of magnitude higher compared to the classical single beam optical trap (optical tweezers). This would mean that such interferometric optical traps may have uses for force measurements within biological systems. Using the Bessel beam standing waves created from two independent counterpropagating Bessel beams, they were able to confine polystyrene particles of radius 100 nanometers, and arrange them into a one-dimensional chain over a length one millimeter (Cizmar et al., 2006c). In related studies, such a moving interferometric pattern created by the same Bessel laser beam modes were used to horizontally transport ultracold atoms over macroscopic distances of up to 20 centimeter while ensuring uniform trapping conditions. The stability of the interference traps allows rapid transport with velocities of up to 6 m/s. (Schmid et al., 2006).

With co-propagating BB, periodic oscillations of the on-axis intensity maxima (termed as self-imaging) can be changed by having two Bessel beams of different conical angles (slightly different wavevector)

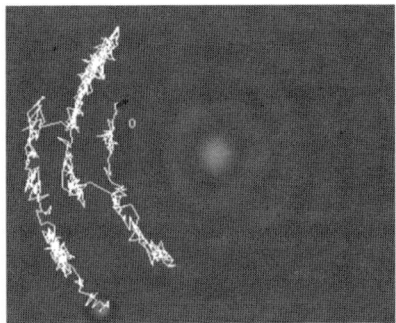

FIGURE 27 Brownian motion of particles confined within the concentric rings of a bessel beam. Typical transversal trajectory of radius =1.15 micrometer silica sphere in a BB. Reprinted figure with permission from (Milne et al., 2007a) © 2007, by Optical Society of America

interfering with one another (Chávez-Cerda et al., 1998; Courtial et al., 2006; Cizmar et al., 2006a). By changing the path difference of one of the Bessel beams, particles can be selectively trapped and manipulated. Through numerical simulation, researchers have showed that by simply increasing the number of interfering beams allows one to select the types of particles that would be confined. Experimentally, they trapped particles with radii from 100 nanometers up to 300 nanometers but required the assistance of fluid flow against the beam propagation. Particle hopping from one trap to the neighboring-occupied traps was observed to generate a domino-like effect propagating with constant velocity over the subsequent occupied traps (Cizmar et al., 2006a).

Tatarkova et al. (2003) examined an axially misalignment (diverging) of the Gaussian beam that is used to illuminate the axicon: in turn this created an asymmetry in the potential wells created by any one ring of the Bessel profile. This causes a preferential lateral guide that brings particle hopping toward the beam center. Subsequent studies explored these issues in more detail where Milne et al. (2007a) used both theory and experiment to look at the motion of silica microspheres by video tracking when placed within a BB as shown in Figure 27. Two different computational models, Mie scattering and geometrical ray optics, gave analogous data and predicted the existence of a distinct size dependence of particle dynamics and equilibria positions within a BB.

Studies with such BBs of higher order ($l > 0$) may complement those performed with LG modes described earlier. In 2002, Garces-Chavez et al. (Garces-Chavez et al., 2003) explored angular momentum transfer from a higher order Bessel beam to high index particles: in this study the particle rotation rates were accurately determined and supported with the theoretical ray optics model, showing good agreement and the expected linear relationship between power and rotation rate. The same team also explored the transfer of orbital angular momentum to low index particles (Garces-Chavez et al., 2002b) where a low index particle was trapped between two of the bright concentric rings of the Bessel beam, and that was the first transfer of orbital angular momentum to low index particles. Here, the authors observed that angular momentum can be transferred via scattering and in the absence of any on-axis scattering light. A further observation of rotation in a birefringent particle around the concentric rings of the higher order Bessel beam produced a conclusive demonstration of the intrinsic and extrinsic nature of optical angular momentum. The birefringent particles can help map the optical angular momentum density in the light field with the particle acting as the probe. Further theoretical studies explored in more detail the dynamics and equilibrium positions of particles in a Bessel beam (Volke-Sepulveda et al., 2004, 2002). The Mathieu beams, described earlier, have been recently used in optical micromanipulation exploring their orbital angular momentum properties (Lopez-Mariscal et al., 2006) and showing non-uniform velocities of particle motion around the beam profile.

Paterson et al. (2005) looked at the motion of red and white blood cells, with no markers attached, across the optical potential energy landscape created by the Bessel beam. This work showed an important result that red and white blood cells could be separated in this purely passive manner by the Bessel beam. White cells were accumulated within the central core whilst red cells, depending upon the exact power and beam parameters, were trapped in the concentric rings where they oriented themselves (flipped) by ninety degrees. The behavior was attributed to the difference between red and white blood cells: lymphocytes are near spherical and have a nucleus whereas erythrocytes are more like a bi-concave disk. This study was recently extended to see if passive sorting could be applied to stem cells and human cancer cells. However the intrinsic differences between the given cell types were typically not sufficient to initiate sorting and the authors used dielectric tagging of the cells to enhance the sorting process: this involved using antibody–antigen binding and treated microspheres to preferentially attach them to certain cells. Subsequently these 'tagged' cells responded more strongly to the Bessel mode (Paterson et al., 2007), as shown in Figure 28. In Figure 28(a) the white and black arrows point to lymphocytes that have been attached to streptavidin-coated microspheres using a mouse, anti-CD2 monoclonal antibody, and a biotinylated anti-mouse antibody. The surrounding cells are other mononuclear cells isolated from venous blood. The mixed ensemble of cells is at the bottom of a sample chamber and imaged using a 60X microscope objective. In (b) and (c) the sphere-labeled cells travel into the central maximum of the Bessel beam and are guided vertically within the center until they reach the top of the sample chamber where they form

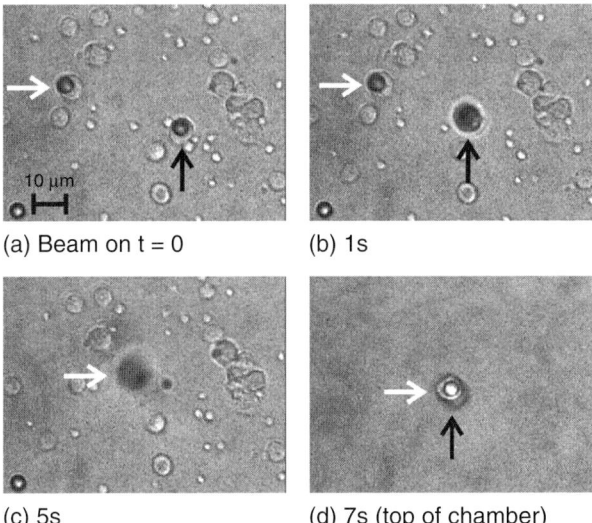

FIGURE 28 Use of bessel beam for the sorting and isolation of CD2 T-lymphocytes from a sample of mononuclear cells. Reprinted figure with permission from (Paterson et al., 2005) © 2005, from American Institute of Physics

a vertical stack as shown in (d). Over the time scale required for isolation of the CD2 T-cells in seconds, the untagged cells do not move significantly.

5. OPTICAL BINDING

Trapping of many particles in desired configurations, be it in interferometric, timed shared, holographic, or other light modes, is an important topic. A rather less well-discussed but fundamentally significant issue is that multiple microparticles may self-organize themselves into arrays by their interaction with light. Indeed when we consider the creation by any means of multiple, optical trap sites and our optical landscapes, to date we have ignored how the very scattering or redistribution of light by one given trapped particle placed within this light field may influence the equilibrium position of its nearest trapped particle (neglecting taking into account any hydrodynamic interactions). This leads one to the emergent and key area of *optical binding*.

Light forces interacting between particles may act to optically bind themselves. Particle organization through interactions of optical scattering in the beam propagation direction has been recently observed and, allow interactions between micron-sized particles separated by distances and order of magnitude larger than their individual diameters. It may manifest itself in a form either transverse to or in the direction of (longitudinal) the beam propagation. There are various types of 'optical binding': Burns et al. (1990, 1989) investigated systems where the interaction of coherently induced dipole moments of the spheres was said to interact to binding matter. It took another decade before new experimental results revived the seminal work of binding matter with light interaction arising from the particles themselves. Tatarkova et al. (2002) and Singer et al. (2003) conducted separate studies in exploring particles in a dual counterpropagating trap and found a very interesting assembly of particles in this trapping geometry, which is now understood as a longitudinal optical binding effect (Metzger et al., 2006a,c).

Rich dynamics in such systems, including recent observations of bistability (Metzger et al., 2007a) and interesting inter-particle correlations (Metzger et al., 2007a) were also reported. Substantial theoretical work has been performed on such systems (Ng et al., 2005; Grzegorczyk et al., 2006; Karasek et al., 2006; Metzger et al., 2006b,c, 2007b) and other manifestations of binding have been observed in near-field optical geometries (Reece et al., 2006; Mellor and Bain, 2006; Garces-Chavez et al., 2006). This is a young area and is likely to have impact across the biological and colloidal sciences and indeed may present a very powerful future method for creation of large scale extended arrays in up to three dimensions. It is noteworthy to emphasize that such 'optical binding' is radically different from individually trapping each micro-object: the very interaction between an object and its nearest neighbors creates a self-consistent and homogeneous solution. This could be the very technique that would allow a straightforward optical geometry in principle, to create a large scale colloidal array.

6. CONCLUSIONS

Optical micromanipulation remains a powerful and versatile tool for the natural sciences. The weak forces it exerts may be calibrated and used to measure the dynamics and forces exerted by motor proteins and other biological macromolecules. The system may be readily calibrated to incredible accuracy and resolve femtonewton forces. An upcoming theme within this field has been the use of structured light fields: these can take the form of multiple optical traps or more formal non-zero order laser modes such as Laguerre-Gaussian or Bessel beams. These latter modes have made major headway into fundamental science, biomedical and chemical applications as well as capturing the imagination of the atomic and optical physics communities with their unusual propagation characteristics. We have summarized in this article the experiments and theory behind such light patterns and given a flavor of some of their impact within the ever growing area of optical micromanipulation.

ACKNOWLEDGMENTS

The authors would like to thank the UK Science and Engineering Research Council for funding and Michael Mazilu, Tomas Cizmar, Joerg Baumgartl and David Stevenson for proof reading and useful comments.

REFERENCES

Abbondanzieri, E.A., Greenleaf, W.J., Shaevitz, J.W., Landick, R. and Block, S.M. (2005). Direct Observation of Base-Pair Stepping by RNA Polymerase. *Nature*, **438**, 460.

Agate, B., Brown, C.T.A., Sibbett, W. and Dholakia, K. (2004). Femtosecond Optical Tweezers for in-situ Control of Two-Photon Fluorescence. *Optics Express*, **12**, 3011.

Akselrod, G.M., Timp, W., Mirsaidov, U., Zhao, Q., Li, C., Timp, R., Timp, K., Matsudaira, P. and Timp, G. (2006). Laser-Guided Assembly of Heterotypic Three-Dimensional Living Cell Microarrays. *Biophysical Journal*, **91**, 3465.

Allen, L., Beijersbergen, M.W., Spreeuw, R.J.C. and Woerdman, J.P. (1992). Orbital Angular-Momentum of light and the Transformation of Laguerre–Gaussian Laser Modes. *Physical Review A*, **45**, 8185.

Allen, L., Padgett, M.J. and Babiker, M. (1999). *Progress in Optics, Vol XXXIX*. pp. 291.

Allersma, M.W., Gittes, F., deCastro, M.J., Stewart, R.J. and Schmidt, C.F. (1998). Two-Dimensional Tracking of ncd Motility by Back Focal Plane Interferometry. *Biophysical Journal*, **74**, 1074.

Andersen, M.F., Ryu, C., Clade, P., Natarajan, V., Vaziri, A., Helmerson, K. and Phillips, W.D. (2006). Quantized Rotation of Atoms from Photons with Orbital Angular Momentum. *Physical Review Letters*, **97**, 170406.

Anderson, M.H., Ensher, J.R., Matthews, M.R., Wieman, C.E. and Cornell, E.A. (1995). Observation of Bose–Einstein Condensation in a Dilute Atomic Vapor. *Science*, **269**, 198.

Appleyard, D.C., Vandermeulen, K.Y., Lee, H. and Lang, M.J. (2007). Optical Trapping for Undergraduates. *American Journal of Physics*, **75**, 5.

Arlt, J. and Dholakia, K. (2000). Generation of High-Order Bessel Beams by Use of an Axicon. *Optics Communications*, **177**, 297.

Arlt, J., Dholakia, K., Soneson, J. and Wright, E.M. (2001a). Optical Dipole Traps and Atomic Waveguides Based on Bessel Light Beams. *Physical Review A*, **63**, 063602.

Arlt, J., Garces-Chavez, V., Sibbett, W. and Dholakia, K. (2001b). Optical Micromanipulation Using a Bessel Light Beam. *Optics Communications*, **197**, 239.

Arlt, J., MacDonald, M., Paterson, L., Sibbett, W., Dholakia, K. and Volke-Sepulveda, K. (2002). Moving Interference Patterns Created Using the Angular Doppler-Effect. *Optics Express*, **10**, 844.

Ashkin, A. (1970). Acceleration and Trapping of Particles by Radiation Pressure. *Physical Review Letters*, **24**, 156.

Ashkin, A. (1992). Forces of a Single-Beam Gradient Laser Trap on a Dielectric Sphere in the Ray Optics Regime. *Biophysical Journal*, **61**, 569.

Ashkin, A. (1997). Optical Trapping and Manipulation of Neutral Particles Using Lasers. *Proceedings of the National Academy of Sciences of the United States of America*, **94**, 4853.

Ashkin, A. and Dziedzic, J.M. (1974). Stability of Optical Levitation by Radiation Pressure. *Applied Physics Letters*, **24**, 586.

Ashkin, A. and Dziedzic, J.M. (1987). Optical Trapping and Manipulation of Viruses and Bacteria. *Science*, **235**, 1517.

Ashkin, A. and Dziedzic, J.M. (1989a). Internal Cell Manipulation Using Infrared-Laser Traps. *Proceedings of the National Academy of Sciences of the United States of America*, **86**, 7914.

Ashkin, A. and Dziedzic, J.M. (1989b). Optical Trapping and Manipulation of Single Living Cells Using Infrared-Laser Beams. *Berichte Der Bunsen-Gesellschaft-Physical Chemistry Chemical Physics*, **93**, 254.

Ashkin, A., Dziedzic, J.M., Bjorkholm, J.E. and Chu, S. (1986). Observation of a Single-Beam Gradient Force Optical Trap for Dielectric Particles. *Optics Letters*, **11**, 288.

Ashkin, A., Schutze, K., Dziedzic, J.M., Euteneuer, U. and Schliwa, M. (1990). Force Generation of Organelle Transport Measured In vivo by an Infrared-Laser Trap. *Nature*, **348**, 346.

Barnett, S.M. (2002). Optical Angular-Momentum Flux. *Journal of Optics B-Quantum and Semiclassical Optics*, **4**, S7.

Barton, J.P. and Alexander, D.R. (1989). Fifth-Order Corrected Electromagnetic Field Components for a Fundamental Gaussian Beam. *Journal of Applied Physics*, **66**, 2800.

Barton, J.P., Alexander, D.R. and Schaub, S.A. (1989). Theoretical Determination of Net-Radiation Force and Torque for a Spherical-Particle Illuminated by a Focused Laser-Beam. *Journal of Applied Physics*, **66**, 4594.

Bechinger, C., Brunner, M. and Leiderer, P. (2001). Phase Behavior of Two-Dimensional Colloidal Systems in the Presence of Periodic Light Fields. *Physical Review Letters*, **86**, 930.

Beijersbergen, M.W., Allen, L., Vanderveen, H. and Woerdman, J.P. (1993). Astigmatic Laser Mode Converters and Transfer of Orbital Angular-Momentum. *Optics Communications*, **96**, 123.

Beijersbergen, M.W., Coerwinkel, R.P.C., Kristensen, M. and Woerdman, J.P. (1994). Helical-Wave-Front Laser-Beams Produced with a Spiral Phaseplate. *Optics Communications*, **112**, 321.

Berg-Sorensen, K. and Flyvbjerg, H. (2004). Power Spectrum Analysis for Optical Tweezers. *Review of Scientific Instruments*, **75**, 594.

Berns, M.W., Tadir, Y., Liang, H. and Tromberg, B. (1998). *Methods in Cell Biology, Vol 55*. pp. 71.

Beth, R.A. (1936). Mechanical Detection and Measurement of the Angular Momentum of Light. *Physical Review*, **50**, 115.

Bin, Z. and Zhu, L. (1998). Diffraction Property of an Axicon in Oblique Illumination. *Applied Optics*, **37**, 2563.

Birkbeck, A.L., Flynn, R.A., Ozkan, M., Song, D.Q., Gross, M. and Esener, S.C. (2003). Vcsel Arrays as Micromanipulators in Chip-Based Biosystems. *Biomedical Microdevices*, **5**, 47.

Bishop, A.I., Nieminen, T.A., Heckenberg, N.R. and Rubinsztein-Dunlop, H. (2003a). Optical Application and Measurement of Torque on Microparticles of Isotropic Nonabsorbing Material. *Physical Review A*, **68**, 033802.

Bishop, A.I., Nieminen, T.A., Heckenberg, N.R. and Rubinsztein-Dunlop, H. (2003b). Optical Application and Measurement of Torque on Microparticles of Isotropic Nonabsorbing Material. *Physical Review A*, **68**, 033802.

Bishop, A.I., Nieminen, T.A., Heckenberg, N.R. and Rubinsztein-Dunlop, H. (2004). Optical Microrheology Using Rotating Laser-Trapped Particles. *Physical Review Letters*, **92**, 198104.

Block, S.M. (1996). Nanometers and Piconewtons: Using Optical Tweezers to Study Biological Motors. *Progress in Biophysics & Molecular Biology*, **65**, SH301.

Block, S.M., Goldstein, L.S.B. and Schnapp, B.J. (1990). Bead Movement by Single Kinesin Molecules Studied with Optical Tweezers. *Nature*, **348**, 348.

Block, S.M., Larson, M.H., Greenleaf, W.J., Herbert, K.M., Guydosh, N.R. and Anthony, P.C. (2007). Molecule by Molecule, the Physics and Chemistry of Life: Smb 2007. *Nat Chem Biol*, **3**, 193.

Brau, R.R., Ferrer, J.M., Lee, H., Castro, C.E., Tam, B.K., Tarsa, P.B., Matsudaira, P., Boyce, M.C., Kamm, R.D. and Lang, M.J. (2007). Passive and Active Microrheology with Optical Tweezers. *Journal of Optics A: Pure and Applied Optics*, **9**, S103.

Brunner, M. and Bechinger, C. (2002). Phase Behavior of Colloidal Molecular Crystals on Triangular Light Lattices. *Physical Review Letters*, **88**, 248302.

Burns, M.M., Fournier, J.M. and Golovchenko, J.A. (1989). Optical Binding. *Physical Review Letters*, **63**, 1233.

Burns, M.M., Fournier, J.M. and Golovchenko, J.A. (1990). Optical Matter — Crystallization and Binding in Intense Optical-Fields. *Science*, **249**, 749.

Bustamante, C., Bryant, Z. and Smith, S.B. (2003). Ten Years of Tension: Single-Molecule DNA Mechanics. *Nature*, **421**, 423.
Bustamante, C., Liphardt, J. and Ritort, F. (2005). The Nonequilibrium Thermodynamics of Small Systems. *Physics Today*, **58**, 43.
Cai, L.Z., Yang, X.L. and Wang, Y.R. (2002). All Fourteen Bravais Lattices can be Formed by Interference of Four Noncoplanar Beams. *Optics Letters*, **27**, 900.
Chavez-Cerda, S., Gutierrez-Vega, J.C. and New, G.H.C. (2001). Elliptic Vortices of Electromagnetic Wave Fields. *Optics Letters*, **26**, 1803.
Chávez-Cerda, S., Meneses-Nava, M.A. and Hickmann, J.M. (1998). Interference of Traveling Nondiffracting Beams. *Optics Letters*, **23**, 1871.
Cheong, W.C., Ahluwalia, B.P.S., Yuan, X.C., Zhang, L.S., Wang, H., Niu, H.B. and Peng, X. (2005). Fabrication of Efficient Microaxicon by Direct Electron-Beam Lithography for Long Nondiffracting Distance of Bessel Beams for Optical Manipulation. *Applied Physics Letters*, **87**.
Cheong, W.C., Lee, W.M., Yuan, X.C., Zhang, L.S., Dholakia, K. and Wang, H. (2004). Direct Electron-Beam Writing of Continuous Spiral Phase Plates in Negative Resist with High Power Efficiency for Optical Manipulation. *Applied Physics Letters*, **85**, 5784.
Chiou, A.E., Wang, W., Sonek, G.J., Hong, J. and Berns, M.W. (1997). Interferometric Optical Tweezers. *Optics Communications*, **133**, 7.
Chiou, P.Y., Ohta, A.T. and Wu, M.C. (2005). Massively Parallel Manipulation of Single Cells and Microparticles Using Optical Images. *Nature*, **436**, 370.
Chowdhury, A., Ackerson, B.J. and Clark, N.A. (1985). Laser-Induced Freezing. *Physical Review Letters*, **55**, 833.
Chu, S. (1998). The Manipulation of Neutral Particles. *Reviews of Modern Physics*, **70**, 685.
Cizmar, T. 2006. Optical Traps Generated by Non-Traditional Beams, *Ph.D. Thesis, Masaryk University, Faculty of Science. Brno.*.
Cizmar, T., Garces-Chavez, V., Dholakia, K. and Zemanek, P. (2005). Optical Conveyor Belt for Delivery of Submicron Objects. *Applied Physics Letters*, **86**.
Cizmar, T., Kollarova, E., Bouchal, Z. and Zemanek, P. (2006a). Sub-Micron Particle Organization by Self-Imaging of Non-Diffracting Beams. *New Journal of Physics*, **8**, 43.
Cizmar, T., Siler, M., Sery, M., Zemanek, P., Garces-Chavez, V. and Dholakia, K. (2006b). Optical Sorting and Detection of Submicrometer Objects in a Motional Standing Wave. *Physical Review B*, **74**, 035105.
Cizmar, T., Siler, M. and Zemanek, P. (2006c). An Optical Nanotrap Array Movable Over a Milimetre Range. *Applied Physics B-Lasers and Optics*, **84**, 197.
Cizmar, T. and Zemanek, P. (2007). Optical Tracking of Spherical Micro-Objects in Spatially Periodic Interference Fields. *Optics Express*, **15**, 2260.
Clapp, A.R., Ruta, A.G. and Dickinson, R.B. (1999). Three-Dimensional Optical Trapping and Evanescent Wave Light Scattering for Direct Measurement of Long Range Forces Between a Colloidal Particle and a Surface. *Review of Scientific Instruments*, **70**, 2627.
Cohen-Tannoudji, C.N. (1998). Manipulating Atoms with Photons. *Reviews of Modern Physics*, **70**, 707.
Constable, A., Kim, J., Mervis, J., Zarinetchi, F. and Prentiss, M. (1993). Demonstration of a Fiberoptic Light-Force Trap. *Optics Letters*, **18**, 1867.
Courtial, J., Whyte, G., Bouchal, Z. and Wagner, J. (2006). Iterative Algorithms for Holographic Shaping of Non-Diffracting and Self-Imaging Light Beams. *Optics Express*, **14**, 2108.
Creely, C.M., Singh, G.P. and Petrov, D. (2005). Dual Wavelength Optical Tweezers for Confocal Raman Spectroscopy. *Optics Communications*, **245**, 465.
Crocker, J.C. (1997). Measurement of the Hydrodynamic Corrections to the Brownian Motion of Two Colloidal Spheres. *Journal of Chemical Physics*, **106**, 2837.
Crocker, J.C. and Grier, D.G. (1996). Methods of Digital Video Microscopy for Colloidal Studies. *Journal of Colloid and Interface Science*, **179**, 298.
Crocker, J.C., Matteo, J.A., Dinsmore, A.D. and Yodh, A.G. (1999). Entropic Attraction and Repulsion in Binary Colloids Probed with a Line Optical Tweezer. *Physical Review Letters*, **82**, 4352.

Crocker, J.C., Valentine, M.T., Weeks, E.R., Gisler, T., Kaplan, P.D., Yodh, A.G. and Weitz, D.A. (2000). Two-point Microrheology of Inhomogeneous Soft Materials. *Physical Review Letters*, **85**, 888.

Curtis, J.E., Koss, B.A. and Grier, D.G. (2002). Dynamic Holographic Optical Tweezers. *Optics Communications*, **207**, 169.

Dao, M., Lim, C.T. and Suresh, S. (2005). Mechanics of the Human Red Blood Cell Deformed by Optical Tweezers. *Journal of the Mechanics and Physics of Solids*, **53**, 493.

Davis, J.A., Carcole, E. and Cottrell, D.M. (1996a). Intensity and Phase Measurements of Nondiffracting Beams Generated with a Magneto-Optic Spatial Light Modulator. *Applied Optics*, **35**, 593.

Davis, J.A., Carcole, E. and Cottrell, D.M. (1996b). Nondiffracting Interference Patterns Generated with Programmable Spatial Light Modulators. *Applied Optics*, **35**, 599.

Davis, K.B., Mewes, M.O., Andrews, M.R., Vandruten, N.J., Durfee, D.S., Kurn, D.M. and Ketterle, W. (1995). Bose–Einstein Condensation in a Gas of Sodium Atoms. *Physical Review Letters*, **75**, 3969.

Derenyi, I. and Vicsek, T. (1995). Cooperative Transport of Brownian Particles. *Physical Review Letters*, **75**, 374.

Dholakia, K. (2008). Optics: Against the Spread of the Light. *Nature*, **451**, 413.

Dholakia, K., Lee, W.M., Paterson, L., MacDonald, M.P., McDonald, R., Andreev, I., Mthunzi, P., Brown, C.T.A., Marchington, R.F. and Riches, A.C. (2007a). Optical Separation of Cells on Potential Energy Landscapes: Enhancement with Dielectric Tagging. *IEEE Journal of Selected Topics in Quantum Electronics*, **13**, 1646.

Dholakia, K., Little, H., Brown, C.T.A., Agate, B., McGloin, D., Paterson, L. and Sibbett, W. (2004). Imaging in Optical Micromanipulation Using Two-Photon Excitation. *New Journal of Physics*, **6**, 136.

Dholakia, K., MacDonald, M.P., Zemanek, P. and Cizmar, T. (2007b). *Laser Manipulation of Cells and Tissues*. p. 467.

Dholakia, K. and Reece, P. (2006). Optical Micromanipulation Takes Hold. *Nano Today*, **1**, 18.

Dholakia, K., Reece, P. and Gu, M. (2008). Optical Micromanipulation. *Chem. Soc. Rev*, **37**, 42.

Di Leonardo, R., Keen, S., Leach, J., Saunter, C.D., Love, G.D., Ruocco, G. and Padgett, M.J. (2007). Eigenmodes of a Hydrodynamically Coupled Micron-Size Multiple-Particle Ring. *Physical Review E (Statistical, Nonlinear, and Soft Matter Physics)*, **76**, 061402.

Di Leonardo, R., Leach, J., Mushfique, H., Cooper, J.M., Ruocco, G. and Padgett, M.J. (2006). Multipoint Holographic Optical Velocimetry in Microfluidic Systems. *Physical Review Letters*, **96**.

Dienerowitz, M., Mazilu, M., Reece, P.J., Krauss, T.F. and Dholakia, K. (2008). Optical Vortex Trap for Resonant Confinement of Metal Nanoparticles. *Optics Express*, **16**, 4991.

Dillen, T.V., Blaaderen, A.V. and Polman, A. (2004). Shaping Colloidal Assemblies. *Materials Today*, **7**, 40.

Dufresne, E.R., Spalding, G.C., Dearing, M.T., Sheets, S.A. and Grier, D.G. (2001). Computer-Generated Holographic Optical Tweezer Arrays. *Review of Scientific Instruments*, **72**, 1810.

Durnin, J., Miceli, J.J. and Eberly, J.H. (1987). Diffraction-Free Beams. *Physical Review Letters*, **58**, 1499.

Durnin, J., Miceli, J.J. and Eberly, J.H. (1988). Comparison of Bessel and Gaussian Beams. *Optics Letters*, **13**, 79.

Ehrlicher, A., Betz, T., Stuhrmann, B., Koch, D., Milner, V., Raizen, M.G. and Kas, J. (2002). Guiding Neuronal Growth with Light. *Proceedings of the National Academy of Sciences of the United States of America*, **99**, 16024.

Eriksen, R.L., Mogensen, P.C. and Gluckstad, J. (2002). Multiple-Beam Optical Tweezers Generated by the Generalized Phase-Contrast Method. *Optics Letters*, **27**, 267.

Fallman, E. and Axner, O. (1997). Design for Fully Steerable Dual-Trap Optical Tweezers. *Applied Optics*, **36**, 2107.

Faucheux, L.P., Bourdieu, L.S., Kaplan, P.D. and Libchaber, A.J. (1995a). Optical Thermal Ratchet. *Physical Review Letters*, **74**, 1504.

Faucheux, L.P., Stolovitzky, G. and Libchaber, A. (1995b). Periodic Forcing of a Brownian Particle. *Physical Review E*, **51**, 5239.
Finer, J.T., Mehta, A.D. and Spudich, J.A. (1995). Characterization of Single Actin–Myosin Interactions. *Biophysical Journal*, **68**, S291.
Fischer, P., Brown, C.T.A., Morris, J.E., Lopez-Mariscal, C., Wright, E.M., Sibbett, W. and Dholakia, K. (2005). White Light Propagation Invariant Beams. *Optics Express*, **13**, 6657.
Flynn, R.A., Birkbeck, A.L., Gross, M., Ozkan, M., Shao, B., Wang, M.M. and Esener, S.C. (2002). Parallel Transport of Biological Cells Using Individually Addressable VCSEL Arrays as Optical Tweezers. *Sensors and Actuators B-Chemical*, **87**, 239.
Fournier, J.M.R., Burns, M.M. and Golovchenko, J.A. (1995). Writing Diffractive Structures by Optical Trapping. *Proc. SPIE*, **2406**, 101.
Franze, K., Grosche, J., Skatchkov, S.N., Schinkinger, S., Foja, C., Schild, D., Uckermann, O., Travis, K., Reichenbach, A. and Guck, J. (2007). From the Cover: Muller Cells are Living Optical Fibers in the Vertebrate Retina. *Proceedings of the National Academy of Sciences*, **104**, 8287.
Friese, M.E.J., Enger, J., RubinszteinDunlop, H. and Heckenberg, N.R. (1996). Optical Angular-Momentum Transfer to Trapped Absorbing Particles. *Physical Review A*, **54**, 1593.
Friese, M.E.J., Nieminen, T.A., Heckenberg, N.R. and Rubinsztein-Dunlop, H. (1998). Optical Alignment and Spinning of Laser-Trapped Microscopic Particles. *Nature*, **394**, 348.
Friese, M.E.J., Rubinsztein-Dunlop, H., Gold, J., Hagberg, P. and Hanstorp, D. (2001). Optically Driven Micromachine Elements. *Applied Physics Letters*, **78**, 547.
Gabor, D. (1951). Microscopy by Reconstructed Wave Fronts: Ii. *Proceedings of the Physical Society. Section B*, **64**, 449.
Gahagan, K.T. and Swartzlander, G.A. (1999). Simultaneous Trapping of Low-Index and High-Index Microparticles Observed with an Optical-Vortex Trap. *Journal of the Optical Society of America B-Optical Physics*, **16**, 533.
Galajda, P. and Ormos, P. (2001). Complex Micromachines Produced and Driven by Light. *Applied Physics Letters*, **78**, 249.
Garces-Chavez, V., Dholakia, K. and Spalding, G.C. (2005). Extended-Area Optically Induced Organization of Microparticles on a Surface. *Applied Physics Letters*, **86**, 031106.
Garces-Chavez, V., McGloin, D., Melville, H., Sibbett, W. and Dholakia, K. (2002a). Simultaneous Micromanipulation in Multiple Planes Using a Self-Reconstructing Light Beam. *Nature*, **419**, 145.
Garces-Chavez, V., McGloin, D., Padgett, M.J., Dultz, W., Schmitzer, H. and Dholakia, K. (2003). Observation of the Transfer of the Local Angular Momentum Density of a Multiringed Light Beam to an Optically Trapped Particle. *Physical Review Letters*, **91**, 093602.
Garces-Chavez, V., Quidant, R., Reece, P.J., Badenes, G., Torner, L. and Dholakia, K. (2006). Extended Organization of Colloidal Microparticles by Surface Plasmon Polariton Excitation. *Physical Review B*, **73**, 085417.
Garces-Chavez, V., Volke-Sepulveda, K., Chavez-Cerda, S., Sibbett, W. and Dholakia, K. (2002b). Transfer of Orbital Angular Momentum to an Optically Trapped Low-Index Particle. *Physical Review A*, **66**, 063402.
Grandbois, M., Beyer, M., Rief, M., Clausen-Schaumann, H. and Gaub, H.E. (1999). How Strong is a Covalent Bond? *Science*, **283**, 1727.
Greenleaf, W.J., Woodside, M.T., Abbondanzieri, E.A. and Block, S.M. (2005). Passive All-Optical Force Clamp for High-Resolution Laser Trapping. *Physical Review Letters*, **95**, 208102.
Greenleaf, W.J., Woodside, M.T. and Block, S.M. (2007). High-Resolution, Single-Molecule Measurements of Biomolecular Motion. *Annual Review of Biophysics and Biomolecular Structure*, **36**, 171.
Grzegorczyk, T.M., Kemp, B.A. and Kong, J.A. (2006). Stable Optical Trapping Based on Optical Binding Forces. *Physical Review Letters*, **96**, 113903.
Grier, D.G. (2003). A Revolution in Optical Manipulation. *Nature*, **424**, 810.
Grier, D.G. and Roichman, Y. (2006). Holographic Optical Trapping. *Applied Optics*, **45**, 880.

Grimm, R., Weidemuller, M. and Ovchinnikov, Y.B. (2000). *Advances in Atomic Molecular, and Optical Physics*, Vol. 42. p. 95.

Guck, J., Ananthakrishnan, R., Cunningham, C.C. and Kas, J. (2002). Stretching Biological Cells with Light. *Journal of Physics-Condensed Matter*, **14**, 4843.

Guck, J., Ananthakrishnan, R., Mahmood, H., Moon, T.J., Cunningham, C.C. and Kas, J. (2001). The Optical Stretcher: A Novel Laser Tool to Micromanipulate Cells. *Biophysical Journal*, **81**, 767.

Guck, J., Ananthakrishnan, R., Moon, T.J., Cunningham, C.C. and Kas, J. (2000). Optical Deformability of Soft Biological Dielectrics. *Physical Review Letters*, **84**, 5451.

Guck, J., Schinkinger, S., Lincoln, B., Wottawah, F., Ebert, S., Romeyke, M., Lenz, D., Erickson, H.M., Ananthakrishnan, R., Mitchell, D., Kas, J., Ulvick, S. and Bilby, C. (2005). Optical Deformability as an Inherent Cell Marker for Testing Malignant Transformation and Metastatic Competence. *Biophysical Journal*, **88**, 3689.

Hale, G.M. and Querry, M.R. (1973). Optical Constants of Water in 200 nm to 200 μm Wavelength Region. *Applied Optics*, **12**, 555.

Hansen, P.M., Bhatia, V.K., Harrit, N. and Oddershede, L. (2005). Expanding the Optical Trapping Range of Gold Nanoparticles. *Nano Letters*, **5**, 1937.

Harada, T. and Yoshikawa, K. (2004). Fluctuation-Response Relation in a Rocking Ratchet. *Physical Review E*, **69**.

Harada, Y. and Asakura, T. (1996). Radiation Forces on a Dielectric Sphere in the Rayleigh Scattering Regime. *Optics Communications*, **124**, 529.

He, H., Friese, M.E.J., Heckenberg, N.R. and Rubinszteindunlop, H. (1995a). Direct Observation of Transfer of Angular-Momentum to Absorptive Particles from a Laser-Beam with a Phase Singularity. *Physical Review Letters*, **75**, 826.

He, H., Heckenberg, N.R. and Rubinszteindunlop, H. (1995b). Optical-Particle Trapping with Higher-Order Doughnut Beams Produced Using High-Efficiency Computer-Generated Holograms. *Journal of Modern Optics*, **42**, 217.

Hertlein, C., Helden, L., Gambassi, A., Dietrich, S. and Bechinger, C. (2008). Direct Measurement of Critical Casimir Forces. *Nature*, **451**, 172.

Hossack, W.J., Theofanidou, E., Crain, J., Heggarty, K. and Birch, M. (2003). High-Speed Holographic Optical Tweezers Using a Ferroelectric Liquid Crystal Microdisplay. *Optics Express*, **11**, 2053.

Howard, R.P. (2007). Fluorescence Microscopy: Established and Emerging Methods, Experimental Strategies, and Applications in Immunology. *Microscopy Research and Technique*, **70**, 687.

Huisstede, J.H.G., van der Werf, K.O., Bennink, M.L. and Subramaniam, V. (2005). Force Detection in Optical Tweezers Using Backscattered Light. *Optics Express*, **13**, 1113.

Indebetouw, G. (1989). Nondiffracting Optical-Fields — Some Remarks on their Analysis and Synthesis. *Journal of the Optical Society of America a-Optics Image Science and Vision*, **6**, 150.

Jarzynski, C. (1997). Nonequilibrium Equality for Free Energy Differences. *Physical Review Letters*, **78**, 2690.

Jesacher, A., Furhapter, S., Bernet, S. and Ritsch-Marte, M. (2004). Diffractive Optical Tweezers in the Fresnel Regime. *Optics Express*, **12**, 2243.

Jesacher, A., Furhapter, S., Maurer, C., Bernet, S. and Ritsch-Marte, M. (2006). Holographic Optical Tweezers for Object Manipulations at an Air–Liquid Surface. *Optics Express*, **14**, 6342.

Jess, P.R.T., Garces-Chavez, V., Smith, D., Mazilu, M., Paterson, L., Riches, A., Herrington, C.S., Sibbett, W. and Dholakia, K. (2006). Dual Beam Fibre Trap for Raman Microspectroscopy of Single Cells. *Optics Express*, **14**, 5779.

Jordan, P., Clare, H., Flendrig, L., Leach, J., Cooper, J. and Padgett, M. (2004). Permanent 3d Microstructures in a Polymeric Host Created Using Holographic Optical Tweezers. *Journal of Modern Optics*, **51**, 627.

Jordan, P., Leach, J., Padgett, M., Blackburn, P., Isaacs, N., Goksor, M., Hanstorp, D., Wright, A., Girkin, J. and Cooper, J. (2005). Creating Permanent 3d Arrangements of Isolated Cells Using Holographic Optical Tweezers. *Lab On A Chip*, **5**, 1224.

Karasek, V., Dholakia, K. and Zemanek, P. (2006). Analysis of Optical Binding in one Dimension. *Applied Physics B-Lasers And Optics*, **84**, 149.

Keen, S., Leach, J., Gibson, G. and Padgett, M.J. (2007). Comparison of a High-Speed Camera and a Quadrant Detector for Measuring Displacements in Optical Tweezers. *Journal of Optics a-Pure and Applied Optics*, **9**, S264.

Kelemen, L., Valkai, S. and Ormos, P. (2007). Parallel Photopolymerisation with Complex Light Patterns Generated by Diffractive Optical Elements. *Optics Express*, **15**, 14488.

Kogelnik, H. and Li, T. (1966). Laser Beams and Resonators. *Applied Optics*, **5**, 1550.

Konig, K., Liang, H., Berns, M.W. and Tromberg, B.J. (1996). Cell Damage in Near-Infrared Multimode Optical Traps as a Result of Multiphoton Absorption. *Optics Letters*, **21**, 1090.

Korda, P.T., Spalding, G.C. and Grier, D.G. (2002a). Evolution of a Colloidal Critical State in an Optical Pinning Potential Landscape. *Physical Review B*, **66**, 024504.

Korda, P.T., Taylor, M.B. and Grier, D.G. (2002b). Kinetically Locked-In Colloidal Transport in an Array of Optical Tweezers. *Physical Review Letters*, **89**.

Kress, H., Stelzer, E.H.K., Griffiths, G. and Rohrbach, A. (2005). Control of Relative Radiation Pressure in Optical Traps: Application to Phagocytic Membrane Binding Studies. *Physical Review E (Statistical, Nonlinear, and Soft Matter Physics)*, **71**, 061927.

Kress, H., Stelzer, E.H.K., Holzer, D., Buss, F., Griffiths, G. and Rohrbach, A. (2007). Filopodia Act as Phagocytic Tentacles and Pull with Discrete Steps and a Load-Dependent Velocity. *Proceedings of the National Academy of Sciences*, **104**, 11633.

Kuga, T., Torii, Y., Shiokawa, N., Hirano, T., Shimizu, Y. and Sasada, H. (1997). Novel Optical Trap of Atoms with a Doughnut Beam. *Physical Review Letters*, **78**, 4713.

La Porta, A. and Wang, M.D. (2004). Optical Torque Wrench: Angular Trapping, Rotation, and Torque Detection of Quartz Microparticles. *Physical Review Letters*, **92**, 190801.

Lacasta, A.M., Sancho, J.M., Romero, A.H. and Lindenberg, K. (2005). Sorting on Periodic Surfaces. *Physical Review Letters*, **94**.

Ladavac, K. and Grier, D.G. (2004). Microoptomechanical Pumps Assembled and Driven by Holographic Optical Vortex Arrays. *Optics Express*, **12**, 1144.

Ladavac, K., Kasza, K. and Grier, D.G. (2004). Sorting Mesoscopic Objects with Periodic Potential Landscapes: Optical Fractionation. *Physical Review E*, **70**, 010901.

Lang, M.J., Fordyce, P.M., Engh, A.M., Neuman, K.C. and Block, S.M. (2004). Simultaneous, Coincident Optical Trapping and Single-Molecule Fluorescence. *Nature Methods*, **1**, 133.

Leach, J., Gibson, G.M., Padgett, M.J., Esposito, E., McConnell, G., Wright, A.J. and Girkin, J.M. (2006a). Generation of Achromatic Bessel Beams Using a Compensated Spatial Light Modulator. *Optics Express*, **14**, 5581.

Leach, J., Mushfique, H., di Leonardo, R., Padgett, M. and Cooper, J. (2006b). An Optically Driven Pump for Microfluidics. *Lab on a Chip*, **6**, 735.

Leach, J., Sinclair, G., Jordan, P., Courtial, J., Padgett, M.J., Cooper, J. and Laczik, Z.J. (2004). 3d Manipulation of Particles into Crystal Structures Using Holographic Optical Tweezers. *Optics Express*, **12**, 220.

Lee, S.H. and Grier, D.G. (2005). Robustness of Holographic Optical Traps Against Phase Scaling Errors. *Optics Express*, **13**, 7458.

Lee, W.M., Ahluwalia, B.P.S., Yuan, X.C., Cheong, W.C. and Dholakia, K. (2005). Optical Steering of High and Low Index Microparticles by Manipulating an Off-Axis Optical Vortex. *Journal of Optics a-Pure and Applied Optics*, **7**, 1.

Lee, W.M., Reece, P.J., Marchington, R.F., Metzger, N.K. and Dholakia, K. (2007). Construction and Calibration of an Optical Trap on a Fluorescence Optical Microscope. *Nat. Protocols*, **2**, 3226.

Lee, W.M., Yuan, X.C. and Cheong, W.C. (2004). Optical Vortex Beam Shaping by use of Highly Efficient Irregular Spiral Phase Plates for Optical Micromanipulation. *Optics Letters*, **29**, 1796.

Li, P., Shi, K.B. and Liu, Z.W. (2005). Manipulation and Spectroscopy of a Single Particle by use of White-Light Optical Tweezers. *Optics Letters*, **30**, 156.

Liang, H., Wright, W.H., Cheng, S., He, W. and Berns, M.W. (1993). Micromanipulation of Chromosomes in ptk2 Cells Using Laser Microsurgery (Optical Scalpel) in Combination with Laser Induced Optical Forces (Optical Tweezers). *Experimental Cell Research*, **204**, 110.

Liesener, J., Reicherter, M., Haist, T. and Tiziani, H.J. (2000). Multi-Functional Optical Tweezers Using Computer-Generated Holograms. *Optics Communications*, **185**, 77.

Lin, Y., Seka, W., Eberly, J.H., Huang, H. and Brown, D.L. (1992). Experimental Investigation of Bessel Beam Characteristics. *Applied Optics*, **31**, 2708.

Lincoln, B., Erickson, H.M., Schinkinger, S., Wottawah, F., Mitchell, D., Ulvick, S., Bilby, C. and Guck, J. (2004). Deformability-Based Flow Cytometry. *Cytometry Part A*, **59A**, 203.

Lincoln, B., Schinkinger, S., Travis, K., Wottawah, F., Ebert, S., Sauer, F. and Guck, J. (2007). Reconfigurable Microfluidic Integration of a Dual-Beam Laser Trap with Biomedical Applications. *Biomedical Microdevices*, **9**, 703.

Liphardt, J., Dumont, S., Smith, S.B., Tinoco Jr., I. and Bustamante, C. (2002). Equilibrium Information from Nonequilibrium Measurements in an Experimental Test of Jarzynski's Equality. *Science*, **296**, 1832.

Lopez-Mariscal, C., Gutierrez-Vega, J.C., Milne, G. and Dholakia, K. (2006). Orbital Angular Momentum Transfer in Helical Mathieu Beams. *Optics Express*, **14**, 4182.

Lorenz, R.M., Edgar, J.S., Jeffries, G.D.M., Zhao, Y.Q., McGloin, D. and Chiu, D.T. (2007). Vortex-Trap-Induced Fusion of Femtoliter-Volume Aqueous Droplets. *Analytical Chemistry*, **79**, 224.

Lutz, C., Kollmann, M. and Bechinger, C. (2004a). Single-File Diffusion of Colloids in One-Dimensional Channels. *Physical Review Letters*, **93**, 026001.

Lutz, C., Kollmann, M., Leiderer, P. and Bechinger, C. (2004b). Diffusion of Colloids in One-Dimensional Light Channels. *Journal of Physics-Condensed Matter*, **16**, S4075.

Lutz, C., Reichert, M., Stark, H. and Bechinger, C. (2006). Surmounting Barriers: The Benefit of Hydrodynamic Interactions. *Europhysics Letters*, **74**, 719.

MacDonald, M.P., Neale, S., Paterson, L., Richies, A., Dholakia, K. and Spalding, G.C. (2004). Cell Cytometry with a Light Touch: Sorting Microscopic Matter with an Optical Lattice. *Journal of Biological Regulators and Homeostatic Agents*, **18**, 200.

MacDonald, M.P., Paterson, L., Sibbett, W., Dholakia, K. and Bryant, P.E. (2001). Trapping and Manipulation of Low-Index Particles in a Two-Dimensional Interferometric Optical Trap. *Optics Letters*, **26**, 863.

MacDonald, M.P., Paterson, L., Volke-Sepulveda, K., Arlt, J., Sibbett, W. and Dholakia, K. (2002). Creation and Manipulation of Three-Dimensional Optically Trapped Structures. *Science*, **296**, 1101.

MacDonald, M.P., Spalding, G.C. and Dholakia, K. (2003). Microfluidic Sorting in an Optical Lattice. *Nature*, **426**, 421.

Mair, A., Vaziri, A., Weihs, G. and Zeilinger, A. (2001). Entanglement of the Orbital Angular Momentum States of Photons. *Nature*, **412**, 313.

Malagnino, N., Pesce, G., Sasso, A. and Arimondo, E. (2002). Measurements of Trapping Efficiency and Stiffness in Optical Tweezers. *Optics Communications*, **214**, 15.

Martin-Badosa, E., Montes-Usategui, M., Carnicer, A., Andilla, J., Pleguezuelos, E. and Juvells, I. (2007). Design Strategies for Optimizing Holographic Optical Tweezers Set-Ups. *Journal of Optics A: Pure and Applied Optics*, **9**, S267.

McGloin, D., Burnham, D.R., Summers, M.D., Rudd, D., Dewar, N. and Anand, S. (2008). Optical Manipulation of Airborne Particles: Techniques and Applications. *Faraday Discussions*, **137**, 335.

McGloin, D. and Dholakia, K. (2005). Bessel Beams: Diffraction in a New Light. *Contemporary Physics*, **46**, 15.

McLeod, J.H. (1954). The Axicon: A New Type of Optical Element. *Journal of the Optical Society of America*, **44**, 592.

Mehta, A.D., Rief, M., Spudich, J.A., Smith, D.A. and Simmons, R.M. (1999). Single-Molecule Biomechanics with Optical Methods. *Science*, **283**, 1689.

Meiners, J.C. and Quake, S.R. (1999). Direct Measurement of Hydrodynamic Cross Correlations Between two Particles in an External Potential. *Physical Review Letters*, **82**, 2211.

Mellor, C.D. and Bain, C.D. (2006). Array Formation in Evanescent Waves. *Chemphyschem*, **7**, 329.

Melville, H., Milne, G.F., Spalding, G.C., Sibbett, W., Dholakia, K. and McGloin, D. (2003). Optical Trapping of Three-Dimensional Structures Using Dynamic Holograms. *Optics Express*, **11**, 3562.

Metzger, N.K., Dholakia, K. and Wright, E.M. (2006a). Observation of Bistability and Hysteresis in Optical Binding of Two Dielectric Spheres. *Physical Review Letters*, **96**.

Metzger, N.K., Dholakia, K. and Wright, E.M. (2006b). Observation of Bistability and Hysteresis in Optical Binding of Two Dielectric Spheres. *Physical Review Letters*, **96**, 068102.

Metzger, N.K., Marchington, R.F., Mazilu, M., Smith, R.L., Dholakia, K. and Wright, E.M. (2007a). Measurement of the Restoring Forces Acting on Two Optically Bound Particles from Normal Mode Correlations. *Physical Review Letters*, **98**.

Metzger, N.K., Marchington, R.F., Mazilu, M., Smith, R.L., Dholakia, K. and Wright, E.M. (2007b). Measurement of the Restoring Forces Acting on two Optically Bound Particles from Normal Mode Correlations. *Physical Review Letters*, **98**, 068102.

Metzger, N.K., Wright, E.M., Sibbett, W. and Dholakia, K. (2006c). Visualization of Optical Binding of Microparticles Using a Femtosecond Fiber Optical Trap. *Optics Express*, **14**, 3677.

Milne, G., Dholakia, K., McGloin, D., Volke-Sepulveda, K. and Zemanek, P. (2007a). Transverse Particle Dynamics in a Bessel Beam. *Optics Express*, **15**, 13972.

Milne, G., Rhodes, D., MacDonald, M. and Dholakia, K. (2007b). Fractionation of Polydisperse Colloid with Acousto-Optically Generated Potential Energy Landscapes. *Optics Letters*, **32**, 1144.

Mio, C., Gong, T., Terray, A. and Marr, D.W.M. (2000). Design of a Scanning Laser Optical Trap for Multiparticle Manipulation. *Review of Scientific Instruments*, **71**, 2196.

Mohanty, S.K., Mohanty, K.S. and Gupta, P.K. (2005). Dynamics of Interaction of rbc with Optical Tweezers. *Optics Express*, **13**, 4745.

Molloy, J.E., Burns, J.E., Kendrickjones, J., Tregear, R.T. and White, D.C.S. (1995). Movement and Force Produced by a Single Myosin Head. *Nature*, **378**, 209.

Molloy, J.E. and Padgett, M.J. (2002). Lights, Action: Optical Tweezers. *Contemporary Physics*, **43**, 241.

Neale, S.L., Macdonald, M.P., Dholakia, K. and Krauss, T.F. (2005). All-Optical Control of Microfluidic Components Using form Birefringence. *Nature Materials*, **4**, 530.

Nemet, B.A., Shabtai, Y. and Cronin-Golomb, M. (2002). Imaging Microscopic Viscosity with Confocal Scanning Optical Tweezers. *Optics Letters*, **27**, 264.

Neuman, K.C. and Block, S.M. (2004). Optical Trapping. *Review of Scientific Instruments*, **75**, 2787.

Neuman, K.C. and Nagy, A. (2008). Single-molecule Force Spectroscopy: Optical Tweezers, Magnetic Tweezers and Atomic Force Microscopy. *Nature Methods*, **5** (6), 491–505.

Neuman, K.C., Chadd, E.H., Liou, G.F., Bergman, K. and Block, S.M. (1999). Characterization of Photodamage to Escherichia Coli in Optical Traps. *Biophysical Journal*, **77**, 2856.

Ng, J., Lin, Z.F., Chan, C.T. and Sheng, P. (2005). Photonic Clusters Formed by Dielectric Microspheres: Numerical Simulations. *Physical Review B*, **72**, 085130.

O'Hara, K.M., Gehm, M.E., Granada, S.R., Bali, S. and Thomas, J.E. (2000). Stable, Strongly Attractive, Two-State Mixture of Lithium Fermions in an Optical Trap. *Physical Review Letters*, **85**, 2092.

O'Neil, A.T., MacVicar, I., Allen, L. and Padgett, M.J. (2002). Intrinsic and Extrinsic Nature of the Orbital Angular Momentum of a Light Beam. *Physical Review Letters*, **88**, 053601.

O'Neil, A.T. and Padgett, M.J. (2000). Three-Dimensional Optical Confinement of Micron-Sized Metal Particles and the Decoupling of the Spin and Orbital Angular Momentum within an Optical Spanner. *Optics Communications*, **185**, 139.

Oemrawsingh, S.S.R., Van Houwelingen, J.A.W., Eliel, E.R, Woerdman, J.P., Verstegen, E.J.K, Kloosterboer, G.W and t Hooft (2004). Production and Characterization of Spiral Phase Plates for Optical Wavelengths. *Appl. Opt.*, **43**, 688.

Okida, M.O.T., Itoh, M. and Yatagai, T. (2007). Direct Generation of High Power Laguerre–Gaussian Output from a Diode- Pumped nd:Yvo4 1.3-μm Bounce Laser. *Optics Express*, **15**, 7616.

Oroszi, L., Galajda, P., Kirei, H., Bottka, S. and Ormos, P. (2006). Direct Measurement of Torque in an Optical Trap and its Application to Double-Strand DNA. *Physical Review Letters*, **97**, 058301.

Parkin, S.J., Knoner, G., Nieminen, T.A., Heckenberg, N.R. and Rubinsztein-Dunlop, H. (2007). Picoliter Viscometry Using Optically Rotated Particles. *Physical Review E*, **76**.

Paterson, L., MacDonald, M.P., Arlt, J., Sibbett, W., Bryant, P.E. and Dholakia, K. (2001). Controlled Rotation of Optically Trapped Microscopic Particles. *Science*, **292**, 912.

Paterson, L., Papagiakoumou, E., Milne, G., Garces-Chavez, V., Briscoe, T., Sibbett, W., Dholakia, K. and Riches, A.C. (2007). Passive Optical Separation within a 'Nondiffracting' Light Beam. *Journal of Biomedical Optics*, **12**, 054017.

Paterson, L., Papagiakoumou, E., Milne, G., Garces-Chavez, V., Tatarkova, S.A., Sibbett, W., Gunn-Moore, F.J., Bryant, P.E., Riches, A.C. and Dholakia, K. (2005). Light-Induced Cell Separation in a Tailored Optical Landscape. *Applied Physics Letters*, **87**, 123901.

Perch-Nielsen, I.R., Rodrigo, P.J. and Gluckstad, J. (2005). Real-Time Interactive 3d Manipulation of Particles Viewed in Two Orthogonal Observation Planes. *Optics Express*, **13**, 2852.

Perkins, T.T., Quake, S.R., Smith, D.E. and Chu, S. (1994). Relaxation of a Single DNA Molecule Observed by Optical Microscopy. *Science*, **264**, 822.

Pesce, G., Sasso, A. and Fusco, S. (2005). Viscosity Measurements on Micron-Size Scale Using Optical Tweezers. *Review of Scientific Instruments*, **76**, 115105.

Phillips, W.D. (1998). Laser Cooling and Trapping of Neutral Atoms. *Reviews of Modern Physics*, **70**, 721.

Polin, M., Grier, D.G. and Quake, S.R. (2006). Anomalous Vibrational Dispersion in Holographically Trapped Colloidal Arrays. *Physical Review Letters*, **96**.

Polin, M., Ladavac, K., Lee, S.H., Roichman, Y. and Grier, D.G. (2005). Optimized Holographic Optical Traps. *Optics Express*, **13**, 5831.

Pontecorvo, G. (1963). The Leeuwenhoek Lecture: Microbial Genetics: Retrospect and Prospect. *Proceedings of the Royal Society of London. Series B, Biological Sciences (1934-1990)*, **158**, 1.

Prentice, P., Cuschierp, A., Dholakia, K., Prausnitz, M. and Campbell, P. (2005). Membrane Disruption by Optically Controlled Microbubble Cavitation. *Nature Physics*, **1**, 107.

Prentice, P.A., MacDonald, M.P., Frank, T.G., Cuschieri, A., Spalding, G.C., Sibbett, W., Campbell, P.A. and Dholakia, K. (2004). Manipulation and Filtration of Low Index Particles with Holographic Laguerre–Gaussian Optical Trap Arrays. *Optics Express*, **12**, 593.

Ramser, K., Enger, J., Goksor, M., Hanstorp, D., Logg, K. and Kall, M. (2005). A Microfluidic System Enabling Raman Measurements of the Oxygenation Cycle in Single Optically Trapped Red Blood Cells. *Lab on a Chip*, **5**, 431.

Reece, P.J., Garces-Chavez, V. and Dholakia, K. (2006). Near-Field Optical Micromanipulation with Cavity Enhanced Evanescent Waves. *Applied Physics Letters*, **88**, 221116.

Reihani, S.N.S. and Oddershede, L.B. (2007). Optimizing Immersion Media Refractive Index Improves Optical Trapping by Compensating Spherical Aberrations. *Optics Letters*, **32**, 1998.

Rhodes, D.P., Lancaster, G.P.T., Livesey, J., McGloin, D., Arlt, J. and Dholakia, K. (2002). Guiding a Cold Atomic Beam Along a Co-Propagating and Oblique Hollow Light Guide. *Optics Communications*, **214**, 247.

Ricardez-Vargas, I., Rodriguez-Montero, P., Ramos-Garcia, R. and Volke-Sepulveda, K. (2006). Modulated Optical Sieve for Sorting of Polydisperse Microparticles. *Applied Physics Letters*, **88**, 121116.

Righini, M., Zelenina, A.S., Girard, C. and Quidant, R. (2007). Parallel and Selective Trapping in a Patterned Plasmonic Landscape. *Nature Physics*, **3**, 477.

Rodrigo, P.J., Eriksen, R.L., Daria, V.R. and Gluckstad, J. (2002). Interactive Light-Driven and Parallel Manipulation of Inhomogeneous Particles. *Optics Express*, **10**, 1550.

Rohrbach, A. (2005). Stiffness of Optical Traps: Quantitative Agreement Between Experiment and Electromagnetic Theory. *Physical Review Letters*, **95**, 168102.

Rohrbach, A. and Stelzer, E.H.K. (2002). Trapping Forces, Force Constants, and Potential Depths for Dielectric Spheres in the Presence of Spherical Aberrations. *Applied Optics*, **41**, 2494.

Rohrbach, A., Tischer, C., Neumayer, D., Florin, E.L. and Stelzer, E.H.K. (2004). Trapping and Tracking a Local Probe with a Photonic Force Microscope. *Review of Scientific Instruments*, **75**, 2197.

Roichman, Y. and Grier, D.G. (2005). Holographic Assembly of Quasicrystalline Photonic Heterostructures. *Optics Express*, **13**, 5434.

Roichman, Y., Grier, D.G. and Zaslavsky, G. (2007). Anomalous Collective Dynamics in Optically Driven Colloidal Rings. *Physical Review E*, **75**, 020401.

Sasaki, K., Koshioka, M., Misawa, H., Kitamura, N. and Masuhara, H. (1991). Pattern-Formation and Flow-Control of Fine Particles by Laser-Scanning Micromanipulation. *Optics Letters*, **16**, 1463.

Sasaki, K., Koshioka, M., Misawa, H., Kitamura, N. and Masuhara, H. (1992). Optical Trapping of a Metal-Particle and a Water Droplet by a Scanning Laser-Beam. *Applied Physics Letters*, **60**, 807.

Sato, S., Ishigure, M. and Inaba, H. (1991). Optical Trapping and Rotational Manipulation of Microscopic Particles and Biological Cells Using Higher-Order Mode Nd-Yag Laser-Beams. *Electronics Letters*, **27**, 1831.

Schiffer, M., Rauner, M., Kuppens, S., Zinner, M., Sengstock, K. and Ertmer, W. (1998). Guiding, Focusing, and Cooling of Atoms in a Strong Dipole Potential. *Applied Physics B-Lasers and Optics*, **67**, 705.

Schmid, S., Thalhammer, G., Winkler, K., Lang, F. and Denschlag, J.H. (2006). Long Distance Transport of Ultracold Atoms Using a 1d Optical Lattice. *New Journal of Physics*, **8**, 159.

Schmitz, C.H.J., Spatz, J.P. and Curtis, J.E. (2005). High-Precision Steering of Multiple Holographic Optical Traps. *Optics Express*, **13**, 8678.

Seeger, S., Monajembashi, S., Hutter, K.J., Futterman, G., Wolfrum, J. and Greulich, K.O. (1991). Application of Laser Optical Tweezers in Immunology and Molecular-Genetics. *Cytometry*, **12**, 497.

Seol, Y., Carpenter, A.E. and Perkins, T.T. (2006). Gold Nanoparticles: Enhanced Optical Trapping and Sensitivity Coupled with Significant Heating. *Optics Letters*, **31**, 2429.

Shaevitz, J.W., Abbondanzieri, E.A., Landick, R. and Block, S.M. (2003). Backtracking by Single RNA Polymerase Molecules Observed at Near-Base-Pair Resolution. *Nature*, **426**, 684.

Shao, B., Zlatanovic, S., Ozkan, M., Birkbeck, A.L. and Esener, S.C. (2006). Manipulation of Microspheres and Biological Cells with Multiple Agile vcsel Traps. *Sensors and Actuators B-Chemical*, **113**, 866.

Siegman, A.E. (1998). How to (Maybe) Measure Laser Beam Quality. *OSA TOPS*, **17**, 184.

Simpson, N.B., Allen, L. and Padgett, M.J. (1996). Optical Tweezers and Optical Spanners with Laguerre–Gaussian Modes. *Journal of Modern Optics*, **43**, 2485.

Simpson, N.B., Dholakia, K., Allen, L. and Padgett, M.J. (1997). Mechanical Equivalence of Spin and Orbital Angular Momentum of Light: An Optical Spanner. *Optics Letters*, **22**, 52.

Simpson, N.B., McGloin, D., Dholakia, K., Allen, L. and Padgett, M.J. (1998). Optical Tweezers with Increased Axial Trapping Efficiency. *Journal of Modern Optics*, **45**, 1943.

Sinclair, G., Jordan, P., Leach, J., Padgett, M.J. and Cooper, J. (2004a). Defining the Trapping Limits of Holographical Optical Tweezers. *Journal of Modern Optics*, **51**, 409.

Sinclair, G., Leach, J., Jordan, P., Gibson, G., Yao, E., Laczik, Z.J., Padgett, M.J. and Courtial, J. (2004b). Interactive Application in Holographic Optical Tweezers of a Multi-Plane Gerchberg-Saxton Algorithm for Three-Dimensional Light Shaping. *Optics Express*, **12**, 1665.

Singer, W., Frick, M., Bernet, S. and Ritsch-Marte, M. (2003). Self-Organized Array of Regularly Spaced Microbeads in a Fiber-Optical Trap. *Journal of the Optical Society of America B-Optical Physics*, **20**, 1568.

Siviloglou, G.A. and Christodoulides, D.N. (2007). Accelerating Finite Energy Airy Beams. *Optics Letters*, **32**, 979.

Smith, S.B., Cui, Y.J. and Bustamante, C. (2003). Optical-Trap Force Transducer that Operates by Direct Measurement of Light Momentum. in: Marriott, G. and Parker, I. (Eds.), *Methods in Enzymology, 361, Biophotonics, Part B*. Academic Press, San Diego, California, p. 134.

Squires, T.M. and Quake, S.R. (2005). Microfluidics: Fluid Physics at the Nanoliter Scale. *Reviews of Modern Physics*, **77**, 977.

Stevenson, D.J., Lake, T.K., Agate, B., Gárcés-Chávez, V., Dholakia, K. and Gunn-Moore, F. (2006). Optically Guided Neuronal Growth at Near Infrared Wavelengths. *Optics Express*, **14**, 9786.

Svoboda, K. and Block, S.M. (1994a). Biological Applications of Optical Forces. *Annual Review of Biophysics and Biomolecular Structure*, **23**, 247.

Svoboda, K. and Block, S.M. (1994b). Optical Trapping of Metallic Rayleigh Particles. *Optics Letters*, **19**, 930.

Tanaka, T. and Yamamoto, S. (2000). Comparison of Aberration between Axicon and Lens. *Optics Communications*, **184**, 113.

Tatarkova, S.A., Carruthers, A.E. and Dholakia, K. (2002). One-Dimensional Optically Bound Arrays of Microscopic Particles. *Physical Review Letters*, **89**, 283901.

Tatarkova, S.A., Sibbett, W. and Dholakia, K. (2003). Brownian Particle in an Optical Potential of the Washboard Type. *Physical Review Letters*, **91**, 038101.

Terray, A., Oakey, J. and Marr, D.W.M. (2002). Microfluidic Control Using Colloidal Devices. *Science*, **296**, 1841.

Thaning, A., Jaroszewicz, Z. and Friberg, A.T. (2003). Diffractive Axicons in Oblique Illumination: Analysis and Experiments and Comparison with Elliptical Axicons. *Applied Optics*, **42**, 9.

Theofanidou, E., Wilson, L., Hossack, W.J. and Arlt, J. (2004). Spherical Aberration Correction for Optical Tweezers. *Optics Communications*, **236**, 145.

Townes Anderson, E., St Jules, R.S., Sherry, D.M., Lichtenberger, J. and Hassanain, M. (1997). Micromanipulation of Retinal Cells by Optical Tweezers. *Investigative Ophthalmology & Visual Science*, **38**, 130.

Tricoles, G. (1987). Computer Generated Holograms: An Historical Review. *Appl. Opt.*, **26**, 4351.

Tsampoula, X., Garces-Chavez, V., Comrie, M., Stevenson, D.J., Agate, B., Brown, C.T.A., Gunn-Moore, F. and Dholakia, K. (2007). Femtosecond Cellular Transfection Using a Nondiffracting Light Beam. *Applied Physics Letters*, **91**, 053902.

Valentine, M.T., Guydosh, N.R., Gutiérrez-Medina, B., Fehr, A.N., Andreasson, J.O. and Block, S.M. (2008). Precision Steering of an Optical Trap by Electro-Optic Deflection. *Opt. Lett.*, **33**, 599.

Vasara, A., Turunen, J. and Friberg, A.T. (1989). Realization of General Nondiffracting Beams with Computer-Generated Holograms. *Journal of the Optical Society of America a-Optics Image Science and Vision*, **6**, 1748.

Vermeulen, K.C., van Mameren, J., Wuite, G.J.L. and Schmidt, C.F. (2005). Dependence of Optical Trap Stiffness on Focusing Depth in the Presence and Absence of Spherical Aberrations. *Biophysical Journal*, **88**, 663A.

Visscher, K., Brakenhoff, G.J. and Krol, J.J. (1993). Micromanipulation by Multiple Optical Traps Created by a Single Fast Scanning Trap Integrated with the Bilateral Confocal Scanning Laser Microscope. *Cytometry*, **14**, 105.

Visscher, K., Gross, S.P. and Block, S.M. (1996). Construction of Multiple-Beam Optical Traps With Nanometer-Resolution Position Sensing. *IEEE Journal of Selected Topics in Quantum Electronics*, **2**, 1066.

Visscher, K., Schnitzer, M.J. and Block, S.M. (1999). Single Kinesin Molecules Studied with a Molecular Force Clamp. *Nature*, **400**, 184.

Volke-Sepulveda, K., Chavez-Cerda, S., Garces-Chavez, V. and Dholakia, K. (2004). Three-Dimensional Optical Forces and Transfer of Orbital Angular Momentum from Multiringed Light Beams to Spherical Microparticles. *Journal of the Optical Society of America B-Optical Physics*, **21**, 1749.

Volke-Sepulveda, K., Garces-Chavez, V., Chavez-Cerda, S., Arlt, J. and Dholakia, K. (2002). Orbital Angular Momentum of a High-Order Bessel Light Beam. *Journal of Optics B-Quantum and Semiclassical Optics*, **4**, S82.

Volpe, G., Kozyreff, G. and Petrov, D. (2007a). Backscattering Position Detection for Photonic Force Microscopy. *Journal of Applied Physics*, **102**, 084701.

Volpe, G., Volpe, G. and Petrov, D. (2007b). Brownian Motion in a Nonhomogeneous Force Field and Photonic Force Microscope. *Physical Review E (Statistical, Nonlinear, and Soft Matter Physics)*, **76**, 061118.

Vossen, D.L.J., van der Horst, A., Dogterom, M. and van Blaaderen, A. (2004). Optical Tweezers and Confocal Microscopy for Simultaneous Three-Dimensional Manipulation and Imaging in Concentrated Colloidal Dispersions. *Review of Scientific Instruments*, **75**, 2960.

Wallin, A.E., Ojala, H., Korsback, A., Haeggstrom, E. and Tuma, R. (2007). Real-Time Control of Optical Tweezers. in: Dholakia, K. and Spalding, G.C. (Eds.), *Optical Trapping and Optical Micromanipulation iv*. p. 66441Y.

Wang, M.M., Tu, E., Raymond, D.E., Yang, J.M., Zhang, H.C., Hagen, N., Dees, B., Mercer, E.M., Forster, A.H., Kariv, I., Marchand, P.J. and Butler, W.F. (2005). Microfluidic Sorting of Mammalian Cells by Optical Force Switching. *Nature Biotechnology*, **23**, 83.

Watson, J.D. and Crick, F.H.C. (1953). A Structure for Deoxyribose Nucleic Acid. *Nature*, **171**, 737.

Whyte, G. and Courtial, J. (2005). Experimental Demonstration of Holographic Three-Dimensional Light Shaping Using a Gerchberg-Saxton Algorithm. *New Journal of Physics*, **7**.

Wright, E.M., Arlt, J. and Dholakia, K. (2000). Toroidal Optical Dipole Traps for Atomic Bose–Einstein Condensates Using Laguerre–Gaussian Beams. *Physical Review A*, **63**, 013608.

Wuite, G.J.L., Smith, S.B., Young, M., Keller, D. and Bustamante, C. (2000). Single-Molecule Studies of the Effect of Template Tension on t7 DNA Polymerase Activity. *Nature*, **404**, 103.

Wulff, K.D., Cole, D.G., Clark, R.L., DiLeonardo, R., Leach, J., Cooper, J., Gibson, G. and Padgett, M.J. (2006). Aberration Correction in Holographic Optical Tweezers. *Optics Express*, **14**, 4169.

Xie, C.G., Dinno, M.A. and Li, Y.Q. (2002). Near-Infrared Raman Spectroscopy of Single Optically Trapped Biological Cells. *Optics Letters*, **27**, 249.

Zernike, F. (1955). How I Discovered Phase Contrast. *Science*, **121**, 345.

Index

Anderson localization, 120, 127, 133, 143
antimatter, 14
antiparticles, 14
argon, 6, 7, 9, 13–15, 17–19, 21, 22, 24–27
atomic fragments, 51, 53, 83, 84, 86–88, 91, 93, 96, 98, 112, 113
atomic mixture, 121, 152
Aubry–André model, 137, 138, 145

beam shaping Techniques, 312
bichromatic lattice, 125, 137, 145, 148
biophysical techniques (research methods), 302
Bloch oscillations, 143
Bogoliubov spectrum, 142
Bose glass, 142, 147–149, 151, 152
Bose–Einstein condensates, 121, 128, 130
Bose–Hubbard Hamiltonian, 146

coherent backscattering, 128
coherent control, 95, 113
collisions, 8, 14, 27, 28, 33, 38
complex fluids and colloidal systems, 262
cross-sections, 2, 5, 6, 8–15, 18, 20, 22, 23, 25–30, 32–35

differential cross-sections, 1, 15, 17, 18, 21, 36
dipole blockade, 162, 200–205
dipole potential, 122
dipole–dipole, 161
dipole–dipole induced motion, 166
dipole–dipole interactions, 52
direct ionization, 1, 5, 11, 12, 14, 16
disordered potential, 121
dynamic Stark shift, 237

electron-capture-to-the-continuum, 36

electron-loss-to-the-continuum, 36
energy transfer, 161–167, 180, 189–192, 194–198, 207, 214
energy transfer collision, 162, 164, 215
exponential localization, 128, 137, 138

Feshbach resonance, 127, 130, 143
four wave mixing, 51, 53, 56, 113
Fourier optics, 289
fragmentation, 1, 30–32, 34–38
fragmented BEC, 133, 142
Franck–Condon region, 84, 102
frozen Rydberg gas, 165, 166, 189, 197, 198, 216

glassy states, 120, 130, 147, 148
Gross–Pitaevskii equation, 129, 139

Hanbury Brown & Twiss effect, 151
Harper model, 137
healing length, 137, 139
helium, 2, 9, 11, 13, 14, 19, 20, 24, 25, 28–30, 32, 33, 36–38
holographic optical elements, 270

idler wave, 57
incommensurate lattice, 121, 138, 145, 149
integrated cross-sections, 1, 6, 11, 21, 32
Ioffe–Regel criterion, 134
ionization, 1, 4–6, 8–10, 12–18, 20–22, 29, 31–36

krypton, 9, 13–15, 17, 19, 24, 26, 27

laser chirp, 62
Lifshitz states, 141
line broadening, 198, 199
localization length, 128
localization, effects of interactions, 136

localized state, 127, 139, 145
long range molecules, 163

many body dipole interactions, 53, 69, 70, 82, 112
many body interactions, 162, 189, 194, 195
mechanical effects of light on atoms and molecules, 224
metal–insulator transition, 119, 127
microtrap, 127, 142
mobility edge, 128
molecular autoionization, 166, 214, 215
molecular dissociation, 51, 53, 54, 82–86, 112, 113
Mott insulator, 121, 146, 149
multi-chromatic lattices, 121, 124
multiphoton ionization and excitation, 225
multiple ionization, 5, 10, 14, 19

neon, 8, 9, 13, 14, 24
noise correlations, 151

optical lattices, 121, 139, 145
optical potentials, 121
optical trapping, 262–265, 268–270, 273, 274, 281–283, 285, 291, 293, 300, 303, 304, 307, 308, 317, 318

PFWM, 53, 56, 57, 59, 61, 63, 64, 68–70, 73, 75, 77, 79, 83, 85, 86, 95, 96, 112, 113
phase matching, 63
photoassociation, 50, 52
positron, 1–8, 13, 15, 17–22, 26, 30, 31, 35–39
positron beam, 8, 9, 25
positronium, 1–4, 8, 17, 25–27, 29, 31, 32, 38, 39
positronium beam, 31
positronium formation, 1, 3, 9, 21, 23, 25, 32

PSWM, 53, 57, 63, 69, 70

quantum beating, 53
quantum scattering, 136
quasi-periodic potentials, 126

radiative collision, 182
Ramsey fringe, 196
random-field induced order, 143
Rb, 53
Rb_2, 53
Rydberg atom, 164, 199–202, 209, 212

scattering, 2–5, 7–9, 13, 14, 18, 19, 22, 26–30, 32–36, 38
scattering length, 129
signal wave, 53, 56–59, 63, 64, 68, 70, 83, 113
single-molecule techniques, 275–277, 301
speckle autocorrelation length, 137, 144
speckle patterns, 121, 122, 134
stationary phase approximation, 100, 102
strong localization, 128
strong-field interactions, 220
Stuckelburg oscillation, 187
superfluid-insulator transition, 120, 147

total ionization, 4–6, 9, 10, 15, 16
transfer ionization, 5, 21

ultracold plasma, 162, 209
ultrafast laser spectroscopy, 52
ultrafast x-rays, 222

wavepackets, 55–57, 64, 66, 68, 70, 71, 77, 83–86, 91, 94, 95, 98, 112, 113
weak localization, 128

x-ray photoionization, 220
xenon, 2, 6, 7, 9, 13–15, 24, 26, 28, 30, 32, 34–36, 38

CONTENTS OF VOLUMES IN THIS SERIAL

Volume 1

Molecular Orbital Theory of the Spin Properties of Conjugated Molecules, *G.G. Hall and A.T. Amos*

Electron Affinities of Atoms and Molecules, *B.L. Moiseiwitsch*

Atomic Rearrangement Collisions, *B.H. Bransden*

The Production of Rotational and Vibrational Transitions in Encounters between Molecules, *K. Takayanagi*

The Study of Intermolecular Potentials with Molecular Beams at Thermal Energies, *H. Pauly and J.P. Toennies*

High-Intensity and High-Energy Molecular Beams, *J.B. Anderson, R.P. Anders and J.B. Fen*

Volume 2

The Calculation of van der Waals Interactions, *A. Dalgarno and W.D. Davison*

Thermal Diffusion in Gases, *E.A. Mason, R.J. Munn and Francis J. Smith*

Spectroscopy in the Vacuum Ultraviolet, *W.R.S. Garton*

The Measurement of the Photoionization Cross Sections of the Atomic Gases, *James A.R. Samson*

The Theory of Electron–Atom Collisions, *R. Peterkop and V. Veldre*

Experimental Studies of Excitation in Collisions between Atomic and Ionic Systems, *F.J. de Heer*

Mass Spectrometry of Free Radicals, *S.N. Foner*

Volume 3

The Quantal Calculation of Photoionization Cross Sections, *A.L. Stewart*

Radiofrequency Spectroscopy of Stored Ions I: Storage, *H.G. Dehmelt*

Optical Pumping Methods in Atomic Spectroscopy, *B. Budick*

Energy Transfer in Organic Molecular Crystals: A Survey of Experiments, *H.C. Wolf*

Atomic and Molecular Scattering from Solid Surfaces, *Robert E. Stickney*

Quantum, Mechanics in Gas Crystal-Surface van der Waals Scattering, *E. Chanoch Beder*

Reactive Collisions between Gas and Surface Atoms, *Henry Wise and Bernard J. Wood*

Volume 4

H.S.W. Massey—A Sixtieth Birthday Tribute, *E.H.S. Burhop*

Electronic Eigenenergies of the Hydrogen Molecular Ion, *D.R. Bates and R.H.G. Reid*

Applications of Quantum Theory to the Viscosity of Dilute Gases, *R.A. Buckingham and E. Gal*

Positrons and Positronium in Gases, *P.A. Fraser*

Classical Theory of Atomic Scattering, *A. Burgess and I.C. Percival*

Born Expansions, *A.R. Holt and B.L. Moiseiwitsch*

Resonances in Electron Scattering by Atoms and Molecules, *P.G. Burke*

Relativistic Inner Shell Ionizations, C.B.O. Mohr
Recent Measurements on Charge Transfer, J.B. Hasted
Measurements of Electron Excitation Functions, D.W.O. Heddle and R.G.W. Keesing
Some New Experimental Methods in Collision Physics, R.F. Stebbings
Atomic Collision Processes in Gaseous Nebulae, M.J. Seaton
Collisions in the Ionosphere, A. Dalgarno
The Direct Study of Ionization in Space, R.L.F. Boyd

Volume 5

Flowing Afterglow Measurements of Ion-Neutral Reactions, E.E. Ferguson, F.C. Fehsenfeld and A.L. Schmeltekopf
Experiments with Merging Beams, Roy H. Neynaber
Radiofrequency Spectroscopy of Stored Ions II: Spectroscopy, H.G. Dehmelt
The Spectra of Molecular Solids, O. Schnepp
The Meaning of Collision Broadening of Spectral Lines: The Classical Oscillator Analog, A. Ben-Reuven
The Calculation of Atomic Transition Probabilities, R.J.S. Crossley
Tables of One- and Two-Particle Coefficients of Fractional Parentage for Configurations $s^\lambda s^{tu} p^q$, C.D.H. Chisholm, A. Dalgarno and F.R. Innes
Relativistic Z-Dependent Corrections to Atomic Energy Levels, Holly Thomis Doyle

Volume 6

Dissociative Recombination, J.N. Bardsley and M.A. Biondi
Analysis of the Velocity Field in Plasmas from the Doppler Broadening of Spectral Emission Lines, A.S. Kaufman
The Rotational Excitation of Molecules by Slow Electrons, Kazuo Takayanagi and Yukikazu Itikawa
The Diffusion of Atoms and Molecules, E.A. Mason and T.R. Marrero
Theory and Application of Sturmian Functions, Manuel Rotenberg
Use of Classical Mechanics in the Treatment of Collisions between Massive Systems, D.R. Bates and A.E. Kingston

Volume 7

Physics of the Hydrogen Maser, C. Audoin, J.P. Schermann and P. Grivet
Molecular Wave Functions: Calculations and Use in Atomic and Molecular Process, J.C. Browne
Localized Molecular Orbitals, Harel Weinstein, Ruben Pauncz and Maurice Cohen
General Theory of Spin-Coupled Wave Functions for Atoms and Molecules, J. Gerratt
Diabatic States of Molecules—Quasi Stationary Electronic States, Thomas F. O'Malley
Selection Rules within Atomic Shells, B.R. Judd
Green's Function Technique in Atomic and Molecular Physics, Gy. Csanak, H.S. Taylor and Robert Yaris
A Review of Pseudo-Potentials with Emphasis on Their Application to Liquid Metals, Nathan Wiser and A.J. Greenfield

Volume 8

Interstellar Molecules: Their Formation and Destruction, D. McNally
Monte Carlo Trajectory Calculations of Atomic and Molecular Excitation in Thermal Systems, James C. Keck
Nonrelativistic Off-Shell Two-Body Coulomb Amplitudes, Joseph C.Y. Chen and Augustine C. Chen
Photoionization with Molecular Beams, R.B. Cairns, Halstead Harrison and R.I. Schoen
The Auger Effect, E.H.S. Burhop and W.N. Asaad

Volume 9

Correlation in Excited States of Atoms, A.W. Weiss

The Calculation of Electron–Atom
 Excitation Cross Section, *M.R.H. Rudge*
Collision-Induced Transitions between
 Rotational Levels, *Takeshi Oka*
The Differential Cross Section of
 Low-Energy Electron–Atom Collisions,
 D. Andrick
Molecular Beam Electric Resonance
 Spectroscopy, *Jens C. Zorn and Thomas
 C. English*
Atomic and Molecular Processes in the
 Martian Atmosphere, *Michael
 B. McElroy*

Recent Progress in the Theory of Atomic
 Isotope Shift, *J. Bauche and
 R.-J. Champeau*
Topics on Multiphoton Processes in Atoms,
 P. Lambropoulos
Optical Pumping of Molecules, *M. Broyer,
 G. Goudedard, J.C. Lehmann and
 J. Vigué*
Highly Ionized Ions, *Ivan A. Sellin*
Time-of-Flight Scattering Spectroscopy,
 Wilhelm Raith
Ion Chemistry in the D Region, *George
 C. Reid*

Volume 10

Relativistic Effects in the Many-Electron
 Atom, *Lloyd Armstrong Jr. and Serge
 Feneuille*
The First Born Approximation, *K.L. Bell
 and A.E. Kingston*
Photoelectron Spectroscopy, *W.C. Price*
Dye Lasers in Atomic Spectroscopy,
 W. Lange, J. Luther and A. Steudel
Recent Progress in the Classification of the
 Spectra of Highly Ionized Atoms,
 B.C. Fawcett
A Review of Jovian Ionospheric Chemistry,
 Wesley T. Huntress Jr.

Volume 11

The Theory of Collisions between Charged
 Particles and Highly Excited Atoms,
 I.C. Percival and D. Richards
Electron Impact Excitation of Positive
 Ions, *M.J. Seaton*
The *R*-Matrix Theory of Atomic Process,
 P.G. Burke and W.D. Robb
Role of Energy in Reactive Molecular
 Scattering: An Information-Theoretic
 Approach, *R.B. Bernstein and
 R.D. Levine*
Inner Shell Ionization by Incident Nuclei,
 Johannes M. Hansteen
Stark Broadening, *Hans R. Griem*
Chemiluminescence in Gases, *M.F. Golde
 and B.A. Thrush*

Volume 12

Nonadiabatic Transitions between Ionic
 and Covalent States, *R.K. Janev*

Volume 13

Atomic and Molecular Polarizabilities—
 Review of Recent Advances, *Thomas
 M. Miller and Benjamin Bederson*
Study of Collisions by Laser Spectroscopy,
 Paul R. Berman
Collision Experiments with Laser-Excited
 Atoms in Crossed Beams, *I.V. Hertel and
 W. Stoll*
Scattering Studies of Rotational and
 Vibrational Excitation of Molecules,
 Manfred Faubel and J. Peter Toennies
Low-Energy Electron Scattering by
 Complex Atoms: Theory and
 Calculations, *R.K. Nesbet*
Microwave Transitions of Interstellar
 Atoms and Molecules, *W.B. Somerville*

Volume 14

Resonances in Electron Atom and
 Molecule Scattering, *D.E. Golden*
The Accurate Calculation of Atomic
 Properties by Numerical Methods, *Brain
 C. Webster, Michael J. Jamieson and
 Ronald F. Stewart*
(e, 2e) Collisions, *Erich Weigold and Ian
 E. McCarthy*
Forbidden Transitions in One- and
 Two-Electron Atoms, *Richard Marrus
 and Peter J. Mohr*
Semiclassical Effects in Heavy-Particle
 Collisions, *M.S. Child*
Atomic Physics Tests of the Basic
 Concepts in Quantum Mechanics,
 Francies M. Pipkin

Quasi-Molecular Interference Effects in
Ion–Atom Collisions, *S.V. Bobashev*
Rydberg Atoms, *S.A. Edelstein and
T.F. Gallagher*
UV and X-Ray Spectroscopy in
Astrophysics, *A.K. Dupree*

Volume 15

Negative Ions, *H.S.W. Massey*
Atomic Physics from Atmospheric and
Astrophysical, *A. Dalgarno*
Collisions of Highly Excited Atoms,
R.F. Stebbings
Theoretical Aspects of Positron Collisions
in Gases, *J.W. Humberston*
Experimental Aspects of Positron
Collisions in Gases, *T.C. Griffith*
Reactive Scattering: Recent Advances in
Theory and Experiment, *Richard
B. Bernstein*
Ion–Atom Charge Transfer Collisions at
Low Energies, *J.B. Hasted*
Aspects of Recombination, *D.R. Bates*
The Theory of Fast Heavy Particle
Collisions, *B.H. Bransden*
Atomic Collision Processes in Controlled
Thermonuclear Fusion Research,
H.B. Gilbody
Inner-Shell Ionization, *E.H.S. Burhop*
Excitation of Atoms by Electron Impact,
D.W.O. Heddle
Coherence and Correlation in Atomic
Collisions, *H. Kleinpoppen*
Theory of Low Energy Electron–Molecule
Collisions, *P.O. Burke*

Volume 16

Atomic Hartree–Fock Theory, *M. Cohen
and R.P. McEachran*
Experiments and Model Calculations to
Determine Interatomic Potentials,
R. Düren
Sources of Polarized Electrons,
R.J. Celotta and D.T. Pierce
Theory of Atomic Processes in Strong
Resonant Electromagnetic Fields,
S. Swain
Spectroscopy of Laser-Produced Plasmas,
M.H. Key and R.J. Hutcheon
Relativistic Effects in Atomic Collisions
Theory, *B.L. Moiseiwitsch*
Parity Nonconservation in Atoms: Status of
Theory and Experiment, *E.N. Fortson
and L. Wilets*

Volume 17

Collective Effects in Photoionization of
Atoms, *M.Ya. Amusia*
Nonadiabatic Charge Transfer,
D.S.F. Crothers
Atomic Rydberg States, *Serge Feneuille
and Pierre Jacquinot*
Superfluorescence, *M.F.H. Schuurmans,
Q.H.F. Vrehen, D. Polder and
H.M. Gibbs*
Applications of Resonance Ionization
Spectroscopy in Atomic and Molecular
Physics, *M.G. Payne, C.H. Chen,
G.S. Hurst and G.W. Foltz*
Inner-Shell Vacancy Production in
Ion–Atom Collisions, *C.D. Lin and
Patrick Richard*
Atomic Processes in the Sun, *P.L. Dufton
and A.E. Kingston*

Volume 18

Theory of Electron–Atom Scattering in a
Radiation Field, *Leonard Rosenberg*
Positron–Gas Scattering Experiments,
Talbert S. Stein and Walter E. Kauppila
Nonresonant Multiphoton Ionization of
Atoms, *J. Morellec, D. Normand and
G. Petite*
Classical and Semiclassical Methods in
Inelastic Heavy-Particle Collisions,
A.S. Dickinson and D. Richards
Recent Computational Developments in the
Use of Complex Scaling in Resonance
Phenomena, *B.R. Junker*
Direct Excitation in Atomic Collisions:
Studies of Quasi-One-Electron Systems,
N. Andersen and S.E. Nielsen
Model Potentials in Atomic Structure,
A. Hibbert
Recent Developments in the Theory of
Electron Scattering by Highly Polar
Molecules, *D.W. Norcross and
L.A. Collins*

Quantum Electrodynamic Effects in Few-Electron Atomic Systems, *G.W.F. Drake*

Volume 19

Electron Capture in Collisions of Hydrogen Atoms with Fully Stripped Ions, *B.H. Bransden and R.K. Janev*
Interactions of Simple Ion Atom Systems, *J.T. Park*
High-Resolution Spectroscopy of Stored Ions, *D.J. Wineland, Wayne M. Itano and R.S. Van Dyck Jr.*
Spin-Dependent Phenomena in Inelastic Electron–Atom Collisions, *K. Blum and H. Kleinpoppen*
The Reduced Potential Curve Method for Diatomic Molecules and Its Applications, *F. Jenč*
The Vibrational Excitation of Molecules by Electron Impact, *D.G. Thompson*
Vibrational and Rotational Excitation in Molecular Collisions, *Manfred Faubel*
Spin Polarization of Atomic and Molecular Photoelectrons, *N.A. Cherepkov*

Volume 20

Ion–Ion Recombination in an Ambient Gas, *D.R. Bates*
Atomic Charges within Molecules, *G.G. Hall*
Experimental Studies on Cluster Ions, *T.D. Mark and A.W. Castleman Jr.*
Nuclear Reaction Effects on Atomic Inner-Shell Ionization, *W.E. Meyerhof and J.-F. Chemin*
Numerical Calculations on Electron-Impact Ionization, *Christopher Bottcher*
Electron and Ion Mobilities, *Gordon R. Freeman and David A. Armstrong*
On the Problem of Extreme UV and X-Ray Lasers, *I.I. Sobel'man and A.V. Vinogradov*
Radiative Properties of Rydberg States in Resonant Cavities, *S. Haroche and J.M. Raimond*
Rydberg Atoms: High-Resolution Spectroscopy and Radiation Interaction—Rydberg Molecules, *J.A.C. Gallas, G. Leuchs, H. Walther, and H. Figger*

Volume 21

Subnatural Linewidths in Atomic Spectroscopy, *Dennis P. O'Brien, Pierre Meystre and Herbert Walther*
Molecular Applications of Quantum Defect Theory, *Chris H. Greene and Ch. Jungen*
Theory of Dielectronic Recombination, *Yukap Hahn*
Recent Developments in Semiclassical Floquet Theories for Intense-Field Multiphoton Processes, *Shih-I Chu*
Scattering in Strong Magnetic Fields, *M.R.C. McDowell and M. Zarcone*
Pressure Ionization, Resonances and the Continuity of Bound and Free States, *R.M. More*

Volume 22

Positronium—Its Formation and Interaction with Simple Systems, *J.W. Humberston*
Experimental Aspects of Positron and Positronium Physics, *T.C. Griffith*
Doubly Excited States, Including New Classification Schemes, *C.D. Lin*
Measurements of Charge Transfer and Ionization in Collisions Involving Hydrogen Atoms, *H.B. Gilbody*
Electron Ion and Ion–Ion Collisions with Intersecting Beams, *K. Dolder and B. Peart*
Electron Capture by Simple Ions, *Edward Pollack and Yukap Hahn*
Relativistic Heavy-Ion–Atom Collisions, *R. Anholt and Harvey Gould*
Continued-Fraction Methods in Atomic Physics, *S. Swain*

Volume 23

Vacuum Ultraviolet Laser Spectroscopy of Small Molecules, *C.R. Vidal*
Foundations of the Relativistic Theory of Atomic and Molecular Structure, *Ian P. Grant and Harry M. Quiney*
Point-Charge Models for Molecules Derived from Least-Squares Fitting of

the Electric Potential, *D.E. Williams and Ji-Min Yan*
Transition Arrays in the Spectra of Ionized Atoms, *J. Bauche, C. Bauche-Arnoult and M. Klapisch*
Photoionization and Collisional Ionization of Excited Atoms Using Synchrotron and Laser Radiation, *F.J. Wuilleumier, D.L. Ederer and J.L. Picqué*

Volume 24

The Selected Ion Flow Tube (SIDT): Studies of Ion-Neutral Reactions, *D. Smith and N.G. Adams*
Near-Threshold Electron–Molecule Scattering, *Michael A. Morrison*
Angular Correlation in Multiphoton Ionization of Atoms, *S.J. Smith and G. Leuchs*
Optical Pumping and Spin Exchange in Gas Cells, *R.J. Knize, Z. Wu and W. Happer*
Correlations in Electron–Atom Scattering, *A. Crowe*

Volume 25

Alexander Dalgarno: Life and Personality, *David R. Bates and George A. Victor*
Alexander Dalgarno: Contributions to Atomic and Molecular Physics, *Neal Lane*
Alexander Dalgarno: Contributions to Aeronomy, *Michael B. McElroy*
Alexander Dalgarno: Contributions to Astrophysics, *David A. Williams*
Dipole Polarizability Measurements, *Thomas M. Miller and Benjamin Bederson*
Flow Tube Studies of Ion–Molecule Reactions, *Eldon Ferguson*
Differential Scattering in He–He and He^+–He Collisions at keV Energies, *R.F. Stebbings*
Atomic Excitation in Dense Plasmas, *Jon C. Weisheit*
Pressure Broadening and Laser-Induced Spectral Line Shapes, *Kenneth M. Sando and Shih-I. Chu*
Model-Potential Methods, *C. Laughlin and G.A. Victor*

Z-Expansion Methods, *M. Cohen*
Schwinger Variational Methods, *Deborah Kay Watson*
Fine-Structure Transitions in Proton–Ion Collisions, *R.H.G. Reid*
Electron Impact Excitation, *R.J.W. Henry and A.E. Kingston*
Recent Advances in the Numerical Calculation of Ionization Amplitudes, *Christopher Bottcher*
The Numerical Solution of the Equations of Molecular Scattering, *A.C. Allison*
High Energy Charge Transfer, *B.H. Bransden and D.P. Dewangan*
Relativistic Random-Phase Approximation, *W.R. Johnson*
Relativistic Sturmian and Finite Basis Set Methods in Atomic Physics, *G.W.F. Drake and S.P. Goldman*
Dissociation Dynamics of Polyatomic Molecules, *T. Uzer*
Photodissociation Processes in Diatomic Molecules of Astrophysical Interest, *Kate P. Kirby and Ewine F. van Dishoeck*
The Abundances and Excitation of Interstellar Molecules, *John H. Black*

Volume 26

Comparisons of Positrons and Electron Scattering by Gases, *Walter E. Kauppila and Talbert S. Stein*
Electron Capture at Relativistic Energies, *B.L. Moiseiwitsch*
The Low-Energy, Heavy Particle Collisions—A Close-Coupling Treatment, *Mineo Kimura and Neal F. Lane*
Vibronic Phenomena in Collisions of Atomic and Molecular Species, *V. Sidis*
Associative Ionization: Experiments, Potentials and Dynamics, *John Weiner Françoise Masnou-Seeuws and Annick Giusti-Suzor*
On the β Decay of [loc=pre]187Re: An Interface of Atomic and Nuclear Physics and Cosmochronology, *Zonghau Chen, Leonard Rosenberg and Larry Spruch*
Progress in Low Pressure Mercury-Rare Gas Discharge Research, *J. Maya and R. Lagushenko*

Volume 27

Negative Ions: Structure and Spectra, *David R. Bates*

Electron Polarization Phenomena in Electron–Atom Collisions, *Joachim Kessler*

Electron–Atom Scattering, *I.E. McCarthy and E. Weigold*

Electron–Atom Ionization, *I.E. McCarthy and E. Weigold*

Role of Autoionizing States in Multiphoton Ionization of Complex Atoms, *V.I. Lengyel and M.I. Haysak*

Multiphoton Ionization of Atomic Hydrogen Using Perturbation Theory, *E. Karule*

Volume 28

The Theory of Fast Ion–Atom Collisions, *J.S. Briggs and J.H. Macek*

Some Recent Developments in the Fundamental Theory of Light, *Peter W. Milonni and Surendra Singh*

Squeezed States of the Radiation Field, *Khalid Zaheer and M. Suhail Zubairy*

Cavity Quantum Electrodynamics, *E.A. Hinds*

Volume 29

Studies of Electron Excitation of Rare-Gas Atoms into and out of Metastable Levels Using Optical and Laser Techniques, *Chun C. Lin and L.W. Anderson*

Cross Sections for Direct Multiphoton Ionization of Atoms, *M.V. Ammosov, N.B. Delone, M.Yu. Ivanov, I.I. Bandar and A.V. Masalov*

Collision-Induced Coherences in Optical Physics, *G.S. Agarwal*

Muon-Catalyzed Fusion, *Johann Rafelski and Helga E. Rafelski*

Cooperative Effects in Atomic Physics, *J.P. Connerade*

Multiple Electron Excitation, Ionization, and Transfer in High-Velocity Atomic and Molecular Collisions, *J.H. McGuire*

Volume 30

Differential Cross Sections for Excitation of Helium Atoms and Helium-Like Ions by Electron Impact, *Shinobu Nakazaki*

Cross-Section Measurements for Electron Impact on Excited Atomic Species, *S. Trajmar and J.C. Nickel*

The Dissociative Ionization of Simple Molecules by Fast Ions, *Colin J. Latimer*

Theory of Collisions between Laser Cooled Atoms, *P.S. Julienne, A.M. Smith and K. Burnett*

Light-Induced Drift, *E.R. Eliel*

Continuum Distorted Wave Methods in Ion–Atom Collisions, *Derrick S.F. Crothers and Louis J. Dube*

Volume 31

Energies and Asymptotic Analysis for Helium Rydberg States, *G.W.F. Drake*

Spectroscopy of Trapped Ions, *R.C. Thompson*

Phase Transitions of Stored Laser-Cooled Ions, *H. Walther*

Selection of Electronic States in Atomic Beams with Lasers, *Jacques Baudon, Rudalf Dülren and Jacques Robert*

Atomic Physics and Non-Maxwellian Plasmas, *Michèle Lamoureux*

Volume 32

Photoionization of Atomic Oxygen and Atomic Nitrogen, *K.L. Bell and A.E. Kingston*

Positronium Formation by Positron Impact on Atoms at Intermediate Energies, *B.H. Bransden and C.J. Noble*

Electron–Atom Scattering Theory and Calculations, *P.G. Burke*

Terrestrial and Extraterrestrial H_3^+, *Alexander Dalgarno*

Indirect Ionization of Positive Atomic Ions, *K. Dolder*

Quantum Defect Theory and Analysis of High-Precision Helium Term Energies, *G.W.F. Drake*

Electron–Ion and Ion–Ion Recombination Processes, *M.R. Flannery*

Studies of State-Selective Electron Capture in Atomic Hydrogen by Translational Energy Spectroscopy, *H.B. Gilbody*

Relativistic Electronic Structure of Atoms and Molecules, *I.P. Grant*

The Chemistry of Stellar Environments, *D.A. Howe, J.M.C. Rawlings and D.A. Williams*

Positron and Positronium Scattering at Low Energies, *J.W. Humberston*

How Perfect are Complete Atomic Collision Experiments?, *H. Kleinpoppen and H. Handy*

Adiabatic Expansions and Nonadiabatic Effects, *R. McCarroll and D.S.F. Crothers*

Electron Capture to the Continuum, *B.L. Moiseiwitsch*

How Opaque Is a Star?, *M.T. Seaton*

Studies of Electron Attachment at Thermal Energies Using the Flowing Afterglow–Langmuir Technique, *David Smith and Patrik Španěl*

Exact and Approximate Rate Equations in Atom–Field Interactions, *S. Swain*

Atoms in Cavities and Traps, *H. Walther*

Some Recent Advances in Electron-Impact Excitation of $n = 3$ States of Atomic Hydrogen and Helium, *J.F. Williams and J.B. Wang*

Volume 33

Principles and Methods for Measurement of Electron Impact Excitation Cross Sections for Atoms and Molecules by Optical Techniques, *A.R. Filippelli, Chun C. Lin, L.W. Andersen and J.W. McConkey*

Benchmark Measurements of Cross Sections for Electron Collisions: Analysis of Scattered Electrons, *S. Trajmar and J.W. McConkey*

Benchmark Measurements of Cross Sections for Electron Collisions: Electron Swarm Methods, *R.W. Crompton*

Some Benchmark Measurements of Cross Sections for Collisions of Simple Heavy Particles, *H.B. Gilbody*

The Role of Theory in the Evaluation and Interpretation of Cross-Section Data, *Barry I. Schneider*

Analytic Representation of Cross-Section Data, *Mitio Inokuti, Mineo Kimura, M.A. Dillon, Isao Shimamura*

Electron Collisions with N_2, O_2 and O: What We Do and Do Not Know, *Yukikazu Itikawa*

Need for Cross Sections in Fusion Plasma Research, *Hugh P. Summers*

Need for Cross Sections in Plasma Chemistry, *M. Capitelli, R. Celiberto and M. Cacciatore*

Guide for Users of Data Resources, *Jean W. Gallagher*

Guide to Bibliographies, Books, Reviews and Compendia of Data on Atomic Collisions, *E.W. McDaniel and E.J. Mansky*

Volume 34

Atom Interferometry, *C.S. Adams, O. Carnal and J. Mlynek*

Optical Tests of Quantum Mechanics, *R.Y. Chiao, P.G. Kwiat and A.M. Steinberg*

Classical and Quantum Chaos in Atomic Systems, *Dominique Delande and Andreas Buchleitner*

Measurements of Collisions between Laser-Cooled Atoms, *Thad Walker and Paul Feng*

The Measurement and Analysis of Electric Fields in Glow Discharge Plasmas, *J.E. Lawler and D.A. Doughty*

Polarization and Orientation Phenomena in Photoionization of Molecules, *N.A. Cherepkov*

Role of Two-Center Electron–Electron Interaction in Projectile Electron Excitation and Loss, *E.C. Montenegro, W.E. Meyerhof and J.H. McGuire*

Indirect Processes in Electron Impact Ionization of Positive Ions, *D.L. Moores and K.J. Reed*

Dissociative Recombination: Crossing and Tunneling Modes, *David R. Bates*

Volume 35

Laser Manipulation of Atoms, *K. Sengstock and W. Ertmer*
Advances in Ultracold Collisions: Experiment and Theory, *J. Weiner*
Ionization Dynamics in Strong Laser Fields, *L.F. DiMauro and P. Agostini*
Infrared Spectroscopy of Size Selected Molecular Clusters, *U. Buck*
Fermosecond Spectroscopy of Molecules and Clusters, *T. Baumer and G. Gerber*
Calculation of Electron Scattering on Hydrogenic Targets, *I. Bray and A.T. Stelbovics*
Relativistic Calculations of Transition Amplitudes in the Helium Isoelectronic Sequence, *W.R. Johnson, D.R. Plante and J. Sapirstein*
Rotational Energy Transfer in Small Polyatomic Molecules, *H.O. Everitt and F.C. De Lucia*

Volume 36

Complete Experiments in Electron–Atom Collisions, *Nils Overgaard Andersen and Klaus Bartschat*
Stimulated Rayleigh Resonances and Recoil-Induced Effects, *J.-Y. Courtois and G. Grynberg*
Precision Laser Spectroscopy Using Acousto-Optic Modulators, *W.A. van Wijngaarden*
Highly Parallel Computational Techniques for Electron–Molecule Collisions, *Carl Winstead and Vincent McKoy*
Quantum Field Theory of Atoms and Photons, *Maciej Lewenstein and Li You*

Volume 37

Evanescent Light-Wave Atom Mirrors, Resonators, Waveguides, and Traps, *Jonathan P. Dowling and Julio Gea-Banacloche*
Optical Lattices, *P.S. Jessen and I.H. Deutsch*
Channeling Heavy Ions through Crystalline Lattices, *Herbert F. Krause and Sheldon Datz*
Evaporative Cooling of Trapped Atoms, *Wolfgang Ketterle and N.J. van Druten*
Nonclassical States of Motion in Ion Traps, *J.I. Cirac, A.S. Parkins, R. Blatt and P. Zoller*
The Physics of Highly-Charged Heavy Ions Revealed by Storage/Cooler Rings, *P.H. Mokler and Th. Stöhlker*

Volume 38

Electronic Wavepackets, *Robert R. Jones and L.D. Noordam*
Chiral Effects in Electron Scattering by Molecules, *K. Blum and D.G. Thompson*
Optical and Magneto-Optical Spectroscopy of Point Defects in Condensed Helium, *Serguei I. Kanorsky and Antoine Weis*
Rydberg Ionization: From Field to Photon, *G.M. Lankhuijzen and L.D. Noordam*
Studies of Negative Ions in Storage Rings, *L.H. Andersen, T. Andersen and P. Hvelplund*
Single-Molecule Spectroscopy and Quantum Optics in Solids, *W.E. Moerner, R.M. Dickson and D.J. Norris*

Volume 39

Author and Subject Cumulative Index Volumes 1–38
Author Index
Subject Index
Appendix: Tables of Contents of Volumes 1–38 and Supplements

Volume 40

Electric Dipole Moments of Leptons, *Eugene D. Commins*
High-Precision Calculations for the Ground and Excited States of the Lithium Atom, *Frederick W. King*
Storage Ring Laser Spectroscopy, *Thomas U. Kühl*
Laser Cooling of Solids, *Carl E. Mangan and Timothy R. Gosnell*
Optical Pattern Formation, *L.A. Lugiato, M. Brambilla and A. Gatti*

Volume 41

Two-Photon Entanglement and Quantum Reality, *Yanhua Shih*

Quantum Chaos with Cold Atoms, *Mark G. Raizen*
Study of the Spatial and Temporal Coherence of High-Order Harmonics, *Pascal Salières, Anne L'Huillier, Philippe Antoine and Maciej Lewenstein*
Atom Optics in Quantized Light Fields, *Matthias Freyburger, Alois M. Herkommer, Daniel S. Krähmer, Erwin Mayr and Wolfgang P. Schleich*
Atom Waveguides, *Victor I. Balykin*
Atomic Matter Wave Amplification by Optical Pumping, *Ulf Janicke and Martin Wikens*

Volume 42

Fundamental Tests of Quantum Mechanics, *Edward S. Fry and Thomas Walther*
Wave-Particle Duality in an Atom Interferometer, *Stephan Dürr and Gerhard Rempe*
Atom Holography, *Fujio Shimizu*
Optical Dipole Traps for Neutral Atoms, *Rudolf Grimm, Matthias Weidemüller and Yurii B. Ovchinnikov*
Formation of Cold ($T \leq 1$ K) Molecules, *J.T. Bahns, P.L. Gould and W.C. Stwalley*
High-Intensity Laser-Atom Physics, *C.J. Joachain, M. Dorr and N.J. Kylstra*
Coherent Control of Atomic, Molecular and Electronic Processes, *Moshe Shapiro and Paul Brumer*
Resonant Nonlinear Optics in Phase Coherent Media, *M.D. Lukin, P. Hemmer and M.O. Scully*
The Characterization of Liquid and Solid Surfaces with Metastable Helium Atoms, *H. Morgner*
Quantum Communication with Entangled Photons, *Herald Weinfurter*

Volume 43

Plasma Processing of Materials and Atomic, Molecular, and Optical Physics: An Introduction, *Hiroshi Tanaka and Mitio Inokuti*
The Boltzmann Equation and Transport Coefficients of Electrons in Weakly Ionized Plasmas, *R. Winkler*
Electron Collision Data for Plasma Chemistry Modeling, *W.L. Morgan*
Electron–Molecule Collisions in Low-Temperature Plasmas: The Role of Theory, *Carl Winstead and Vincent McKoy*
Electron Impact Ionization of Organic Silicon Compounds, *Ralf Basner, Kurt Becker, Hans Deutsch and Martin Schmidt*
Kinetic Energy Dependence of Ion–Molecule Reactions Related to Plasma Chemistry, *P.B. Armentrout*
Physicochemical Aspects of Atomic and Molecular Processes in Reactive Plasmas, *Yoshihiko Hatano*
Ion–Molecule Reactions, *Werner Lindinger, Armin Hansel and Zdenek Herman*
Uses of High-Sensitivity White-Light Absorption Spectroscopy in Chemical Vapor Deposition and Plasma Processing, *L.W. Anderson, A.N. Goyette and J.E. Lawler*
Fundamental Processes of Plasma–Surface Interactions, *Rainer Hippler*
Recent Applications of Gaseous Discharges: Dusty Plasmas and Upward-Directed Lightning, *Ara Chutjian*
Opportunities and Challenges for Atomic, Molecular and Optical Physics in Plasma Chemistry, *Kurl Becker Hans Deutsch and Mitio Inokuti*

Volume 44

Mechanisms of Electron Transport in Electrical Discharges and Electron Collision Cross Sections, *Hiroshi Tanaka and Osamu Sueoka*
Theoretical Consideration of Plasma-Processing Processes, *Mineo Kimura*
Electron Collision Data for Plasma-Processing Gases, *Loucas G. Christophorou and James K. Olthoff*
Radical Measurements in Plasma Processing, *Toshio Goto*

Radio-Frequency Plasma Modeling for Low-Temperature Processing, *Toshiaki Makabe*
Electron Interactions with Excited Atoms and Molecules, *Loucas G. Christophorou and James K. Olthoff*

Volume 45

Comparing the Antiproton and Proton, and Opening the Way to Cold Antihydrogen, *G. Gabrielse*
Medical Imaging with Laser-Polarized Noble Gases, *Timothy Chupp and Scott Swanson*
Polarization and Coherence Analysis of the Optical Two-Photon Radiation from the Metastable $2^2Si_{1/2}$ State of Atomic Hydrogen, *Alan J. Duncan, Hans Kleinpoppen and Marian O. Scully*
Laser Spectroscopy of Small Molecules, *W. Demtröder, M. Keil and H. Wenz*
Coulomb Explosion Imaging of Molecules, *Z. Vager*

Volume 46

Femtosecond Quantum Control, *T. Brixner, N.H. Damrauer and G. Gerber*
Coherent Manipulation of Atoms and Molecules by Sequential Laser Pulses, *N.V. Vitanov, M. Fleischhauer, B.W. Shore and K. Bergmann*
Slow, Ultraslow, Stored, and Frozen Light, *Andrey B. Matsko, Olga Kocharovskaya, Yuri Rostovtsev, George R. Welch, Alexander S. Zibrov and Marlan O. Scully*
Longitudinal Interferometry with Atomic Beams, *S. Gupta, D.A. Kokorowski, R.A. Rubenstein, and W.W. Smith*

Volume 47

Nonlinear Optics of de Broglie Waves, *P. Meystre*
Formation of Ultracold Molecules ($T \leq 200$ μK) via Photoassociation in a Gas of Laser-Cooled Atoms, *Françoise Masnou-Seeuws and Pierre Pillet*
Molecular Emissions from the Atmospheres of Giant Planets and Comets: Needs for Spectroscopic and Collision Data, *Yukikazu Itikawa, Sang Joon Kim, Yong Ha Kim and Y.C. Minh*
Studies of Electron-Excited Targets Using Recoil Momentum Spectroscopy with Laser Probing of the Excited State, *Andrew James Murray and Peter Hammond*
Quantum Noise of Small Lasers, *J.P. Woerdman, N.J. van Druten and M.P. van Exter*

Volume 48

Multiple Ionization in Strong Laser Fields, *R. Dörner Th. Weber, M. Weckenbrock, A. Staudte, M. Hattass, R. Moshammer, J. Ullrich and H. Schmidt-Böcking*
Above-Threshold Ionization: From Classical Features to Quantum Effects, *W. Becker, F. Grasbon, R. Kapold, D.B. Milošević, G.G. Paulus and H. Walther*
Dark Optical Traps for Cold Atoms, *Nir Friedman, Ariel Kaplan and Nir Davidson*
Manipulation of Cold Atoms in Hollow Laser Beams, *Heung-Ryoul Noh, Xenye Xu and Wonho Jhe*
Continuous Stern–Gerlach Effect on Atomic Ions, *Günther Werth, Hartmut Haffner and Wolfgang Quint*
The Chirality of Biomolecules, *Robert N. Compton and Richard M. Pagni*
Microscopic Atom Optics: From Wires to an Atom Chip, *Ron Folman, Peter Krüger, Jörg Schmiedmayer, Johannes Denschlag and Carsten Henkel*
Methods of Measuring Electron–Atom Collision Cross Sections with an Atom Trap, *R.S. Schappe, M.L. Keeler, T.A. Zimmerman, M. Larsen, P. Feng, R.C. Nesnidal, J.B. Boffard, T.G. Walker, L.W. Anderson and C.C. Lin*

Volume 49

Applications of Optical Cavities in Modern Atomic, Molecular, and Optical Physics, *Jun Ye and Theresa W. Lynn*
Resonance and Threshold Phenomena in Low-Energy Electron Collisions with

Molecules and Clusters, *H. Hotop, M.-W. Ruf, M. Allan and I.I. Fabrikant*
Coherence Analysis and Tensor Polarization Parameters of $(\gamma, e\gamma)$ Photoionization Processes in Atomic Coincidence Measurements, *B. Lohmann, B. Zimmermann, H. Kleinpoppen and U. Becker*
Quantum Measurements and New Concepts for Experiments with Trapped Ions, *Ch. Wunderlich and Ch. Balzer*
Scattering and Reaction Processes in Powerful Laser Fields, *Dejan B. Milošević and Fritz Ehlotzky*
Hot Atoms in the Terrestrial Atmosphere, *Vijay Kumar and E. Krishnakumar*

Volume 50

Assessment of the Ozone Isotope Effect, *K. Mauersberger, D. Krankowsky, C. Janssen and R. Schinke*
Atom Optics, Guided Atoms, and Atom Interferometry, *J. Arlt, G. Birkl, E. Rasel and W. Ertmet*
Atom–Wall Interaction, *D. Bloch and M. Ducloy*
Atoms Made Entirely of Antimatter: Two Methods Produce Slow Antihydrogen, *G. Gabrielse*
Ultrafast Excitation, Ionization, and Fragmentation of C_{60}, *I.V. Hertel, T. Laarmann and C.P. Schulz*

Volume 51

Introduction, *Henry H. Stroke*
Appreciation of Ben Bederson as Editor of Advances in Atomic, Molecular, and Optical Physics
Benjamin Bederson Curriculum Vitae
Research Publications of Benjamin Bederson
A Proper Homage to Our Ben, *H. Lustig*
Benjamin Bederson in the Army, World War II, *Val L. Fitch*
Physics Needs Heroes Too, *C. Duncan Rice*
Two Civic Scientists—Benjamin Bederson and the other Benjamin, *Neal Lane*
An Editor *Par Excellence*, *Eugen Merzbacher*
Ben as APS Editor, *Bernd Crasemann*
Ben Bederson: Physicist–Historian, *Roger H. Stuewer*
Pedagogical Notes on Classical Casimir Effects, *Larry Spruch*
Polarizabilities of ^3P Atoms and van der Waals Coefficients for Their Interaction with Helium Atoms, *X. Chu and A. Dalgarno*
The Two Electron Molecular Bonds Revisited: From Bohr Orbits to Two-Center Orbitals, *Goong Chen, Siu A. Chin, Yusheng Dou, Kishore T. Kapale, Moochan Kim, Anatoly A. Svidzinsky, Kerim Urtekin, Han Xiong and Marlan O. Scully*
Resonance Fluorescence of Two-Level Atoms, *H. Walther*
Atomic Physics with Radioactive Atoms, *Jacques Pinard and H. Henry Stroke*
Thermal Electron Attachment and Detachment in Gases, *Thomas M. Miller*
Recent Developments in the Measurement of Static Electric Dipole Polarizabilities, *Harvey Gould and Thomas M. Miller*
Trapping and Moving Atoms on Surfaces, *Robert J. Celotta and Joseph A. Stroscio*
Electron-Impact Excitation Cross Sections of Sodium, *Chun C. Lin and John B. Boffard*
Atomic and Ionic Collisions, *Edward Pollack*
Atomic Interactions in Weakly Ionized Gas: Ionizing Shock Waves in Neon, *Leposava Vušković and Svetozar Popović*
Approaches to Perfect/Complete Scattering Experiments in Atomic and Molecular Physics, *H. Kleinpoppen, B. Lohmann, A. Grum-Grzhimailo and U. Becker*
Reflections on Teaching, *Richard E. Collins*

Volume 52

Exploring Quantum Matter with Ultracold Atoms in Optical Lattices, *Immanuel Bloch and Markus Greiner*
The Kicked Rydberg Atom, *F.B. Dunning, J.C. Lancaster, C.O. Reinhold, S. Yoshida and J. Burgdörfer*
Photonic State Tomography, *J.B. Altepeter, E.R. Jeffrey and P.G. Kwiat*

Fine Structure in High-*L* Rydberg States: A Path to Properties of Positive Ions, *Stephen R. Lundeen*

A Storage Ring for Neutral Molecules, *Floris M.H. Crompvoets, Hendrick L. Bethlem and Gerard Meijer*

Nonadiabatic Alignment by Intense Pulses. Concepts, Theory, and Directions, *Tamar Seideman and Edward Hamilton*

Relativistic Nonlinear Optics, *Donald Umstadter, Scott Sepke and Shouyuan Chen*

Coupled-State Treatment of Charge Transfer, *Thomas G. Winter*

Volume 53

Non-Classical Light from Artificial Atoms, *Thomas Aichele, Matthias Scholz, Sven Ramelow and Oliver Benson*

Quantum Chaos, Transport, and Control—in Quantum Optics, *Javier Madroñero, Alexey Ponomarev, Andrí R.R. Carvalho, Sandro Wimberger, Carlos Viviescas, Andrey Kolovsky, Klaus Hornberger, Peter Schlagheck, Andreas Krug and Andreas Buchleitner*

Manipulating Single Atoms, *Dieter Meschede and Arno Rauschenbeutel*

Spatial Imaging with Wavefront Coding and Optical Coherence Tomography, *Thomas Hellmuth*

The Quantum Properties of Multimode Optical Amplifiers Revisited, *G. Leuchs, U.L. Andersen and C. Fabre*

Quantum Optics of Ultra-Cold Molecules, *D. Meiser, T. Miyakawa, H. Uys and P. Meystre*

Atom Manipulation in Optical Lattices, *Georg Raithel and Natalya Morrow*

Femtosecond Laser Interaction with Solid Surfaces: Explosive Ablation and Self-Assembly of Ordered Nanostructures, *Juergen Reif and Florenta Costache*

Characterization of Single Photons Using Two-Photon Interference, *T. Legero, T. Wilk, A. Kuhn and G. Rempe*

Fluctuations in Ideal and Interacting Bose–Einstein Condensates: From the Laser Phase Transition Analogy to Squeezed States and Bogoliubov Quasiparticles, *Vitaly V. Kocharovsky, Vladimir V. Kocharovsky, Martin Holthaus, C.H. Raymond Ooi, Anatoly Svidzinsky, Wolfgang Ketterle and Marlan O. Scully*

LIDAR-Monitoring of the Air with Femtosecond Plasma Channels, *Ludger Wöste, Steffen Frey and Jean-Pierre Wolf*

Volume 54

Experimental Realization of the BCS-BEC Crossover with a Fermi Gas of Atoms, *C.A. Regal and D.S. Jin*

Deterministic Atom–Light Quantum Interface, *Jacob Sherson, Brian Julsgaard and Eugene S. Polzik*

Cold Rydberg Atoms, *J.-H. Choi, B. Knuffman, T. Cubel Liebisch, A. Reinhard and G. Raithel*

Non-Perturbative Quantal Methods for Electron–Atom Scattering Processes, *D.C. Griffin and M.S. Pindzola*

R-Matrix Theory of Atomic, Molecular and Optical Processes, *P.G. Burke, C.J. Noble and V.M. Burke*

Electron-Impact Excitation of Rare-Gas Atoms from the Ground Level and Metastable Levels, *John B. Boffard, R.O. Jung, L.W. Anderson and C.C. Lin*

Internal Rotation in Symmetric Tops, *I. Ozier and N. Moazzen-Ahmadi*

Attosecond and Angstrom Science, *Hiromichi Niikura and P.B. Corkum*

Atomic Processing of Optically Carried RF Signals, *Jean-Louis Le Gouët, Fabien Bretenaker and Ivan Lorgeré*

Controlling Optical Chaos, Spatio-Temporal Dynamics, and Patterns, *Lucas Illing, Daniel J. Gauthier and Rajarshi Roy*

Volume 55

Direct Frequency Comb Spectroscopy, *Matthew C. Stowe, Michael J. Thorpe, Avi Pe'er, Jun Ye, Jason E. Stalnaker, Vladislav Gerginov and Scott A. Diddams*

Collisions, Correlations, and Integrability in Atom Waveguides, *Vladimir A.*

Yurovsky, Maxim Olshanii and David S. Weiss

MOTRIMS: Magneto–Optical Trap Recoil Ion Momentum Spectroscopy, *Brett D. DePaola, Reinhard Morgenstern and Nils Andersen*

All-Order Methods for Relativistic Atomic Structure Calculations, *Marianna S. Safronova and Walter R. Johnson*

B-Splines in Variational Atomic Structure Calculations, *Charlotte Froese Fischer*

Electron–Ion Collisions: Fundamental Processes in the Focus of Applied Research, *Alfred Müller*

Robust Probabilistic Quantum Information Processing with Atoms, Photons, and Atomic Ensembles, *Luming Duan and Christopher R. Monroe*

Supplements

Atoms in Intense Laser Fields, edited by Mihai Gavrila (1992)

Multiphoton Ionization, *H.G. Muller, P. Agostini and G. Petite*

Photoionization with Ultra-Short Laser Pulses, *R.R. Freeman, P.H. Bucksbaum, W.E. Cooke, G. Gibson, T.J. McIlrath and L.D. van Woerkom*

Rydberg Atoms in Strong Microwave Fields, *T.F. Gallagher*

Muiltiphoton Ionization in Large Ponderomotive Potentials, *P.B. Corkum, N.H. Burnett and F. Brunel*

High Order Harmonic Generation in Rare Gases, *Anne L'Huillier, Louis-André Lompré, Gerard Manfrey and Claude Manus*

Mechanisms of Short-Wavelength Generation, *T.S. Luk, A. McPherson, K. Boyer and C.K. Rhodes*

Time-Dependent Studies of Multiphoton Processes, *Kenneth C. Kulander, Kenneth J. Schafer and Jeffrey L. Krause*

Numerical Experiments in Strong and Super-Strong Fields, *J.H. Eberly, R. Grobe, C.K. Law and Q. Su*

Resonances in Multiphoton Ionization, *P. Lambropoulos and X. Tang*

Nonperturbative Treatment of Multiphoton Ionization within the Floquet Framework, *R.M. Potvliege and Robin Shakeshaft*

Atomic Structure and Decay in High Frequency Fields, *Mihai Gavrila*

Cavity Quantum Electrodynamics, edited by Paul R. Berman (1994)

Perturbative Cavity Quantum Electrodynamics, *E.A. Hinds*

The Micromaser: A Proving Ground for Quantum Physics, *Georg Raithel, Christian Wagner, Herbert Walther, Lorenzo M. Narducci and Marlan O. Scully*

Manipulation of Nonclassical Field States in a Cavity by Atom Interferometry, *S. Haroche and J.M. Raimond*

Quantum Optics of Driven Atoms in Colored Vacua, *Thomas W. Mossberg and Maciej Lewenstein*

Structure and Dynamics in Cavity Quantum Electrodynamics, *H.J. Kimble*

One Electron in a Cavity, *G. Gabrielse and J. Tan*

Spontaneous Emission by Moving Atoms, *Pierre Meystre and Martin Wilkens*

Single Atom Emission in an Optical Resonator, *James J. Childs, Kyungwon An, Ramanchandra R. Dasari and Michael S. Feld*

Nonperturbative Atom–Photon Interactions in an Optical Cavity, *H.J. Carmichael, L. Tian, W. Ren and P. Alsing*

New Aspects of the Casimir Effect: Fluctuations and Radiative Reaction, *G. Barton*